P9-AFQ-434

The Use and Abuse
of Nature

WITHDRAWN
UTSA Libraries

RENEWALS 458-4574

DATE DUE

| | | | |
|---|---|---|---|
| | | | |
| | | | |
| | | | |
| | | | |
| | | | |
| | | | |
| | | | |
| | | | |
| | | | |
| | | | |
| | | | |
| | | | |
| | | | |
| | | | |
| | | | |
| | | | |
| | | | |
| | | | |
| | | | |
| GAYLORD | | | PRINTED IN U.S.A. |

Douglas and Alaska

WITHDRAWN
UTSA Libraries

MADHAV GADGIL
AND
RAMACHANDRA GUHA

# The Use and Abuse
# of Nature

incorporating

❧ This Fissured Land
An Ecological History of India

and

❧ Ecology and Equity

OXFORD
UNIVERSITY PRESS

# OXFORD

UNIVERSITY PRESS

YMCA Library Building, Jai Singh Road, New Delhi 110 001

Oxford University Press is a department of the University of Oxford. It furthers the
University's objective of excellence in research, scholarship, and education
by publishing worldwide in

Oxford   New York

Auckland   Bangkok   Buenos Aires   Cape Town   Chennai
Dar es Salaam   Delhi   Hong Kong   Istanbul   Karachi   Kolkata
Kuala Lumpur   Madrid   Melbourne   Mexico City   Mumbai   Nairobi
São Paulo   Shanghai   Taipei   Tokyo   Toronto

Oxford is a registered trademark of Oxford University Press
in the UK and in certain other countries

Published in India
By Oxford University Press, New Delhi

© this edition Oxford University Press 2000
*This Fissured Land* © Oxford University Press 1992
Oxford India Paperbacks 1993

*Ecology and Equity* © United Nations Research Institute
for Social Development 1995
Published by arrangement with the United Nations
Research Institute for Social Development

The moral rights of the author have been asserted
Database right Oxford University Press (maker)

First published 2000
Oxford India Paperbacks 2004

All rights reserved. No part of this publication may be reproduced,
or transmitted in any form or by any means, electronic or mechanical,
including photocopying, recording or by any information storage and
retrieval system, without permission in writing from Oxford University Press.
Enquiries concerning reproduction outside the scope of the above should be
sent to the Rights Department, Oxford University Press, at the address above

You must not circulate this book in any other binding or cover
and you must impose this same condition on any acquirer

ISBN   019 567174 0

Printed in India by Saurabh Print-O-Pack, NOIDA, U.P.
Published by Manzar Khan, Oxford University Press
YMCA Library Building, Jai Singh Road, New Delhi 110 001

Library
University
at San Antonio

# Introduction to the Omnibus Edition of
*This Fissured Land* and *Ecology and Equity*

These two books are principally about the human use and abuse of nature: past, present, and future. Of the species on earth, humans are singular in their capacity for environmental destructiveness and, less visibly, for taking remedial action. The fabrication of artefacts and ingenious deployment of external sources of energy have allowed man to dominate nature. We impose demands on natural resources that are often completely unrelated to our biological needs. These demands frequently escalate, primarily as a result of competition amongst individuals and groups; these then create severe and unsustainable pressure on land, waters, forests, and the atmosphere, as well as on other living species, an increasing number of which have been wiped off the face of this planet by the intended and unintended consequences of human action.

Paradoxically, humans are also the only species that can be prudent in their use of nature. There is a long-running debate in biology on whether animals ever display deliberate restraint in their harvesting of prey, on whether they seek to maintain their resource base on a long-term basis. The consensus among scholars is that they do not. Animals seem to hunt either to maximize their immediate net energy gain, or to minimize their risk. Only humans sometimes set about refraining from harvesting readily accessible resources for the sake of a long-term interest, as when communities desist from fishing in sacred ponds, or declare a season closed for hunting.

The first of the books here, *This Fissured Land*, looks at this patchwork of prudence and profligacy through human history, in particular through the history of India. The book was in part inspired—or should one say provoked—by the near-total neglect of environmental history among the country's professional historians. Although we, the two authors of *This Fissured Land*, were trained in ecology and sociology respectively, both of us had a long interest in historical research. Fortunately, we were not burdened with the biases and fashions common among professional historians, who, in India at any rate, had tended to underplay and often ignore the reciprocal interaction between humans on the one side and nature and natural resources on the other.

When it first appeared, *This Fissured Land* sought to throw down the gauntlet to the community of historians, challenging them to enlarge their frame of analysis by including the environmental dimension. Its theoretical ambition was two-fold. Ecologically, it outlined the conditions under which humans exercised prudence or profligacy in their use of nature. It also identified and enumerated the types of human action that could justifiably be termed prudent or resource-conserving. Sociologically, it presented a new framework of what we called 'modes of resource use'. This classified human societies in terms of their interaction with nature and natural resources. The new framework was required, we argued, because the classical Marxian concept of 'modes of production' was restricted to the realms of field and factory—it had not considered the realms of water, forests, and minerals. For saying this we were thanked neither by the Marxists, who were upset that we dared point out a weakness in their received framework, nor by the non-and-anti-Marxists, for whom it was heresy enough for us to have been inspired in some ways by Marxism.

Readers will notice that the empirical sections of *This Fissured Land* focus on the use and abuse of forests. When we first met each other, in the summer of 1982, we had both been independently drawn into a most contentious debate on forest policy in India. On one side, popular movements such as the Chipko and Jharkhand Andolans had sought to loosen state control over forest areas, and move towards a more inclusively participatory system of resource management. On the other, the state had sought to further consolidate and strengthen the legal and political control it already exercised over that 22 per cent of India's land area legally designated as 'forest'. In 1982 the Government of India proposed a draft act, modelled on but going far beyond the notorious colonial forest act of 1878. This proposal was immediately met by protests from tribal leaders and human rights groups. The debate was also fuelled by the growing evidence from sattelite imagery and ground surveys of massive deforestation and habitat loss.

*This Fissured Land* provided an opportunity for us to pool our researches on Indian forestry. We focused on the colonial period of Indian history, when the state first started intervening, systematically and on a wide scale, in the management of forests. The sociologist's interest was in the responses to state forest management of peasants, tribals, pastoralists, and artisans affected by it. The ecologist's interest was in the formal techniques of forestry, its choice of species and harvesting techniques, its grand but ultimately failed attempt to replicate in the tropics certain methods of sustained-yield silviculture that were originally meant for and developed in the temperate zone. The citizens' interest was in why and how both institutional and (pseudo) scientific underpinnings of colonial forestry were so wholeheartedly adopted by the governments of independent India.

These distinct but complementary perspectives were brought together in our analysis of Indian forest history. Alas, this story could be read as one of doom and disaster, of the deprivation of rural communities and of the impoverishment of our natural endòwment. It is possibly intrinsic to environmental and ecological history for its practitioners to be seen as the bearers of bad tidings. Some readers and reviewers, especially in India, complained that we offered a counsel of despair. Looking back over what we had written, we saw that, though unintended, this was certainly one possible reading of *This Fissured Land*.

*Ecology and Equity* was written in part to answer the criticism that we were, in a manner of speaking, postmen carrying the dreaded telegram to the bereaved family. In part it was a response to what we saw as the negativism of the environmental movement in India. The book is in two sections; a history of resource use and abuse in post-independent India is followed by a programmatic agenda for social and environmental renewal. Here too we were bold enough to offer a theoretical framework of our own. This was based on the idea of a 'resource catchment', the area from which individuals and communities garner the natural resources they require or desire. We posit a basic opposition between 'ecosystem people', who depend almost wholly on natural resources in their immediate vicinity, and 'omnivores', who have the political and economic power to avail themselves of the resources of the entire country, and indeed the globe.

One colleague, Purnendu S. Kavoori, pointed out to us that while the idea of 'modes of resource use' is a theory of ecological *production*, the idea of 'resource catchment' is a theory of ecological *consumption*. However, both seek essentially to understand the dialectic of prudence and profligacy in human history. *Ecology and Equity* thus explores the history of Indian economic development in terms of the contest over resources between socio-ecological classes. We show how 'omnivores' have successfully monopolized the benefits of development, with an 'iron triangle' among them making sure that state incentives, subsidies and technological interventions are designed principally to further their interests. We argue that biases in the development process have allowed these omnivores to successfully pass off costs such as resource depletion, habitat fragmentation, and species loss, to the ecosystem people. This process has created a third class of people, whom we have called 'ecological refugees'.

There are periods in human history when concern for the careful husbanding of resources acquires social visibility and political influence. At other periods the imperatives of intensive and destructive use of nature are paramount, as for example over the first few decades of Indian freedom, when development was promoted at all costs. But worldviews change with every generation, perhaps every twenty-five years or so. A generation passed, and the consequences of the course the nation had taken in 1947 became more evident. The worldview began to shift, and ecological concerns came to the fore.

So it is no accident that India's environmental movement began in the early seventies. This movement, we argue in *Ecology and Equity*, is largely a response of disaffected ecosystem people to the egregious excesses of omnivores. Of course the movement is not homogeneous or unified, and we identify the dominant ideological trends within it—Gandhian, Marxist, Appropriate Technologist, Scientific Conservationist, and Wilderness Lover. Each of these trends seems to us valuable but one-sided, offering acute insights into the environmental crisis but suggesting inadequate or unworkable means to overcome it. Part II of *Ecology and Equity* offers an original synthesis of these different ideologies, a charter of sustainable development which seeks to foster ecological prudence by strengthening the political power of ecosystem people and ecological refugees.

When it began in the 1970s, the Indian environmental debate was dominated by forest-related conflicts. In the 1980s and 1990s, however, movements such as the Narmada Bachao Andolan and the Kerala fisherfolk struggle brought the question of appropriate uses of water and fish to centrestage. Likewise, the opposition to bauxite mining in Orissa and to limestone mining in Garhwal and Kumaun highlighted iniquities of access to mineral resources.

The forestry sector looms large in *Ecology and Equity* but it does not dominate this book—as it did our previous one. We provide a mapping of resource conflicts in contemporary India: conflicts over water, fish, forests, minerals, and so on. Moving beyond protest, the book explores the activities of groups and communities that have been engaged in ecological restoration. Some of these 'invisible environmentalists' (invisible to the media, that is) work on afforestation, others on water conservation, soil replenishment, and energy management. These groups have provided, albeit in a modest way, a prudent counterpart to the profligate trends that have dominated India's experience with development.

Both *This Fissured Land* and *Ecology and Equity* are ecumenical in their choice of research strategies and theoretical approaches. We have used archival material, field work, survey data, ecological experiments—whatever seemed appropriate to the context. We have taken our cues from the literature in sociology, anthropology, ecology, political science, history, and energetics. The intention always was to further our understanding of resource use and abuse in India. In the pursuit of this over-riding goal we were not going to be constrained by the limitations of particular methodologies and the sectarianism of scholarly disciplines.

The response to our work, both within the academic community and outside, has been a source of encouragement and at times consolation. There have been criticisms, occasionally sharp or sharply put. One of these was aimed at some sections of *This Fissured Land*: we were charged with promoting the caste system as an 'eco-friendly' alternative to present patterns of resource use. We were even accused of a Brahmanical bias. This, we submit, is a criticism that owes everything to political correctness and nothing to reason or accuracy. In prefacing our account of caste as a system of resource use, we state categorically that caste society 'was a sharply stratified society, with the terms of exchange between different caste groups weighted strongly in favour of the higher status castes.' Perhaps we should have italicized these words, or have had them set in bold type. Not that that would have helped: in India's politicized climate today, wilful misinterpretations of scholarly work are nearly inevitable.

The analysis of caste in *This Fissured Land* showed that at the lower, so to say productive, levels of the system, there had been elaborated a method of resource diversification, based on endogamy and territoriality, that had fostered ecological prudence. The great historian E. P. Thompson, in his review of our book, thought this the most innovative section, for complementing, as he said, political and ideological explanations of the origins and persistence of caste. Despite his robust activist and left-wing credentials, Thompson was a scholar notably unburdened by political correctness. He responded to our analysis as a historian rather than propagandist. For we had offered it as an interpretation of the past, not as a prescription for the present. Critics who somewhat absurdly took the latter view need only have awaited *Ecology and Equity*. That book was written as an extended brief on behalf of India's ecosystem people, who, as we all know, come from the lower castes. All its prescriptive sections aimed at reducing the power of the omnivores, who are largely though not wholly of upper-caste origin. The authors of *Ecology and Equity* were both born into Brahman households, but this did not impede them from dedicating the book to two Muslims, two Christians, and a Bahujan. This decision was dictated neither by political correctness nor guilty conscience, but by our desire to honour the purely secular contributions to ecology and equity by the four men and one woman to whose memory the book was offered.

Another criticism, aimed this time at our theoretical framework, was that it was too 'simplistic'. This came from Marxists whose own view of capitalist development generously allowed for all of two economic classes—the bourgeoisie and the proletariat. Our interpretation of Indian development was structured around relations between three classes: omnivores, ecosystem people, and ecological refugees. Naturally there are Indians who do not fit into these classes. And naturally these categories admit of different shades. There are upper omnivores, such as Forest and Fisheries Ministers and owners of paper mills and trawlers; medium omnivores, such as gazetted government officers and farmers of large holdings of irrigated land; and lower omnivores, such as agents of pharmaceutical companies organizing the collection of medicinal plants.

We could have divided the population of India into three hundred classes, instead of three. That extension of sane taxonomy into the domain of Beckettian absurdity would have put an end to meaningful analysis. Our three-fold classification must therefore suffice. It seems to us to possess a sufficient virtue in pinpointing key elements of economic development, social conflict, and ecological change in modern India. Its impetus was not so much *inter*disciplinary as *trans*disciplinary. Existing theories and approaches of the social sciences tell us little or nothing of human interactions with nature. Existing theories of ecological change seriously underplay the cognitive distinctiveness of humans. We therefore needed to forge a new framework for ourselves. Now, among the most 'progressive' sections of the Indian scholarly community, it is perfectly acceptable if one adapts the theories of Karl Marx to the Indian context, or those of thinkers such as Gramsci and Foucault whose canonical status is also well established. Our sin was in suggesting that the widely used concepts of 'class' and 'interest group' were inadequate when explaining socio-ecological change and conflict. Our guilt was compounded when we dared to invent new terms of our own (such as 'modes of resource use' and 'omnivores').

Mercifully, not all scholars are restricted within political dogma and academic convention. Where older Marxists continue their reservations over a subject they see as a 'bourgeois fad', scholars of the younger generation have responded enthusiastically to the new challenges of environmental history. We ourselves are General Editors of a series of monographs published by Oxford University Press under the broad head: 'Studies in Social Ecology and Environmental History'. Begun in 1990, this series has so far yielded twelve books, with another half-dozen on their way. The books comprising this series range from historical analyses of pastoralism in Himachal Pradesh, to an account of tribal participation in the Narmada movement, to regionally-focused studies of the history of Indian forestry. In keeping with the temper of the times, several works that foreground water conflicts are to be published. Outside this series, the interest in environmental issues has—and not a day too soon—begun to penetrate the citadels of historical research. Leading journals in the field, such as the *Indian Economic and Social History Review* and *Studies in History*, have started publishing an increasing number of papers on the use and abuse of water, forests, and wildlife.

In the early eighties, when we began collaborative research on environmental history and social ecology, we were provoked simultaneously by scholarly inaction and citizens' action. We were deeply inspired by our own encounters with Indians such as Dr Shivram Karanth, a man with ten professions and a hundred passions, a novelist and social reformer who was also a pioneer of environmentalism in Karnataka; and equally by Chandi Prasad Bhatt, founder of the Chipko Andolan, one of the greatest living Gandhians, a social worker who had made notable contributions both as an agitator of peasant protest and as a motivator of rural reconstruction. Movements such as Chipko had punctured the claim that a developing country like India could not afford to be environmentally prudent. They had shown that, in fact, the converse was true—that environmental degradation had greater social costs in a poor country and generated more intensive social conflicts as well. These signals coming in from civil society were not picked up by the academic community. It was in order to redress this neglect that we initially embarked on the research that resulted in the publication of these two books.

Now, in 1999, we find a curious reversal. The scholarly community, once superciliously silent, is now active, hyper-active even, making up for decades of indifference. The rise of environmental history has already been remarked upon, but beyond this sociologists, economists, and political scientists have begun to take an untypical but not unwelcome interest in questions of natural resource use and abuse.[1] The environmental movement, on the other hand, is going through difficult days. The onset of economic liberalization has presented it with a fresh and formidable challenge. The exhaustion of the old and the cynicism of the young have meant that replenishment in the shape of new activists lacks the desired pace. Finally, and most consequen-

---

[1] For more details see Ramachandra Guha, 'Trends in Social-Ecological Research in India: A Status Report'. *Economic and Political Weekly*, 15 January 1997, and works cited therein.

tially, the movement has not shown itself as especially innovative in the realm of ideas.

One area of reform that has by and large escaped the attention of the environmental movement is the question of *efficiency*. We produce less grain per tonne of nitrogen added to croplands and less national product per unit of energy consumed than most other countries of the world. In particular, we allow incredible inefficiencies in the transmission of power while turning a blind eye to large scale thefts of it. The contrast with other Asian countries, such as Korea and Japan, is striking. Technological innovation has allowed those countries to greatly increase fuel efficiencies and greatly reduce polluting emissions. Indian environmentalists, however, are beset with an almost atavistic hatred of modern science; this precludes or at least inhibits the search for prudent alternatives to prevailing technologies.

Again, the environmental movement in India focuses too much on the *products*, and not enough on the *processes*, of development. Dams, even big dams, thermal power plants, and highways are not good or bad in themselves. They are neither the temples nor the hell-holes of India. The problem lies, rather, in the centralized and closed circle of decision-making. The broader masses of people in India have no role to play in deciding whether a dam should be built, how its catchments should be treated, how the populations displaced by it should be properly compensated, how its waters should be distributed and used. As things stand, the dam becomes an evil chiefly because the process of planning and management is distorted by a corrupt, wasteful and unjust machinery of decision-makers.

Indian democracy, as our books show and everyone generally recognizes, is grossly flawed. However, nobody is completely powerless within this system: there are always avenues for the expression of popular protest, even for the most disadvantaged. *Ecology and Equity* notes moves towards the decentralization of political authority, moves that have gathered force since that book was published. Environmentalists must participate with vigour in this deepening of Indian democracy. They must lobby for freedom of information, insist on public hearings, work to ensure accountability at all levels of government. They must urge upon different state governments the necessity for greater democratic control of natural resources through the formation of village forest protection committees or water users' associations.

xiv • *The Use and Abuse of Nature*

Efficiency and decentralization are vital if the environmental movement, and the country as a whole, are to reap a full and just harvest from ongoing attempts to open out the Indian economy. Contrary to the kneejerk response of groups such as the Swadeshi Jagran Manch, there is no turning back from globalization. We must come to terms with it and bend it as best we can to our own interests. Our current surrender to global forces is a consequence only of our own creation of a high-cost, low-quality economy and our establishment and maintenance of a bloated, parasitic bureaucracy. These have kept our rate of economic growth low, and have meant that industry is not competitive on the global scene. In turn, this has meant a negative balance of payments, which has forced us to borrow from the IMF and accept its terms. These terms include the acceptance of intellectual property rights, including rights over biological products and varieties of cultivated plants.

There is little to be gained by decrying the injustice of it all; we must instead realistically confront these challenges. In doing so we can take advantage of the provisions of the Convention on Biological Diversity that confer on countries sovereign rights over their genetic resources and ask them to share with local communities the profits from commercial applications of their knowledge of biological resources. To this end we must properly catalogue and organize the rich knowledge possessed by India's barefoot ecologists—tribals, herders, peasants and fisherfolk. One such effort, currently under way in parts of Kerala and Karnataka, is the creation of 'People's Biodiversity Registers' through a decentralized effort involving teachers and students who work with local communities. This might be broadened into a participatory programme which monitors the status of environmental resources, a proper prelude to a decentralized, environment-friendly, resource-efficient process of development.

These criticisms of the environmental movement come from two scholars who count themselves among its supporters. The voices of Chipko and Narmada, not the research of academics or the foresight of business, have emphatically placed the environmental question at the forefront of public discourse. We have keenly studied the evolution of India's environmental movement over the past twenty-five years, and look forward to studying its further evolution over the next twenty-five. But social movements, like governments, are not immaculate. If we are obliged by our vocation to criticize the second, we cannot shirk from criticizing the first.

The separate prefaces to the two books that are here published together record the important debts incurred when we first wrote them. Two other debts, not previously mentioned, must now be acknowledged. One is to the Forest Research Institute in Dehra Dun, vanguard for over a century of the 'scientific' forestry that comes in for bitter and sustained criticism in these pages. Ironically, it was at the FRI that the present authors first met, in 1982, when Guha was beginning research on the history of Indian forestry and Gadgil had come to present a seminar on the social behaviour of elephants. It was in the old red-brick rest-house of the FRI that we had the first of our countless and sometimes contentious conversations. Other conversations were conducted in Uttara Kannada, Bellagio, Calcutta, Dharwad, Delhi and Bangalore, in forest, farm and garden, in buses, trains, jeeps, boats, aeroplanes and humdrum academic offices. It was chiefly at the urgings of Oxford University Press that we first thought of making public the contents of those talks. The OUP has never wavered in its consistent support to both of us personally, and also, what matters more, to the publication of serious work on prudence and profligacy in Indian history.

Bangalore, 14 April 1999           Madhav Gadgil
(Dr B. R. Ambedkar's birth anniversary)     Ramachandra Guha

# This Fissured Land
## An Ecological History of India

To the memory of

*Verrier Elwin*
*Irawati Karve*
*D.D. Kosambi*
*Radhakamal Mukerjee*

# Contents

## TABLES

# Acknowledgements

A number of colleagues have contributed ideas and information for this book. We would like especially to thank Bill Burch, Anjan Ghosh, S.R.D. Guha, Kailash Malhotra, Juan Martinez Alier, Narendra Prasad, N.H. Ravindranath, C.V. Subbarao, M.D. Subash Chandran and Romila Thapar. Close readings of a draft by Mike Bell and Jim Scott were of great help in revising the manuscript for publication. Geetha Gadagkar typed numerous drafts with care and precision. Nor can we neglect to mention the support of the Indian Institute of Science, an institution which has successfully withstood the moral and intellectual degeneration of our times. This book was largely written at the Institute's Centre for Ecological Sciences, itself founded with the generous assistance of the Department of Environment, Government of India. Finally, we should note that an earlier version of chapter 5 was published in *Past and Present*, number 123, May 1989.

The dedication indicates our debt to an earlier generation of scholars. Verrier Elwin wrote a series of remarkable ethnographies of tribal communities on the margins of survival. His studies of the cultural meanings of the forest, and of the impact of colonial forestry on tribal life, are of abiding interest.

Through her imaginative work on the pastoral communities of Maharashtra, Irawati Karve laid the foundations of ecological anthropology in India. Her interpretation of the Mahabharata stands out for its insights into the ecological basis of social conflict in ancient India. D.D. Kosambi, mathematician and Sanskrit scholar, broke new ground through his materialist interpretations of Indian history. He also highlighted the importance of different modes of resource use in the mosaic of Indian culture. Radhakamal Mukerjee was an early harbinger of the coming together of ecology and the social sciences. His theoretical studies of 'Social Ecology' were fleshed out through detailed investigations of socio-ecological change in the Indo-Gangetic Plain. These four scholars may justly be regarded as the pioneers of human ecological studies in India; it is our privilege to acknowledge their influence on this work.

# *Prudence and Profligacy*

Contemporary India is a fantastic mosaic of fishing boats and trawlers, of cowherds and milk-processing plants, of paddy fields and rubber estates, of village blacksmiths and steel mills, of handlooms and nuclear reactors. Its 850 million people live in tiny fishing hamlets and camps of nomadic entertainers; in long-settled villages and slowly-decaying old towns; in suburban ghettoes and burgeoning metropolitan cities. Some build their shelters with bamboo and mud, others with cement and steel. Some cook with small twigs on a three-stone hearth, or with coconut husks on a mudstove, some with electricity and gas in modern kitchens.

Naturally, the demands of this remarkable mosaic on the country's resources are exceedingly varied. Thus, bamboo is coveted by rural artisans for weaving baskets and to fashion seed drills, by graziers to feed their cattle, and by industry to convert into paper and polyfibre. Landed peasants want the less fertile lands around villages to graze their cattle, the landless to scratch it to produce some grain, the forest departments to produce marketable timber. Peasants want the mountain valleys to grow paddy, power corporations to construct hydroelectric dams.

The resources so varyingly in demand are regulated in equally varying ways. There are sacred ponds which are not fished at all, and beach seiners recognize customary territorial rights of different fishing villages—even though none of this has any formal legal status. Many tribes in north-eastern India own land communally and put it to shifting cultivation, an ownership pattern recognized by law. Village common lands, used as grazing grounds, and wood-lots were once controlled by village communities; they are now government land under the control of revenue or forest departments. There are tenants who cultivate lands that belong to absentee landowners, though much of the land under cultivation is now owned by the tiller. A good bit of the land cultivated by tribals is, however, legally state-owned; this is part of a vast government estate which covers over a quarter of the country's land surface, mostly designated 'Reserved Forest'. Business corporations control large tracts of land, as tea or rubber estates, and there are moves to permit even larger holdings for forest-based industries.

A whole range of these resources, regulated and utilized in many different ways, is under great stress. There are very few deer and antelope left to hunt for hunter-gatherers such as the Phasepardhis of Maharashtra. A majority of the shepherds in peninsular India have given up keeping sheep for want of pastures to graze them. The shifting cultivators of north-eastern India have drastically shortened their fallow periods from a traditional fifteen to a current five years. All over, peasants have been forced to burn dung in their hearths for want of fuelwood, while there is insufficient manure in fields. Ground-water levels are rapidly going down as commercial farmers sink deeper and deeper bore-wells. There are long shutdowns in industry for want of power and raw material, and every urban centre is groaning under acute shortages of housing, fuel, water, power and transport.

In this ancient land, which harbours what is undoubtedly the most heterogeneous of cultures on earth, these resource shortages have given rise to an amazing range of adjustments, collusions and conflicts. However, the country is living on borrowed time. It is eating, at an accelerating rate, into the capital stock of its renewable resources of soil, water, plant and animal life. Does this mean we are headed for disaster? Or is this a temporary phase before we get back on the path of sustainable resource use? Or perhaps before further technological advances open up an undreamt range of resources? It is obviously true that not *all* resources are being decimated everywhere. In the village of Gopeshwar in the Garhwal Himalaya, for example, there is a nice grove of oak and other broad-leaved species. All families in the village carefully observe traditional regulations on the quantum of plant biomass removed from this grove. But this quantum is quite inadequate for their needs, and for the balance they turn to other hill slopes where exploitation is unregulated and denudation at an advanced stage. This was apparently not the case some decades ago, when the grove was larger and fulfilled the demands of a smaller population in a sustainable fashion.

Human history is, as a whole, precisely such a patchwork of prudence and profligacy, of sustainable and exhaustive resource use. In contemporary India the instances of profligacy clearly outnumber (and outweigh) those of prudence, although this book will argue that such was not always the case. In our own times, acute resource shortages have given rise to a host of social conflicts, and these have significant consequences for what is now happening to the life of India's people and to the health of its land.

The implications of the uses and misuses of India's biological resources are only dimly perceived by the rich and the powerful. But we, in common with the majority of Indians who face the burden of this misuse in their daily lives, believe them to be of tremendous significance: hence this book. We have attempted an ecological history of changing human interactions with living resources, using the Indian case illustratively to explore four themes. The significance of these, in our view, is scarcely restricted to one country, nor even to one continent.

First we ask: under what conditions may we expect human beings to exercise prudence in their use of natural resources?

Second, we investigate both the 'hardware' and 'software' of natural resource use in different historical periods. 'Hardware' refers to the forces and relations of production—namely the technological infrastructure and the systems of property—open access, family, communal, corporate, or state—governing resource use. 'Software' refers to the belief systems (for example, religion, tradition, or science) which legitimize and validate human interactions with nature.

Third, this book analyses the forms of social conflict between different groups of resource users. Here we are especially interested in changes in the intensity of social conflict over time, and in its escalation as one mode of resource use gains ascendancy over another.

Finally, we are interested in the impact of changing patterns of resource use, as well as of social conflict, on the status of living resources. Beginning with the conditions that favour prudence, we come full circle with our analysis of the conditions under which profligacy predominates.

These four themes—i.e. prudence, profligacy, strategies of resource use, and the conflicts to which they give rise—provide the unifying framework for the three parts of this book. The theoretical framework of the study is outlined in Part I. Here we analyse the different forms of restraint on resource use reported from human societies. We locate these practices of resource use in specific cultural and historical contexts. This link between ecology and history is accomplished by means of our typology, which we designate 'modes of resource use'. We put this forward as a supplement to the 'modes of production' scheme which social scientists have traditionally used as a framework for historical periodization.

Part II presents a new interpretation of how the cultural and ecological mosaic of Indian society came together. Given the fragmentary nature of historical evidence on this, our reconstruction may be viewed as *one* plausible scenario of the ebb and flow of different social systems, as well as of different belief systems of resource use, which existed in the Indian subcontinent before British colonialism. Based on our fieldwork in peninsular India, we advance an ecological interpretation of the caste system, thereby complementing standard economic and ideological interpretations of the persistence of caste as the organizing principle of Indian society.

Part III draws on more abundant source material. It presents a socio-ecological analysis of the new modes of resource use which were introduced by the British, and which have continued to operate, with modifications, after independence in 1947. We argue that British colonial rule marks a crucial watershed in the ecological history of India. The country's encounter with a technologically advanced and dynamic culture gave rise to profound dislocations at various levels of Indian society. While sharply critical of colonialism, Indian historians have, in the main, been indifferent to the ecological consequences of British intervention. Our analysis, in contrast, highlights the essential interdependence of ecological and social changes that came in the wake of colonial rule. Furthermore, our study shows that the socio-ecological consequences of European colonialism in India, while significant in themselves, were quite different from those in the New World.

The British did not merely change Indian history, they also changed the writing of it—by providing historians with an unprecedented level of documentation. Consequently, Part III draws upon a far wider range of sources than the earlier two parts; even so, its focus is only on the most important aspect of the ecological encounter between Britain and India—changes in the ownership and management systems of India's forests. For over a century, India has had the benefits—such as these are—of an extensive system of state forestry. With over one-fifth of the country's land area in its charge, the forest department has been, since the late nineteenth century, by far our biggest landlord. Since students of modern Indian history have shown a surprising unawareness of this fact (cf. Sarkar 1983;

Kumar 1983), and also since every member of India's agrarian society had a direct economic relation to forest produce, a historical assessment of colonial forest management is long overdue. The edifice of colonial forestry has been taken over by the government of independent India. Therefore our analysis will shed light on the links between economic development and ecological change in one important Third World country. Finally, the Indian case, as presented in our history, may illuminate several current debates on tropical deforestation.

In this manner, Part I presents a theory of ecological history; Part II provides a fresh interpretive history of pre-modern India; and Part III contains a socio-ecological history of the forest in modern India.

Although our focus is specifically on interactions between humans and living resources, the business of covering the broad sweep of Indian history is daunting. So we have no illusions about the 'definitiveness' of this enterprise. Our generalizations and data are bound to be modified, altered, even overthrown, in the course of time, as is normal with academic ventures in relatively uncharted terrain. Yet it is undeniable that to date historians of India have been almost completely unaware of the ecological dimensions of social life. Their focus has been more or less exclusively on relations around land and within the workplace, never on the ecological fabric within which both field and factory are embedded, and which these in turn transform. Hence the questions which scaffold this book—the hardware and software of natural resource use, social conflicts around nature, and the cumulative impact on ecological health—are, we believe, asked and partially answered for perhaps the first time in relation to ancient, medieval and modern India.

Ultimately, the ecological history of India must be constructed around detailed regional studies, sharply bounded in time and space. Yet there are periods in the development of scholarship when a new interpretation cannot endlessly await the steady accumulation of certified data. Indeed, to plot the pieces of a jigsaw puzzle one must begin determining the shape and structure of the puzzle. In the circumstances, while this book provides new data and new interpretations of old data, it provides above all a new and alternative framework for understanding Indian society and history.

Here we draw inspiration from Marc Bloch, who prefaced his great study of French agriculture—itself a model of ecological analysis—with these words:

> There are moments in the development of a subject when a synthesis, however premature it may appear, can contribute more than a host of analytical studies; in other words, there are times when for once the formulations of problems is more urgent than their solution . . . I could liken myself to an explorer making a rapid survey of the horizon before plunging into thickets from which the wider view is no longer possible. The gaps in my account are naturally enormous. I have done my best not to conceal any deficiencies, whether in the state of our knowledge in general or in my own documentation, which is based partly on first-hand research but to a much greater extent on soundings taken at random . . . When the time comes for my own work to be superseded by studies of deeper penetration, I shall feel well rewarded if confrontation with my false conjectures has made history learn the truth about herself.

# PART ONE

# *A Theory of Ecological History*

CHAPTER ONE

# Habitats in Human History

## Modes of Production and Modes of Resource Use

Many social scientists have found the Marxist concept of modes of production useful when classifying societies according to their technologies and relations of production. Undoubtedly the original scheme, of primitive communism–slavery–feudalism–capitalism, derived largely from the European experience, has been modified by an increasing sophistication within the writing of the histories of non-western cultures. Notwithstanding the problems in applying European models of feudalism to societies such as India and China, as well as the continuing debates around the so-called Asiatic mode of production, the framework itself remains very much in favour. It is strongest while delineating the features of the capitalist mode of production. With the emergence of capitalism as a 'world system', it is currently enjoying a revival in those far-flung corners of the globe where the clash between pre-capitalist relations of production and the capitalist ethos is only now gathering momentum.

Among several important criticisms made of the mode of production scheme, we single out three. The first, made by

Marxists themselves, relates to the relative lack of emphasis in this scheme on political structures and struggles. In his widely-noticed interventions in the 'transition' debate, Robert Brenner argued that the form and intensity of political conflict, rather than changes in production technology or expansion in trade, better explain the nature of the transition from feudalism to capitalism in different parts of Europe (Brenner 1976 and 1978). Other scholars have suggested a supplementary concept, 'mode of power', to more accurately capture the structure of power and domination in different societies (Chatterjee 1983). Second, there are criticisms which, while accepting the relevance of the scheme to European history, express reservations about its application elsewhere. The European model of feudalism does not, for example, fit the Indian experience, and the Asiatic mode is scarcely of any use either, since the state played by no means as important a part in providing public works and irrigation facilities for agriculture as this notion suggests. Finally, there are the criticisms of non-Marxists (and non-economists). These amount to the view that, whatever the merits of the mode of production concept while explaining differences in economic structure, this concept is of little use when interpreting differences in the religious, cultural and ideological attributes of different societies.

While all these criticisms are compelling, from the perspective of this book they do not go far enough. An ecological approach to such questions suggests that the mode of production concept is not adequately materialistic in the first place. This may seem an ironic accusation against a doctrine as supposedly materialist as Marxism, yet a little reflection bears it out. Marxist analyses usually begin with the economic 'infrastructure'—the so-called relations of production and productive forces—without investigating the ecological context, i.e. the soil, water, animal, mineral and vegetative bases of society in which the infrastructure is embedded. As exemplified by recent political and economic histories of modern India, both Marxist and non-Marxist, the most major lacuna in existing scholarship is an inadequate appreciation of the *ecological* infrastructure of human society. We therefore propose to complement the concept of modes of production with the concept of modes of resource use.

While focusing on spheres of production, such as the field and the factory, most analyses of modes of production have ignored the natural contexts in which the field and factory are embedded—the contexts to which they respond, and which they in turn transform. The concept of modes of resource use extends the realm of production to include flora, fauna, water and minerals. It asks very similar questions. With respect to relations of production, for example, it investigates the forms of property, management and control, and of allocation and distribution, which govern the utilization of natural resources in different societies and historical periods. And with respect to productive forces, it analyses the varying technologies of resource exploitation, conversion and transportation that characterize different social orders.

While complementing the mode of production framework, the mode of resource use scheme incorporates two additional dimensions. First, it examines whether one can identify characteristic ideologies that govern different modes. More importantly, it identifies the ecological impact of various modes, and assesses the consequences of these different modes for the pattern, distribution and availability of natural resources.

Three caveats are in order here. First, the mode of resource use concept, like the mode of production concept, is at bottom an 'ideal type'. Hence, the identification of distinct modes does not preclude the existence of more than one mode in any given social (or, more accurately, socio-ecological) formation. Still, it is usually possible to identify the *dominant* mode within a socio-ecological formation. Second, our treatment is largely restricted to human uses of *living* resources—i.e., flora and fauna—both husbanded and in their natural state. This framework can of course be extended to incorporate other natural resources, such as water and minerals. Finally, one important respect in which our scheme differs from the Marxian mode of production scheme is that the industrial mode of resource use, as defined by us, includes both capitalist and socialist societies. While there are significant differences between socialist and capitalist paths of development—for example with respect to property and the role of the market—from an ecological point of view the similarities in these two developmental paths are more significant than the differences. For instance, there are structural similarities in the scale and direction of natural resource flows, the technologies of resource exploitation, the patterns of energy use, the ideologies of human–nature interaction, the specific resource-management practices, and, ultimately, the cumulative impact of all these on the living environment in capitalist and socialist societies. Consequently, it makes sense to treat industrial socialism and industrial capitalism as being, ecologically speaking, simply two variants of one industrial mode of resource use.

## Four Historical Modes

From the long sweep of human history we can distil four distinct modes of resource use: gathering (including shifting cultivation); nomadic pastoralism; settled cultivation; industry. Later we shall examine the distinctive characteristics of each mode across different axes. These shall include:

• aspects of *technology*, such as sources of energy, materials used, and the knowledge base relating to resource use

• aspects of *economy*, such as the spatial scale of resource flows and the modes of resource acquisition

• aspects of *social organization*, such as the size of social group, the division of labour, and mechanisms of control over access to resources

• aspects of *ideology*, including broad perceptions of the man–nature relationship, as well as specific practices promoting resource conservation or destruction

• the nature of the *ecological impact* itself

After presenting the ideal-typical characteristics of different modes of resource use, we outline the characteristic forms of social conflict *between* as well as *within* different modes. The chapter concludes with a recapitulation, while the appendix contains some reflections on the forms of reproductive behaviour typical of different modes.

## Gathering

*Technology*: The largest period of human history has been spent within the gathering mode of resource use, during which the hunting of wild animals and the gathering of vegetable matter were the mainstay of subsistence. Gathering continues to be significant during the phase of shifting cultivation as well (cf. Elwin 1939; von Fürer Haimendorf 1943a), and we may also include societies that practise shifting agriculture under this rubric. In the gathering mode, societies depend almost exclusively on human muscle power and wood fuel as sources of energy, and on naturally available plants, animals and stones to fulfil their material requirements. Their knowledge base is fairly limited, and nature is viewed as almost totally capricious, as something not subject to human control. The ability to store food and other materials is also very limited, as is the ability to transport materials over long distances.

The economy within this mode of resource use is based on resources which are acquired within a small area of, at best, a few hundred square kilometres. Only a very small range of resources, such as shells, peacock feathers and flint tools, may be transported over larger distances. The diversity of plant and animal matter consumed from the social group's immediate surroundings is high; yet, given this restrictive spatial scale as well as the limited ability to process resources, the actual variety of resources used *in toto* is small, while the quantities consumed are restricted to subsistence needs. Societies which pursue the gathering mode of resource use are highly susceptible to variations within resource availability through space and time. They respond to such variations by fine-tuned adaptations to local conditions. In the harsher and more variable environments the people comprising these societies subsist as nomadic bands; in the more productive and stable environments they exist as tribal groups confined to relatively small territories. Such territorial restriction also continues with a swich-over to shifting cultivation.

*Social organization*: The sizes of social groups among hunter gatherer–shifting cultivators are small: kin groups of the order of a few hundred perhaps, largely in face-to-face communication with each other. There are hardly any transactions outside such social groups; the relationship with aliens is largely one of conflict, often over territorial control (Rappaport 1984). The division of labour within these groups is minimal; what exists is primarily based on age and sex, and to some extent upon knowledge and leadership abilities. The women will principally be found involved in gathering plant foods and small animals, while the men will be found hunting the larger animals. As regards the gender equation, men play a greater role in organizing infoimation and taking decisions relating to resource use on behalf of the group as a whole.

In the gathering mode there is little variation among members of a group in terms of access to resources, and notions of private property are extremely poorly developed. Within a group, here, no individual is in a position to dominate and coercc others to any significant degree. While within-group differences thus tend to be low, the differences between groups may or may not be equally low. Till 10,000 years ago, before plants and animals began to be husbanded by humans, between-group differences were, without doubt, at a low level. The human populations of any region might then have been divided amongst a large number of endogamous groups competing with each other for control over land and water. This competition would have been intense in productive and stable environments, such as tropical rain forests, where inter-tribal warfare over territorial rights was probably a routine occurrence. On the other hand, the limits of territories are likely to have been much fuzzier in harsher, highly variable environments, such as tropical deserts or the coniferous forests of higher latitudes. Here, inter-tribal conflicts would have been correspondingly less acute. With the beginnings of cultivation and animal husbandry, and later the industrial revolution, people who pursued the gathering mode came to be increasingly disadvantaged *vis-à-vis* people with access to advanced levels of technology.

*Economy*: We may characterize the economy of gatherers as a *natural* economy, in so far as it draws all its resources directly from nature (Dasmann 1988, Worster 1988). Here, the flows of resources are largely closed on spatial scales of a few hundred, or at best a few thousand, square kilometres, over which each endogamous social group of gatherers might range (Fig. 1.1). While there may be some flows of materials, such as shells or stone tools, over longer distances, these are insignificant (in terms of physical mass) when compared to food and the other resources utilized by the human groups concerned. The material flows would, therefore, not only be closed, they would also be balanced, i.e. with inflows matching outflows on a fairly restricted spatial scale. This balance is apt to be disrupted when gatherers come into contact with people practising more advanced modes of resource use. The latter are then likely to organize net resource outflows from regions inhabited by gatherers, and perhaps also affect such regions through inflows of wastes—of materials they discard.

Gatherer societies, as indeed all human societies, encounter three distinct situations with respect to their resource base. First, their demands on the resource base may be small compared to the overall availability. This may be due to their having colonized a new habitat—as must have happened when some groups first crossed over into North America—or on account of a technological innovation such as the bow and arrow, which opened up a new range of species to prey upon. This is a situation analogous to that of r-strategists in ecological literature, i.e. of populations that increase relatively rapidly for most of the time. The r-strategists include weedy species which rapidly colonize disturbed environments, or micro-organisms like influenza which spread in epidemics. Other societies may be in equilibrium with their resource base, as for instance fishing communities in an isolated coral island. These are analogous to the k-strategists of ecological literature, examples of which include shade-tolerant and timber-tree species in rain forests, or micro-organisms like tubercular bacilli. Finally, societies may be confronted with a shrinking resource base, as may have confronted gatherer societies who were displaced by advancing glaciers, or by technologically more advanced human societies. Ecologists have no specific term for this situation. However, given the fact that some twenty times as many biological species have gone extinct as are alive today, many biological populations must be adapted to dealing with a shrinking resource base.

Under conditions approaching equilibrium with their resource base, gatherers exist in kinship-based, small, viscous groups that are more or less tied to specific localities, depending on the variability of the environment. Such an organization might be broken up when gatherers come in contact with people who work within more advanced modes of resource use, and consequently have to adjust to a shrinking resource base. In the former case, the group's members would tend to be involved in a network of co-operative behaviour, and would be very sensitive to group interests. In the latter case, such co-operative behaviour, as also the perception of group interests, may dissolve.

*Ideology*: To gatherers, with their limited knowledge base, nature follows its own capricious ways, hardly subject to human control. Gatherers typically regard humans as merely part of a community of beings that includes other living creatures, as well as elements of the landscape such as streams and rocks. Especially where gatherers are attached to particular localities, as in productive and stable environments like tropical humid

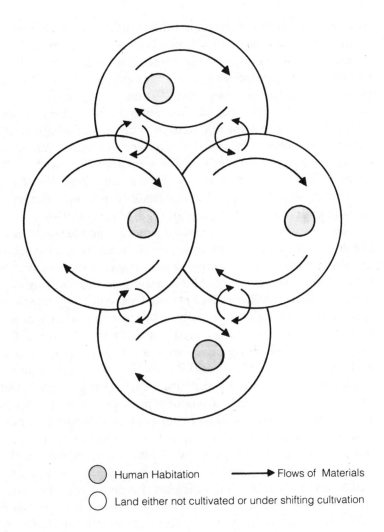

Human Habitation      ⟶ Flows of Materials

Land either not cultivated or under shifting cultivation

**Figure 1.1**

In productive, stable environments, hunter-gatherer cum shifting-cultivators maintain well defined territories. Cycles of materials in such environments are largely closed on the spatial scales of territories with flows of materials across territorial boundaries being much less significant. The thickness of an arrow indicates the intensity of the flow.

forests, they attribute sacred qualities to individual trees, ponds or mountain peaks, or to all members of a plant or animal species, such as *Ficus* trees. They often treat plants, animals or elements of the landscape as kin, or as being in relationships of either mutualism or antagonism. Thus, rivers may be considered as mothers, and totemic animals (like bears or antelopes) as brethren; specific trees may be seen as inhabited by demons who need to be placated. Gatherers therefore enter into a whole range of frequently positive relationships with these other 'beings' of their own locality. By the same token, they have no relationship with plants, animals or landscape elements outside their own locality (Martin 1978; Macleod 1936).

*Restraint—Real and Apparent*: At a more concrete level, these ideologies of nature worship are buttressed by *specific social practices* which orient societies in the gathering mode towards the prudent use of nature. Thus, many gatherer–shifting cultivator societies have a variety of practices which regulate their behaviour towards other members of their community of beings, and which seemingly contribute towards ensuring the long-term sustainability of resource use (Gadgil 1991; Gadgil and Berkes 1991; McNeely and Pitt 1985; Ruddle and Johannes 1985). In the context of ecological debates on prudence and profligacy, it is of interest to examine these practices in order to assess whether they could be better explained in terms of harvesting for short-term gain; and if not, whether they could indeed lead to an enhanced availability of resources to the group as a whole in the long run. These practices, studied by anthropologists and ethnobiologists, involve a variety of restraints on harvesting in terms of quantity, locality, season, and life-history stages. They also involve differential harvests by age, sex, or social class. It is, of course, possible that such apparent restraint may have nothing to do with a long-term conservation of the resource base. A harvester interested in calculating an immediate return may still not use a resource if the net gain obtainable from it is below a certain threshold, which in turn would depend on the net gain obtainable from

alternative resources. We must therefore examine each supposed instance of restraint to assess whether it could involve
such a discontinuation of resource use. This may happen, for
instance, because the cost of harvest increases excessively
(Smith 1983; Borgerhoff Mulder 1988).

The whole range of practices demonstrating the restrained
use of resources by human beings can be classified under ten
broad categories:

• A quantitative restriction on the amount harvested of a given
species, or from a given locality, by harvesters. The imposition
of such quotas implies that the harvest is halted at resource
densities greater than those at which individuals would find
the net gains too low to continue harvesting. As a corollary,
these quotas are likely to enhance total yields over the long-
term, and at the sacrifice of some immediate return. These are
therefore likely to be genuine instances of restraint.

• Harvesting a certain resource may be abandoned when resource densities fall. In parts of New Guinea, for example, the
hunting of Birds of Paradise is temporarily abandoned if their
populations decline (Eaton 1985). This sort of response is expected from harvesters who are attempting to maximize a
short-term net gain, since a fall in the resource density would
progressively increase the costs of harvesting. It is possible,
though not very probable, that harvesting is abandoned well
before this level is reached in the interest of long-term yields.

• Harvesting from a certain habitat patch may be abandoned
if yields from that patch are reduced. Thus, in the Torres Strait,
fishing may be stopped in regions where fish yields are known
to have declined (Nietschmann 1985). This again is a response
which is to be expected from a forager who is attempting to
maximize immediate net returns. It can be related to long-term
resource conservation only if concrete quantitative evidence is
available to indicate that harvesting is abandoned in advance
of the returns reaching a value low enough to justify abandoning harvests.

• Harvesting from a certain species may be abandoned in a

certain season. Illustrative here is the taboo, in many Indian villages, on hunting certain animals between July and October (Gadgil 1985a). Possibly this taboo is a consequence of returns which are too low to justify harvesting for immediate gains in that season. Conversely, if in fact net returns in that season are expected to be relatively high, this is likely to be a conservation measure.

• Harvesting from a certain habitat patch may be abandoned in a certain season. Again, this could be a response to an excessively low level of net gain from that habitat patch in that season. This should be verified by a comparison with net gains in other seasons, and if possible by a quantitative assessment.

• Certain stages of life—by age, sex, size or reproductive status —may immunize an animal or bird from being harvested. For example, in the village of Kokre-Bellur in the state of Karnataka, birds in the breeding within a heronary may be left unmolested, though they may be hunted elsewhere and at other seasons (Gadgil 1985b). If such protected stages appear to be critical to population replenishment, and if they are likely to yield as high or higher net returns than the unprotected stages, it is reasonable to assume that this measure is designed specifically to conserve the resource. On the other hand, if these stages are likely to yield lower net returns in comparison with the unprotected stages, then they might be left unharvested simply in the interest of maximizing the immediate net gain.

• Certain species may never be harvested either because of the relative difficulty of procuring them (risk of injury during the hunt) or because they carry parasites that can affect humans. If these conditions do not operate, conservation can indeed serve the long-term interest of human resource use: if the species thus protected enhances the availability of some other species that are harvested. This is likely for a widely protected species, such as trees belonging to the genus *Ficus*, but much less likely for a wide variety of species protected as totemic by certain tribes (Gadgil 1989).

• Certain habitat patches may either never be harvested, or be subject to very low levels of harvest through strict regulation. It is extremely difficult to arrive at workable prescriptions on quantitative quotas, closed seasons, or protected life-history stages that decisively guard against resource decimation. Providing refugia (sacred groves, sacred ponds, etc.) may then be the most easily perceived and most efficient way of guarding against resource depletion (Gadgil and Vartak 1976).

• Certain methods of resource harvest may be wholly prohibited or strictly regulated. Thus, fishing by poisoning river pools is severely regulated by tradition in many parts of India (Gadgil 1985a). If these methods are likely to provide as high, or higher, net returns than the permitted methods, their regulation may serve the interest of long-term resource conservation.

• Certain age/sex categories or social groups may be banned from employing certain sorts of harvesting methods, or from utilizing certain species or habitat patches. Thus in New Guinea adult males are banned from hunting rodents (Rappaport 1984). This could contribute towards long-term resource conservation by moderating the total amount of harvest. It could also assist in long-term conservation by restricting the access to a limited number of individuals who may more readily come to use the resource in a prudent fashion. It is, of course, quite possible that such restrictions merely benefit segments of the community that are in positions of power, without really serving the interest of long-term conservation.

## Simple Rules of Thumb

Given the complexity of ecological communities, precise prescriptions for the prudent use of living resources are difficult. Detailed quantitative prescriptions seem impossible, given the present state of knowledge (Clark 1985). This is particularly so if the entire prey population is continually subject to being harvested. But certain simple prescriptions to avert a resource collapse seem easy to formulate and should have a significant effect in enabling sustainable use. These prescriptions are five-fold: provide complete protection to certain habitat patches which represent different ecosystems, so that resource populations are always maintained above some threshold; provide complete protection to certain selected species so that community-level interactions are disrupted minimally; protect such life-history stages as appear critical to the maintenance of the resource population; provide complete protection to resource populations at certain times of the year; and organize resource use in such a way that only relatively small groups of people control or have access to a particular resource population (Johannes 1978).

Modern ecological and evolutionary theory suggests that such prescriptions are likely to assist in avoiding an environmental collapse, although they would by no means ensure harvests at maximum sustainable yield levels. In his classic experiments on prey-predator cycles of protozoans, Gause showed that prey extinction could be effectively avoided only by providing the prey a refugium, an area of the experimental arena inaccessible to the predator where the prey could maintain a minimal population and from which other areas could be colonized by it (Gause 1969). Sacred groves, sacred ponds, and stretches of sea coast from which all fishing is prohibited are examples of such refugia. Modern ecological theory also stresses the significance of certain species which serve as keystone resources or mobile links in maintaining the overall functioning of the community (Terborgh 1986). The tree of the genus *Ficus*— to which belong species such as banyan and peepal (widely protected in Asia and Africa)—is one such keystone resource. Contemporary ecological theory also points to the fact that certain stages in a population are of higher 'reproductive value', and therefore more significant in permitting continued population growth. Pregnant does and nesting birds, again often protected by humans, are such stages (Fisher 1958; Slobodkin, 1968). Finally, recent work on the evolution of co-operative behaviour emphasizes that restraint is progressively more likely to evolve if the number of individuals involved in repeated social interactions gets progressively smaller (Joshi 1987 Feldman and Thomas 1986; Berkes and Kence 1987).

It thus appears plausible that, over the course of human history, there have been human groups whose interests were strongly linked to the prudent use of their resource base, and that such groups did indeed evolve conservation practices. Many of the practices described above have in fact been reported from different gatherer societies. These conservation practices were apparently based on several simple rules of thumb that tended to ensure the long-term sustenance of a resource base. These rules of thumb were necessarily approximate and would have been arrived at by trial and error. Practices which seemed to keep the resource base secure may have gradually become stronger; conversely, there could have been a gradual rejection of those practices which appeared to destroy the resource base (Joshi and Gadgil 1991). Conservation practices observed by various other social groups may also be emulated if they appear successful in resource conservation. Such a process is likely to lead to the persistence of a whole range of practices, some beneficial from the point of view of resource conservation, but also others that are neutral; and perhaps some that might once have been beneficial or neutral but later prove harmful on account of changed circumstances.

*Diversity of resources*: In conjunction with their various practices of restrained use, gatherer societies are remarkable for the great diversity of biological resources they utilize. Studies on the American Indians of Amazonia have shown that they utilize several hundred different species of plants and animals for food, and as sources of structural materials and drugs. They have distinct names for as many as 500 to 800 biological species (Berlin 1973). For another thing, different tribal groups may be familiar with and utilize a different set of species. As early as 30,000 years ago, for example, two Neanderthal groups of Dordogne in France had apparently specialized in different prey species, one group concentrating on horses and the other on reindeer (Leakey 1981). We may therefore conclude that primitive gatherer-shifting-cultivator societies valued a very wide range of biological diversity, and evolved cultural practices which promoted the persistence of this diversity over long intervals.

*Deliberate destruction*: While gatherer societies typically have a variety of practices that could help conserve the resource base of their own localities, they may also deliberately destroy the resource bases of aliens. Thus, when New Guinea highlanders defeat and drive away a neighbouring group from its territory, the conquerors do not immediately occupy this territory. They cut valued fruit-yielding trees from the conquered group's territory, thereby rendering it far less desirable for recoloniza-tion by the conquered. The actual territory may be physically occupied only later, if it is not reoccupied by the vanquished group (Rappaport 1984).

*Ecological impact*: Gatherer societies, with their low popula-tion densities, low per capita resource demands, cycles of ma-terials closed on limited spatial scales, and a number of prac-tices that promote sustainable resource use, necessarily have a low level of impact upon the environment. Over long intervals, however, even this can add up to substantial changes. Such changes are especially likely when the resource base changes relatively rapidly, as might have happened with fluctuations during the ice age; or when a gatherer population encounters an entirely new resource base, as during the initial colonization of the Americas. It has, for instance, been suggested that the widespread extinction of large mammalian species during the Pleistocene was a consequence of human overhunting, and that many of the savanna-grassland formations of East and South Africa are a result of fires lit over tens of thousands of years by human populations (Menant *et. al* 1985). Nevertheless, the pace of such impact would be considerably slower than the pace in populations which possess more advanced modes of resource use, as described later in this chapter. Some writers have claimed that hunter-gatherers possess an ecological wisdom far in advance of that shown by modern man (Shepard 1982). Be that as it may, it is indisputable that the ecological impact of this mode of resource use is minimal.

## Pastoralism

*Technology*: The long period of history when human beings were exclusively gatherers began to come to a close with the domestication of plants and animals. This coincided with the withdrawal of glaciers, ten thousand years ago. It is possible that climatic and vegetational change prompted human populations to intensify resource use, and to initiate agriculture and animal husbandry. These processes began in parallel, and have often gone hand-in-hand. While the cultivation of plants has been of greater significance in tracts of moderate-to-high rainfall and moderate-to-high temperatures, animal husbandry has held pride of place in tracts of low rainfall, and at the higher altitudes and latitudes where temperatures are too low to support agriculture (Grigg 1980). Over large tracts where agriculture is not feasible, it is also difficult to maintain herds of domestic animals within a single locality. Animal husbandry is therefore based, in such tracts, on moving herds from place to place often over several hundred kilometres. This requires taking advantage of the seasonal abundance of grazing resources in different parts of a region. Nomadic pastoralism thus evolved as a distinctive mode of resource use, a mode that held sway for several centuries over large regions, particularly in Central Asia and North and Central Africa (Leeds and Vayda 1965; Forde 1963).

Pastorals have access to animal muscle power, an important additional source of energy, especially for transport. The animals also serve as a source of food which can be tapped as required, thus greatly increasing flexibility in the use of different habitats.

*Economy*: Nomadic pastorals move over large distances, and with their access to animal energy, have been critical in creating flows of resources over distance scales that are vastly greater than those which prevail in gatherer societies. The resources they moved have been both high-bulk commodities like salt, and high-value, low-bulk luxury items like precious stones and musk. They also served as carriers of information about resources of distant regions and of technologies elaborated by other societies. Consequently, pastorals not only continued some hunting-gathering while on the move, and produced meat, milk, hide and wool from their animals, but also acquired resources, especially from settled agricultural societies, in exchange for material and information. Even more important, nomadic pastorals could effectively deploy force to usurp resources from societies of cultivators, as Chengiz Khan did with great success over the huge regions of Asia and Europe. Indeed, nomadic pastoral societies, in their heyday, may have behaved much like the r-strategists of ecological literature.

*Social organization*: The social groups of nomadic pastorals remain limited to kin groups of a few thousand; nevertheless, they come in contact with large numbers of other groups over an extensive terrain. Within the social groups of pastorals, the division of labour is fairly limited. It is based on age, sex, and leadership qualities which emerge during inter-group conflict. Women may be more involved in feeding, milking, and tending animals, men in deciding on migration routes and herding animals while on the move.

In the pastoral mode, elements of private property begin to emerge. However, while herds are usually owned by separate households, pastures are invariably common property, with individual herdsmen possessing rights of access and usufruct. However, like gatherer societies, nomadic groups are relatively egalitarian (Khazanov 1984). With this, coercion within groups remains limited; indeed there is considerable premium on cooperation within the group, especially in the context of conflict with other groups, nomadic or otherwise.

*Ideology*: By surviving successfully in harsh and variable environments, and with little attachment to any particular locality, nomadic pastorals were perhaps the first societies to perceive human communities as separate from nature, and therefore in a position to dominate it. Since the usurpation of resources controlled by alien and settled communities constituted a significant part of their strategy of resource acquisition, they were unlikely to evolve strong traditions of careful or restrained resource use. Indeed, ideologies which rejected the attribution of sacred value to living creatures or to natural objects, e.g. religions like Judaism, Christianity and Islam, arose in tracts that were dominated by nomadic pastorals in the Middle East. In fact, as Lynn White (1967) notes, such religions sometimes prescribed the deliberate destruction of sacred trees and sacred groves.

The ritual life of nomads is quite meagre: no pantheon of gods, as in peasant societies, no system of totemism, as in gatherer societies. Ritual importance may be placed on livestock, but almost never on natural locations or specific fields. Equally striking is the relative unimportance here of witchcraft. In comparison with peasants and gatherers, nomadic pastorals have little need to pacify or placate nature; in the event of resource shortages they remove themselves to more resource-abundant areas—something peasants cannot do (Goldschmidt 1979).

Apart from these broad ideologies, in their day-to-day interaction with nature nomads do have practices which reveal a deliberate restraint on resource use. These practices include the complete exclusion of grazing pressure during certain periods within fodder reserves, and its limited use during other periods in terms of the kind and number of animals permitted for grazing—as for example in the system of *ahmias* around Taif in Saudi Arabia (Draz 1985).

*Ecological impact*: It is possible that nomadic pastorals contributed to a gradual overgrazing, and to the expansion of arid regions at their margins, all through their history. This they have certainly done across many regions in modern times. They have also contributed to ecological degradation through the organization of trade and the diffusion of technology over large distances and perhaps most importantly by disseminating the belief in man's mastery over nature.

## Settled Cultivation

*Technology*: Human societies learnt to cultivate plants and domesticate animals around the same time, beginning some 10,000 years ago. In some regions the two developed hand-in-hand, with the traction power of animals and the manurial value of their dung being vital to agricultural operations. This, for instance, was the case in the Middle East, from where the use of cattle and the plough, and the cultivation of wheat and barley, gradually spread over parts of Asia, Europe and Africa. In other areas, domesticated animals played a much less significant part in cultivation, as in the paddy-growing tracts of Asia—or had no role at all—as in the case of maize cultivation in pre-Columbian America (Grigg 1980).

Cultivation involves an intensified production of certain species of plants, and the removal of plant material from a relatively restricted area of land. The plant material so removed, for instance cereal grains, are particularly rich in certain elements, such as nitrogen and phosphorus, and contain a number of micronutrients, like boron and molybdenum, in smaller quantities. The continuation of cultivation on a piece of land therefore depends on returning to the earth what is taken away from it. This happens either through long periods of fallow, as in shifting cultivation, or by the application of river silt, organic manure, or mineral fertilizers if the same piece of land is tilled year after year. Shifting cultivation is, of course, the option followed so long as the amount of land available is large relative to the population. As this ratio declines, the same piece of land has to be used more and more intensively. Almost everywhere, this has called for the extensive use of organic manure derived from natural vegetation in the surrounding areas, gathered either through grazing domestic animals or directly by human effort. This has changed radically only in recent times, when fossil fuel energy began to be used to efficiently mine, transport and synthesize mineral fertilizers to augment agricultural production (Pimentel and Pimentel 1979).

Through most of its history, settled cultivation has thus depended on human muscle power, supplemented in some regions by animal muscle power. In the industrialized world, it has come to depend increasingly on fossil-fuel energy. However, pre-industrial agriculture depends primarily on plant and animal-based materials, along with some control of natural flowing water for irrigation. Consequently, pre-industrial agricultural societies (more properly, peasant societies) have a fairly substantial knowledge base in relation to husbanded plants and animals; they also view nature as being subject to human control to a very significant extent.

*Economy*: In peasant societies, cereal grains can be stored and moved around, especially on the backs of animals or in carts, over long distances. Resources can here flow over much larger distances than in gatherer societies, enabling the concentration in towns of human populations not directly involved in gathering or the production of food. Changes in settlement patterns also correspond to shifts in consumption. Of course, a majority of the agrarian population consumes natural resources largely for subsistence—e.g. for food, clothing, shelter, implements, fodder and manure. However, a small but powerful segment of the population is involved in the large-scale consumption and use of materials not directly related to subsistence—both luxury items such as silk and wine, and instruments of coercion such as the horse and elephant, metal swords and shields. Fig. 1.2 shows the resulting structure of material flows in such a system. There are large-scale exports of materials out of intensively cultivated patches of lands, both to nearby villages and to more distant urban centres. The volume, range of items and distance-scale of such flows steadily increase with technical improvements, especially in animal-based transport. These outflows from agricultural land are balanced by inflows from surrounding non-cultivated lands. In peasant societies there are no counterflows from urban centres back to cultivated lands. However, as Fig. 1.3 shows, in industrial societies there are large flows, back to the land, of materials such as farm machinery and synthetic fertilizers.

Characterized by fairly extensive resource flows, especially of foodgrains and livestock, peasant societies are much less subject to environmental variation in space and time. Nevertheless, the techniques of cultivation and animal husbandry, the choice of plant and animal varieties, and the way that resources of non-cultivated lands are gathered and put to use are all greatly dependent on the local environment. There is therefore a substantial degree of locality-dependent adaptive variation in patterns of resource use within peasant societies. This begins to disappear only with large-scale inputs of fossil-fuel energy and advanced technology into agriculture.

*Social organization*: Cultivation requires intensive inputs of human energy in relatively restricted areas of land—a few hectares per person in the pre-industrial stage. Therefore a small kin group rather than a large band can most effectively organize such inputs; hence the family becomes the basic unit of an agricultural society. Family groups need to co-operate with each other in a variety of ways, including defence against the usurpation of their production (apart from what is surrendered as tax to the state). Several families thus remain banded together in a village, which becomes a social group of a few hundred to a few thousand individuals. This social group also tends to control and manage a territory of non-cultivated land surrounding the cultivated areas, from which come a variety of inputs such as fuel-wood, fodder and leaf manure.

Sex-based division of labour is quite pronounced in the peasant mode. Typically, men confine themselves to operations such as ploughing, which require higher power output. Women take on the burden of more tedious work, such as weeding and transplanting, and, outside cultivation, the collection of fuel, fodder and water.

While cultivated plots are usually (though not always) con-

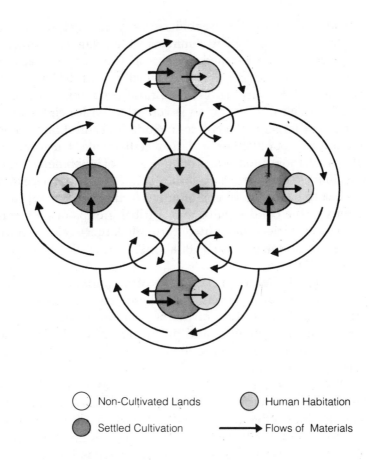

○ Non-Cultivated Lands   ● Human Habitation

● Settled Cultivation   ——→ Flows of Materials

**Figure 1.2**

Material flows in an agrarian society. Settled agriculture makes possible generation of surplus grain and livestock production which can support concentration of non-agricultural populations in towns and cities. This material export from cultivated lands has to be made good by flows from surrounding non-cultivated lands. Material cycles thus become much more open in comparison with the hunter-gatherer shifting cultivator stage. Settlements adjacent to cultivated land represent villages, the larger habitation in the centre, towns. The thickness of the arrow indicates the intensity of the flow.

trolled by individual households, forests, grazing grounds and water are normally held in common by the village (cf. Mukerjee 1926). Several, and sometimes very many, such villages are integrated into a larger chiefdom, which constitutes the terrain over which the surplus of agricultural production is pooled together. The right to do so is contested with neighbouring states. The larger social entities include concentrations of non-agricultural populations in larger settlements, the towns or cities with thousands or tens of thousands of inhabitants. While the villages constitute a face-to-face social group (not necessarily kinship based) of several hundred or thousand people analogous to the bands/endogamous tribal groups of gatherers, the larger society integrated through larger-scale resource flows can no longer deal with all members of the group in a personalized way.

Within the larger social group of an agricultural society there is a great deal of division of labour, made possible by the fact that only a fraction of the population needs to engage directly in the gathering and production of food. Those not directly involved in the production of food take on other occupations. These are—

The processing of materials (e.g. textiles, oilseeds), transportation, the interpretation and dissemination of natural and cultural knowledge (by priests) and coercion (by specialized warrior groups). In this division of labour, men end up monopolizing the more prestigious and skilled jobs, passing on the less skilled and tedious work to the women. There is now a substantial differentiation of coercive abilities within the social group— from peasants who receive very little in return for the surplus production they yield to others, to priests and warriors who provide others little in return for the surplus they manage to get hold of (Service 1975).

In comparison with the gatherer mode, the peasant mode shows a sharp separation between cultivated and non-cultivated land. This separation is significant in directing resource flows (Fig. 1.2), and equally so with regard to differing forms of property and control. At the lowest spatial scale agricultural land may be controlled by a family. Such control may be subject to regulation by the village community, which could reassign plots of land and, further, treat land as a community resource, perhaps for grazing purposes, outside the cropping season (Bloch 1978). The non-cultivated land within village boundaries, typically a few square kilometres in extent, serves to supply fuel, grazing, manure, etc. for the community as a whole. These large chunks of land, different portions of which may be used at different seasons, may be most effectively controlled as community, rather than family, property. (This also applies to water sources such as tanks, rivulets, lakes and springs, typically held in common.) However, with technological advance and the concentration of powers of coercion, village communities may lose control over cultivated lands and become tenant-cultivators. Access to non-cultivated lands may be lost too, for instance with the enclosure of commons by powerful landlords or by the state. In addition to land in the vicinity of villages, the state may lay claim to larger uninhabited tracts, constituted as princely hunting preserves or forests from which the army derives its supply of elephants (Thompson 1975; Trautmann 1982).

The large-scale resource flows of agricultural societies are accompanied by imbalances involving net outflows from rural areas and inflows into urban centres. These imbalances become progressively more acute as technological advances in storage, transport and processing create an effective demand for a larger range of produce from both cultivated and non-cultivated lands, and as the coercive power of the non-agricultural sector increases *vis-à-vis* the agricultural sector. The overall imbalances decrease with large-scale inflows of synthetic chemicals, fertilizers and machinery to agricultural lands in industrial societies; however, it is very likely that, given the larger inflows and outflows, the imbalances pertaining to specific elements such as micro-nutrients, or to larger organic molecules like humic acids, may become all the more acute (Pimentel and Pimentel 1978; Stanhill 1984).

Agricultural societies might be at equilibrium with their resource base; or they might encounter either expanding or shrinking resource bases. Agriculturists who newly colonize lands earlier held by gatherer societies, or who newly benefit from a major resource input such as irrigation, would find themselves with an expanding resource base, analogous to the r-strategists of ecology. On the other hand the base may be shrinking because of an adverse climatic change, or if access is cut off from important inputs like leaf manure and fodder from forests newly taken over by the state. Their resource base may also shrink if agricultural productivity remains stagnant in the face of human population growth. An approximate equilibrium may be maintained if the population grows slowly, if the external demands on agricultural production remain stable, and if technological progress keeps pace with the need to continue increases in agricultural production in consonance with population change.

In the last case—that of approximate equilibrium as with the k-strategists of ecology—the social groups are likely to be highly viscous, with related individuals tending to stay together and tied, perhaps generation after generation, to a given locality. Under these conditions they may exhibit high levels of co-operative behaviour among themselves, as also behaviour which favours long-term group interests. The peasant societies of India, China and South East Asia, in the period before European colonization, perhaps fall into this category. On the other hand, when the resource base is rapidly expanding, especially with new land being brought under cultivation, social groups are likely to be much more fluid and far less tied to any locality. Their level of co-operative behaviour, and especially their willingness to sacrifice individual interests to long-term group interests, would consequently be much lower. This would seem to have been the case with European pioneers in seventeenth and eighteenth century North America. Finally, peasant cultures faced with a shrinking resource base may also lose group coherence and the attachment to a particular locality, as has been reported from several parts of India in recent times (Guha, 1989a).

*Ideology*: In comparison with hunter-gatherer societies, agricultural societies have established substantial control over natural processes; nevertheless, they are still very subject to nature's caprices in the form of droughts, floods, frost, and plagues of locusts. Indeed, certain life-history stages of agricultural crops are themselves especially sensitive to the environment. Hence, agricultural societies continue, in part, to perceive man as one among a community of beings. At the same time, the image of man as a steward of natural resources acquires influence. The restrained use of natural resources could thereby be expected to form one part of the ideology of agricultural societies, especially when they are in a state of near equilibrium with their resource base. On the other hand agricultural societies in the process of encountering an expanding resource base—either through new technologies or, especially, while colonizing lands earlier held by gatherers—are much more likely to view man as separate from nature and with a right to exploit resources as he wishes (cf. Cronon 1983).

In stable peasant societies, practices of restrained use relate to cultivation itself, and are linked to a philosophy of minimizing risk rather than maximizing immediate profit (Scott 1976). The use of a whole variety of different crops and crop rotations, and the careful community-based maintenance of irrigation ponds, may be part of such an approach. This approach would also encompass the non-cultivated lands from which the villagers gather fuel, fodder, small timber, leaf manure, and so on. A variety of practices of restrained resource use with respect to non-cultivated lands have been reported from peasant societies which are at equilibrium with their resource base. For example:
• A quantitative restriction on the amount of harvest from a given locality, e.g. on the amount of wood or grass harvested by a family or their livestock from community lands.

• Restrictions on the harvest in certain seasons. Thus, lopping the green leaves of trees may be permitted only after the rainy season, i.e. after the trees have ceased to put on growth.

• Certain species, e.g. trees belonging to the genus *Ficus*, may be wholly protected.

• Certain habitat patches may never be harvested. In the state of Mizoram, in north-eastern India, the community wood-lots from which regulated harvests are permitted, called supply forests, are complemented by sacred groves, aptly called 'safety forests', from which no harvests are permitted.

• Certain methods of harvest may be completely prohibited. In the Aravalli hills of Rajasthan there are patches of forests called *oraons* where harvesting by using metal tools is prohibited, though wood may be removed by breaking twigs by the hand (Brara 1987).

• Specific age-sex classes or social groups may be banned from employing certain harvesting methods, or from utilizing certain species or habitat patches.

These practices all depend on a high degree of co-operation between the members of a village community. One could say that, in the peasant mode, *custom* and *tradition* provide the overarching framework within which human–nature interactions are carried out. While religion continues to permeate social life, in the realm of resource use it is supplemented to a significant degree by custom. In other words, customary and time-honoured networks govern relationships of reciprocity within peasant society—as, for example between different castes in a village (cf. Chapter 3), between nomadic pastorals and peasant cultivators, or between the village and the state. Clearly, such relations (with respect to resource use, as elsewhere) are often asymmetrical: but they normally fluctuate only within the limits defined by custom.

*Ecological impact*: The ecological impact of the peasant mode may be characterized as *intermediate*. With the march of agriculture a significant proportion of land begins to be converted into artificial grasslands or cropfields, which replace forests, marshes or natural grasslands. Fire, stone axes and metal axes aid in this process of conversion. Cultivation also imposes increasing demands on natural vegetation and a greater removal of forest produce, to be used as fuel, fodder, manure, building-timber and implements. The discovery of iron, which in many areas led to the colonization of the forest by agriculturists, also facilitates the continued felling of individual trees in a forest. In villages where such felling takes place, natural regeneration is relied upon to restore tree cover. At the same time, improvements in weapon technology enable a more flexible hunting strategy. The cumulative impact of these interventions is a striking change in the landscape, which very likely becomes heterogeneous, manifesting a variety of successive stages within a mosaic. It could also result in the local extinction of some species of plants and animals.

Of course, agricultural societies which newly colonize lands held by gatherers have had a dramatic ecological impact even in the short run—transforming the landscape, exterminating certain species and depleting others, introducing weedy species, and so on (Cronon 1983; Crosby 1986). This would also be the case with agricultural societies expanding their resource base through technological innovations such as large-scale irrigation and the use of pesticides. On the other hand, agricultural societies in approximate equilibrium with their environment—dominated by 'local production for local use'—have only moderate levels of impact in transforming landscapes and bringing about gradual changes in the composition of biological communities.

## The Industrial Mode

*Technology*: The latest mode of resource use to appear in human history, large-scale industry, has been with us for just about two hundred years. This is only one-fiftieth of the time that *Homo sapiens* has spent husbanding plants and animals, and one-two hundredths of the time since when hunter-gatherers painted grand hunting scenes in the caves of Lescaux and Altamira. But its ecological impact has been profound, far surpassing all that preceded this revolution. The main reason for this is the quantum jump in the use of energy, with heavy demands on non-renewable sources (coal, oil), coupled with the use of entirely novel sources such as nuclear energy.

If the pattern of energy use in the gatherer mode may be characterized as *passive* (relying only on human-muscle and wood-fuel power) and that of the agricultural mode as *active* (augmenting human power with animal power, wood-fuel and water power), in the industrial mode energy use follows an *extractive* path, wherein natural resources are both *harnessed* (hydro power) and *mined* (fossil fuels) for human consumption. The industrial mode has also brought into use a whole new range of man-made materials; e.g. metals, plastics, silicon chips and synthetic pesticides. These newly-fashioned materials can now be preserved to be used for long periods, and transported for consumption elsewhere. Great improvements in transportation within the industrial mode mean that even bulky, heavy goods—e.g. timber or rocks, can be transported with ease over large distances (Ayers, 1978). 'The windmill gives you the feudal lord, the steam engine the capitalist', says Marx when distinguishing between different modes of production. Distinguishing between different modes of resource use according to their technological infrastructure, we may likewise say: 'the axe and bullock cart give you the peasant mode, the chain-saw and locomotive the industrial mode'.

These abilities *vis-à-vis* materials processing, storage and transport have revolutionized flows of resources, these having now become truly global. Any material, be it animal, vegetable, or mineral, natural or man-made, can now be rapidly transported anywhere, and whenever desired. With this a significant fraction of humanity, albeit still very much a minority, has come to consume vast quantities of a very wide range of resources (IIED and WRI 1987). Consumers in the high centres of industrial civilization can draw upon the natural resources of most parts of the globe, taking for granted the continued supply of teakwood from India, ivory from Africa, and mink from the Arctic. This elite is now scarcely affected by the natural variations within resource availability in space and time; it has developed a lifestyle which could truthfully be called global. In the process it has wiped out myriad locally-adapted lifestyles in different parts of the world.

*Economy*: Over the last three centuries, industrial societies have steadily expanded their resource base. This has been achieved by a growing knowledge about the working of nature—through the hypothetico-deductive method of modern science and by the links established between scientific discovery and practical application—in order to tap additional sources of energy, process materials, and transport goods faster, more economically, and over ever-longer distances. This process of the intensification of resource use has led to the continual over-use and exhaustion of many resources. The typical response to such exhaustion has been to find a substitute, though initially, when the resource being substituted was abundant, the use of the substitute would have involved more effort in terms of energy, material, and labour. In the classic case, when wood became scarce in Europe, coal came to substitute for wood charcoal in the manufacture of iron, while locomotives were invented to replace horse-drawn carts even as there developed serious scarcities of fodder for horses. Industrial societies have gone on to consume resources at an accelerating rate, exhausting them one by one in sequence, going from the economically desirable resource of the moment to less and less desirable resources as the more desirable resources become exhausted. And, alongside, technological innovations now continually extract better and better results from what were once the less desired resources (Wilkinson 1988).

The expansion of the resource base in industrial societies has rested upon access to land and natural resources which were earlier controlled by gatherer and peasant societies. Reeling under an energy and transport crisis brought about by the

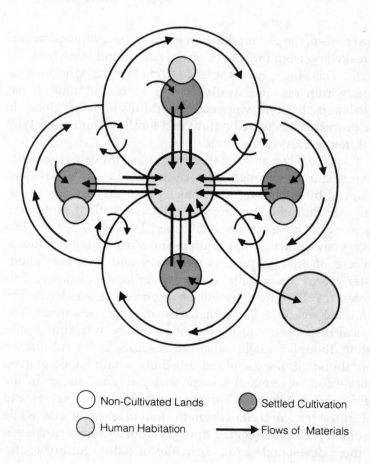

Non-Cultivated Lands     Settled Cultivation

Human Habitation     Flows of Materials

**Figure 1.3**

Material flows characterizing modern Indian society. Such societies not only tap surplus agricultural production, but also a great deal of the produce of non-cultivated lands to meet the requirements of the urban-industrial sector. Thereby the material cycles become totally open, with large outflows from rural hinterlands. These are partially compensated for by the organization of flows of materials such as fertilizers from the urban-industrial sector to the cultivated lands. The large central human habitiation represents an Indian city such as Bombay; the separated habitation in the upper right hand corner, the industrialized countries. The thickness of an arrow indicates the intensity of the flow.

exhaustion of their forests, European colonists laid claim to vast terrains the world over. Wherever the 'portmanteau biota' of these Europeans—wheat and cattle, their weeds and diseases—could establish itself comfortably—as in North America and Argentina, South Africa, Australia and New Zealand—it created neo-Europes (Crosby 1986). Where the ecological setting did not permit such a takeover, as in the older civilizations of the Middle East and Asia, or in the humid tropical forests of the Amazon, the Congo and Malaysia, these Europeans nevertheless established a firm hold on the resources of these regions and organized outflows of what they most desired to their own lands and people. Thus, India became an exporter of teak, cotton, jute, tea, indigo and precious metals; Burma of rice and teak; the West Indies of sugar; Brazil of rubber and coffee; and so on. Although colonialism has formally ended, this process vigorously continues, and India now exports prawns and trained manpower along with tea, rather than teak or cotton; and Brazil exports beef and forest produce instead of rubber.

These flows are highly asymmetric, with industrial societies receiving large volumes of unprocessed resources at low prices and exporting small volumes of processed resources at much higher prices. Simultaneously, this process also implies the production in industrial societies of high volumes of wastes; these are then sought to be disposed of elsewhere, either in the global commons of the oceans and the atmosphere, or through their sale to Third World societies plagued by foreign debt.

*Social organization*: The greatly enhanced scale of resource flows in industrial society goes together with a substantial increase in the number of humans involved in this network. Face-to-face contact is obviously impossible among such large numbers, and they tend to interact through the medium of formally codified transactions. Within such large social groups there is, of course, a rather elaborate division of labour. In modern industrial societies this division largely hinges upon skills in processing, transporting and exchanging materials and information. Smaller social groups based on such division of labour—for instance, car mechanics or teachers—and corporate groups banded together to carry out a task—for instance, car manufacture or schoolteaching—become important. Among these exist relations involving face-to-face communication equivalent to those in hunter-gatherer bands or peasant villages. In these societies, too, men have tended to monopolize the more skilled and prestigious jobs, leaving the bulk of the unskilled and tedious work to women. Such discrimination has persisted despite the rhetoric of sexual equality and the development of technologies which tend to discount the advantages of greater physical strength.

Whereas kinship, locality and region define the forms of association in the pre-industrial world, in industrial societies the impersonal criteria of structural location *vis-à-vis* the means of production define the ways in which individuals come together for collective action. These corporate groups, based on the division of labour, are extremely fluid, the membership being in great flux. Again, the continual expansion of the resource base, both in terms of new extractive techniques and new territories to draw upon, has enhanced the fluidity of social groupings. There is therefore a much greater stress on pursuit of individual interest in these greatly atomized societies (Hawley 1986).

The rise of individualism is also accompanied by a tremendous expansion in the role of the state in regulating individual transactions. In most spheres of social life, the personalized and flexible systems of customary law—typical of agricultural societies—are replaced by impersonal and relatively rigid systems of codified law. The forest codes of modern society are amazingly detailed, often running into several hundred sections and subsections (cf. Guha, R. 1983; Merriman 1975; Linebaugh 1976). While safeguarding private property over land and in the workplace, and taking over the ownership of what was hitherto common property, the modern industrial state completely delegitimizes *community*-based systems of access and control.

Forest management provides a good illustration of how the industrial mode admits only the polarity of individual and state control. In the process of industrialization in Western Europe and the United States, an early phase of anarchic capitalist exploitation was followed by the assertion of state control, wherein the government stepped in to assume responsibility for forest protection and production (Hays 1958; Heske 1937). However, when the situation stabilizes with respect to the supply of raw material for industry, the state may withdraw control over certain forest areas by handing over 'captive plantations' to private industry. Simultaneously, the expansion of the market may encourage the widespread planting of commercially valued trees by individual farmers. Thus, forests in the industrial mode are primarily *state*-owned, supplemented in capitalist societies by a small but often growing *private* sector. The victim in this process is, inevitably, village-based systems of *community* forests and pasture management: the lands in these systems are either sequestered by the state or parcelled out among individuals.

*Ideology*: Developed in a mileu of r-strategists, the ideological underpinnings of industrial society involve the total rejection of the gatherer view of man as part of a community of beings, or even of the agriculturist view of man as a steward of nature. Instead it is emphatically asserted that man is separate from nature, with every right to exploit natural resources to further his own well being. Nature is now desacralized. What has come to be venerated instead is the marketplace; the market is supposed to rationally allocate the use of resources so efficiently that all individuals are as well off as they could possibly be. Among socialists this veneration is transferred to a Central Plan: it is regarded by them as a more efficient allocator than the market.

While the ideology of conquest over nature as well as modern lifestyles have resulted in a radical alteration of the landscape of the globe, industrial societies have, to be fair, made systematic attempts to safeguard their own environments. Such

attempts were triggered off by deforestation and the consequent landslides in the Swiss Alps in the 1860s, and further
strengthened when the western frontier of the United States
was officially declared closed by the 1890 census. Since then
Western conservationists have gradually brought back part of
the forest cover of Europe and North America, set in motion an
effective programme of soil conservation, and set apart wilderness areas (cf. Hays 1958 and 1987; Nash 1982). The problems
of chemical pollution which surfaced after the Second World
War are also being tackled—with the successful cleaning of the
Thames, for instance. But pollution proceeds apace with such
efforts; new environmental issues come up, such as acid rain
and global warming. These require global solutions, so that it
now appears as if industrial societies can no longer restrict
environmental degradation to areas outside their own borders
(WCED 1987).

Despite the increasing scale of environmental problems, the
ideology of industrialism even today rejects as invalid any
concern with physical 'limits to growth', trusting technical
innovation to take care of problems as they arise. It puts its faith
in the efficacy of the marketplace (or of the Central Plan) to
stimulate the discovery of substitutes, and to enhance the efficiency of resource use (Beckerman 1972; Laptev 1977). A variety
of scientifically-based prescriptions on sustainable use of soils,
forests, fisheries and other renewable resources, generated over
the last century, have been interpreted as examples of efficient
resource use, and one of the aims of this book is to examine
whether this is actually so.

*Scientific prescriptions for resource use*: Just as religion and
custom legitimized patterns of resource use in pre-industrial
societies, in the industrial mode science provides the organizing principle for human interactions with nature. However,
scientific prescriptions for the preservation, the sustainable use,
and the deliberate destruction of different species and habitats
parallel, in many ways, the pre-scientific prescriptions typical
of earlier modes. For example:

• Just as there were quantitative restrictions on the amount of fuelwood extracted by a family from a community wood-lot, or on the number of animals to be grazed on a fodder reserve, there are now quantitative prescriptions on the amount of timber to be removed during the course of selection felling from a designated patch of forest, or on the number of deer that may be shot by a licence-holder in a given year.

• Just as New Guinea hunter-gatherers stop hunting a particular species—say the Bird of Paradise, when its population density falls below a certain level—the International Whaling Commission has banned the hunting of certain species of whales whose populations have been severely depleted.

• Just as fishing from certain coastal areas is traditionally stopped if the yields become very low, there have been prescriptions to 'rest' certain forest areas, for instance in India after the excessive harvests of the Second World War.

• Just as there is a traditional ban on hunting in certain seasons, there are closed seasons for mechanized fishing, or for timber extraction.

• Just as certain life-history stages are traditionally given protection—for instance, breeding birds at heronries in many villages in peninsular India—the younger, growing stock in a forest is supposed to be spared all extraction pressure.

• Just as certain methods of harvest, e.g. the poisoning of streams, may be traditionally forbidden, there are regulations forbidding fishing with nets of excessively small mesh size.

• Just as certain habitat patches—for instance, sacred groves or sacred ponds—may be fully protected from harvest, there are prescriptions for keeping certain forest areas totally free from human interference, either for watershed conservation or for the maintenance of biological diversity.

• Just as certain species are totally exempt from being hunted or felled, e.g. monkeys or *Ficus* trees in Indian villages, complete protection may be extended to certain endangered species, such as the California condor or the whooping crane.

In parallel, of course, there exist a whole range of supposedly scientific prescriptions for deliberately destroying certain resources. Thus, just as the entire Khandava forest near present-day New Delhi was supposedly burnt in the time of the Mahabharata (cf. Chapter 2) as an offering to the fire god, massive forest areas in the Narmada valley in Central India are now being destroyed to create reservoirs for hydro-electric power generation and irrigation. And one cannot forget that the large-scale defoliation of forests during the Vietnam war was supported by considerable inputs from the American scientific commuity.

In this manner, scientific prescriptions in the industrial mode closely parallel traditional, 'pre-scientific' prescriptions based on a large but informal base of knowledge and on simple rules of thumb. In part, modern prescriptions are based on a more detailed knowledge of the behaviour of the system, and on a more explicit definition of what is sought to be achieved. With many tree or fish populations, this aim is defined as achieving maximum sustainable yield, i.e. a yield that will not decline in the long run.

However, deciding upon a harvesting regime that would lead to maximum sustainable yields is beset with manifold difficulties (Clark 1985; Beddington and May 1982). First, every plant or animal population is involved in a number of interactions with other members of its community. Thus, harvesting a deer population may, through a change in the composition of the herb layer, affect another resource such as wild cattle. Or harvesting a plant species may result in a change in the availability of honey. In fact, some years ago, biologists warned the fishing industry on the east coast of Canada that scallop fisheries were being over-exploited and were likely to crash. It so happened that just after this there was an outbreak of an epidemic disease among sea urchin populations which prey heavily upon the scallop. With predation pressures reduced, scallop populations increased rather than collapsed, as had been feared by fishery biologists (cf. Berkes 1989). Similarly, wild-life biologists concerned with enhancing the carrying capacity of wetlands within the Bharatpur bird sanctuary in north India recommended the banning of buffalo-grazing in the sanctuary. Following the imposition of this ban it was found that, in the absence of grazing, a grass, *Paspalum*, grew so rapidly that it choked the wetland and reduced its carrying capacity for water fowl (Vijayan 1987).

Because of such complex community interactions, the simultaneous harvesting of more than one resource population from a specific locality can have a variety of unexpected consequences. Theoretical investigations suggest that in some cases the natural tendency of a population to fluctuate might be exaggerated by the simultaneous harvesting of another species, and expose it to the danger of extinction (Pimm 1986; May 1984).

But, aside from such complications which arise out of community interactions, it is difficult to understand the behaviour of harvested populations and to assess whether they run the risk of being over-harvested and becoming extinct. For example, the Elk populations of British Columbia are being simultaneously harvested—for subsistence by Amerindians and for sport by European settlers. Wildlife biologists primarily interested in deciding on regulations for sport hunting contend that the Elk populations are declining on account of subsistence hunting. On the other hand anthropologists looking at the same population data suggest that there is no good evidence of a decline in Elk populations (Freeman 1989). It is even more difficult to assess the fate of fish or whale populations for which the only available information is based on harvests themselves. The level of such harvests is a complex function of the number, age and sex composition of the prey population, its distribution in space and time, and the extent and distribution of the harvesting effort. The parameters which characterize the prey population also depend on the extent of the exploitation to which it has been subjected in the past. It is difficult to predict the responses of such populations to different levels of harvests even on the basis of sophisticated mathematical models because our knowledge of many parameters in the model remains limited. Indeed the International Whaling Commission finds it very difficult to arrive at a consensus on what is actually happening to many whale populations (May 1984).

Given all these difficulties, even today successful prescriptions of how to obtain maximum sustainable yields are possible only in very limited contexts. They work only when community interactions, as well as the population structure, are drastically simplified. This is the case with single-species even-aged populations of forest-tree species such as pines, planted for harvest in temperate latitudes. Here all other plant, insect, and microbial species that might enter the forest naturally are sought to be eliminated. Intra-specific competition is regulated by permiting reproduction only through artificial regeneration, by controlling inter-plant distance, and by harvesting the individual plants which are being suppressed through intra-specific competition. It would not be incorrect to say that scientific prescriptions on maximum sustainable yields have largely failed to work, except in such limited contexts. Even in these contexts, the suspicion is that a gradual exhaustion of soil nutrients is now leading to the decline of forests which have been long invoked as exemplars of sustained-yield management. We shall later see, in much greater detail (Chapter 6), how modern attempts to apply such concepts to the tropical forests in India have led to serious over-harvests (see also FAO 1984).

The prescriptions on giving total protection to certain species and localities, such as those embedded in the US Endangered Species Act or CITES (Convention on International Trade in Endangered Species), or to certain localities—as in programmes setting apart national parks or biosphere reserves—also have rather uncertain scientific foundations. This is inevitable, given our very limited understanding of the extent and distribution of biological diversity, and of the ecosystem processes which govern the survival of the species which make up different biological communities. Heated debates on whether to aim for a few large, several small, or even plentifully patchy nature reserves, as well as the debates on minimum viable population size, show that scientists are still far from finding definitive answers on the appropriate ways of conserving natural populations of living organisms (Soule 1986). Indeed one could argue that scientific prescriptions in industrial societies show little evidence of progress over the simple rule-of-thumb prescriptions for sustainable resource use and the conservation of diversity which characterized gatherer and peasant societies. Equally, the legal and codified procedures which are supposed to ensure the enforcement of scientific prescriptions work little better than earlier procedures based on religion or social convention.

*Ecological impact*: Industrial societies, unlike gatherer or agricultural societies, are no longer directly dependent on the natural resources of their immediate vicinity. At first glance this may suggest that they could effectively conserve their resource base, whereas societies that follow more primitive modes of resource use would perforce suffer a much more adverse impact. This seems borne out by the fact that Japan, the most industrialized Asian country, has the best-preserved forest cover, while the forests of countries like Malaysia and Indonesia —which have large populations dependent on primitive agriculture—are being devastated. But a second look shows that Japan maintains its forest cover, in spite of its enormous per capita consumption of timber, only by shifting the pressure on to Malaysia and Indonesia, precisely the countries which suffer devastation. The Japanese also exploit their coastal fisheries very conservatively, while refusing to give up whaling in international waters. This discrepancy between prudence at home and profligacy abroad also characterizes the behaviour of other industrially advanced countries such as the USA and the (former) USSR.

If one looks at the overall picture, therefore, it is obvious that the enormous resource demands and waste production of industrial nations, and of the industrial segments in less industrialized nations, have the most profound impact on the world's environment (cf. Bahro 1984). This impact includes radical modifications of the landscape, as with the laying of the Alaskan pipeline, or cattle ranching in Amazonian forests; a gradual depletion of forests, as with acid rain in Europe or over-extraction in Malaysia and Kalimantan; drastic reductions in, or extinctions of, populations, as with the passenger pigeon and the bison in the last century, or the African elephant and myriad other species in Amazonia in this century; a wholesale poisoning of the biosphere, as with wide-spectrum pesticides and nuclear waste; a modification of bio-geochemical cycles, as with the increased production of carbon dioxide; and perhaps a long-term adverse modification of the climate as well (cf. Richards 1986).

In assessing the ecological impact of different modes, one is struck by two paradoxes, which we illustrate here with respect to forest use. Spatially, hunter-gatherers live *in* the forest, agriculturists live adjacent to but within *striking distance* of the forest, and urban-industrial men live *away* from the forest. Paradoxically, the more the spatial separation from the forest, the greater the impact on its ecology, and the further removed the actors from the consequences of this impact! The same conditions operate with regard to other resources, such as water.

Second, the faster the development of formal, scientific knowledge about the composition and functioning of forest types, the faster the rate of deforestation. One important reason here is undeniably higher levels of economic activity, but another though less obvious factor is the *idiom* of resource use itself. While enormously enlarging our knowledge about specific physical and chemical processes, modern science has not always displayed the same understanding of the ecological consequences of human interventions which follow the development of scientific knowledge. However, the *belief* that science provides an infallible guide has nonetheless encouraged major interventions in natural ecosystems, and these have had unanticipated and usually unfortunate consequences. The history of both fisheries and forest management are replete with illustrations of the failure of sustained-yield methods to forestall ecological collapse (McEvoy 1988; FAO 1984). While there no longer exist social constraints on levels of intervention, scientists are also inadequately informed on the ecological processes in operation. Ironically, therefore, religion and custom as ideologies of resource use are perhaps better adapted to deal with a situation of imperfect knowledge than a supposedly 'scientific' resource management.

## Conflict Between and Within Modes

*Inter-modal conflict*: As one mode of resource use comes into contact with another mode organized on very different social and ecological principles, we expect the occurrence of substantial social strife. In fact, the clash of two modes has invariably resulted in massive bursts of violent and sometimes genocidal conflict. One of the best documented of such conflicts is the clash between the indigenous hunter-gatherer/shifting-cultivator populations of the New World on the one hand, and the advance guard of European colonists practising an altogether different system of agriculture on the other. Historians of the victorious white race have in recent years depicted, with marked sensitivity, the ecological bases of a conflict that resulted in the extermination of the bulk of the native population and the traumatization of the segment which escaped annihilation (Cronon 1983; Crosby 1986). Yet, as Chapter 2 of this book argues, such episodes are not exactly foreign to the histories of the Old World. For the brutal conflict between the American Indians and the English colonists was anticipated several millenia before—in the conquest of indigenous hunter-gatherer populations within India by invading agriculturists, a clash vividly captured in the sacred text of the conquerors, the Hindu epic Mahabharata.

The environmental and social costs of the encounter between agrarian and industrial modes in Europe have been ably chronicled by a long line of distinguished historians. The rise of industrial capitalism radically altered relations not merely in land and the workplace, but also around the utilization of nature. The conflicts over the enclosure of what was previously common land, and the assertion of state control over forests, while perhaps not as brutal as the clash between hunter-gatherers and the neolithic vanguard, also exacted a heavy human cost (cf. Agulhon 1982). As one American forester, documenting the enclosure of the forests by the state and lords in Germany, laconically put it:

> All these changes from the original communal property conditions did not, of course, take place without friction, the opposition often taking place in peasants' revolts; hundreds of thousands of these being killed in their attempts to preserve their commons, forests and waters free to all, to re-establish their liberty to hunt, fish and cut wood, and to abolish titles, serfdom and duties (Fernow 1907, p. 34).

Even where there was no open revolt, peasants resorted to poaching and the theft of forest produce. These forest 'crimes' were very widespread. In Prussia in 1850, 265,000 wood thefts were reported, as against only 35,000 cases of ordinary theft. In this class struggle over nature, between the peasants on one side and landlords and the state on the other, the wood thieves were 'defending their entire economic system—the family economy which was based on collective usage rights' (Mooser 1986). Indeed until well into the present century, European foresters were unable to fully extinguish the customary rights of user granted as a consequence of the peasant opposition to state forest management (cf. Heske 1937). In the American South too, common rights of hunting and grazing on non-cultivated land withstood landlord and state pressures until the early decades of this century (Hahn 1982).

The European experience is directly comparable to the Indian (detailed in chapters 5 and 7), where the encounter between peasant and industrial modes, mediated partly by colonialism, has also greatly intensified social conflicts over forest resources. In the colonial societies of South East Asia and Africa as well, the takeover of forest land for strategic and commercial purposes fuelled bitter conflicts between the state and the peasantry (Scott 1976; Grove 1990). Here we must note that colonial methods of ecological control also intensified conflicts within the agrarian sector. Two examples come to mind. One is the clash over forests, lands and water in colonial and semi-colonial societies, between large cash-crop plantations and the traditional peasantry (Womack 1969; Pandian 1985). The other is the restrictions placed on subsistence hunting-gathering and grazing by the setting up of game reserves by white colonists (cf. Kjekshus 1977).

While gatherer–peasant and peasant–industrial conflicts are both vividly represented in myth and history, a third inter-modal conflict dominated medieval Europe and medieval Asia, namely between the peasant and pastoral modes. Despite the many instances of nomad–peasant symbiosis, for example the pasturing of livestock on fallow fields in return for manure, this relationship has historically been fraught with friction. Medieval sources in the Near East talk of nomads driving their livestock to graze on *sown* fields—a practice beneficial to the pastoralists but disastrous for cultivators. Conflicts have also arisen when agriculturists shift to crops (e.g. cotton) which do not provide stubble during the dry seasons. In this conflict between two modes of resource use that have, for the most part, overlapped in time and space, the nomads have on occasion greatly expanded their resource base—as with the Mongols of Central Asia in the medieval period—while at other times their own niche has steadily shrunk—as with the reservation of forests and the expansion of irrigated agriculture in modern India (Khazanov 1984; Gadgil and Malhotra 1982).

Our final example of inter-modal conflict comes from the contemporary West. In recent years a distinctively new form of ecological conflict has surfaced, between votaries of the industrial mode of resource use and a mode struggling to be born. Thus, while scientific foresters and industrial users continue to look upon the forest primarily as a resource to be harvested, in the perspective of the Western environmental movement forests are to be preserved as a haven from the workaday world and as a reservoir of biological diversity. Although this conflict has by no means been as violent as those between earlier modes of resource use, the ideological differences between the industrial mode and what environmentalists call the 'post-industrial' mode can hardly be minimized (Nash 1982; Hays 1987).

Two aspects of inter-modal conflicts need to be highlighted. First, apart from massive bursts of social conflict, the encounter between the different modes also signals a spurt in rates of ecological destruction. The deforestation of the Indo-Gangetic plain in the first and second millennium BC, the draining of the fens in eighteenth-century Europe, and the destruction of forest cover in colonial India—all these testify to the enormous environmental costs associated with the advent of a new mode of resource use. (Of course the emergence of a 'post-industrial' mode, if it occurs, may reverse this trend.) As one mode wins out, there is a slow but perceptible diminution in the levels of social conflict and ecological disturbance.

Second, while the conflict between different modes of resource use is, at the most elemental level, a struggle for control over productive resources, it is invariably accompanied by an ideological debate legitimizing the claims of the various modes. Agrarian societies have typically justified their takeover of the lands and resources of gatherer societies in terms of the latters' low productivity and 'wasteful' use of nature, a distinction used by colonists in America for their conquest of Indian territories (Cronon 1983; Prucha 1985). Likewise, votaries of the industrial mode have used the rhetoric of scientific conservation when legitimizing their claims. The scientific model of natural resource management, it is claimed, is a distinctively modern innovation, and inherently superior to the idioms of religion and custom which legitimized human interactions with nature in earlier epochs. And, as the widespread disaffection with excesses within the industrial mode finds expression in modern environmental movements, it is no accident that religion and custom are upheld as being, after all is said and done, more prudent in their use of nature than modern 'scientific' methods.

## Intra-modal Conflict

The ebb and flow of social conflict *within* different modes of resource use is perhaps not as visible in history and myth as the violent struggles *between* modes. Yet such conflict is by no means absent. Most of the primate relatives of human beings are engaged in struggles to hold on to and expand group territories, and it is very likely that hominids were similarly engaged ever since their origins. There is little fossil evidence of violent death in the period before 30,000 b.p., when modern *Homo sapiens*, with its capacity for symbolic communication, finally came into its own. Since that time, conflicts relating to territorial control have undoubtedly been a feature of most hunter-gatherer societies, as documented by anthropologists in such far-flung areas as New Guinea, New Zealand, and the Amazon (cf. Rappaport 1984; Vayda 1974).

Such intra-modal conflicts become more complex as one passess on to more advanced modes of resource use. They become especially acute when the ideal-typical characteristics of a mode, as outlined earlier in this chapter, are perceived as being distorted to subserve the ends of a particular social group. When feudal lords fail to honour the customary codes of the moral economy, for example, or when the workings of forest law and science are seen to be class-specific rather than class-neutral, the ideological basis of the mode of resource use begins to crumble. At such times an idiom of conflict, rather than collaboration, characterizes intra-modal relations.

Coming to the peasant mode, in many periods conflicts between the peasantry and overlords over natural resources has been endemic. In England, for example, conflicts over rights in forests were especially acute in the thirteenth and fourteenth centuries, a period of rising demographic pressure and the expansion of arable land. There were growing numbers of prosecutions for timber theft, with peasants also occasionally invading woodland enclosed by lords and abbots (Faith 1984; Birrel 1987). Conflicts over forest and pasture were an important element in one of the greatest ever anti-feudal revolts, the German Peasant War of 1525 (cf. Engels 1956). And in France from the sixteenth to the eighteenth centuries, peasants repeatedly rose in revolt against attempts by landlords to usurp forests, swamps and grazing grounds earlier held in common (Ladurie 1980; Bloch 1978). Finally, resistance to the enclosure of community pastures by plantations was widespread in early-twentieth-century Mexico (Lewis 1964; Womack 1969).

Another form of intra-modal conflict, characteristic of mature feudalism in Europe, related to peasant rights over pasturage and timber in forests which were reserved exclusively for hunting by the nobility (Thompson 1975; Hay *et al.* 1975). While in normal times peasants were unable to challenge this monopoly (though they continued to breach it on the sly), when the state was vulnerable they quickly and forcefully asserted their rights. Thus, in the rural revolts that accompanied the French Revolution, groups of peasants broke into the hunting preserves of the nobility and 'determinedly hunted the game' (Lefebvre 1982, p. 44 and *passim*). Similar invasions of forests controlled by the nobility were also reported in the wave of peasant strikes during the Russian Revolution of 1905 (Shanin 1986).

With the shift from the peasant to the industrial mode, conflicts within the peasant mode intensify as one class within agrarian society is quicker to adapt to the socio-ecological orientation of the coming mode (e.g. landlords, in the case of the enclosure movement). But the internal conflicts characteristic of the industrial mode—particularly its capitalist variant—are quite different. These are, first, the continuing struggles of individual capitalists with each other, and against the state, for formal proprietorial control of non-cultivated areas, as well as the terms of disposal of living resources. Second, the industrialization of forest usage itself creates a new class of workers whose interests are not always in harmony with those of capitalists in the timber harvesting and processing sectors (Vail 1987).

These forms of intra-modal conflict, and the ways in which they are resolved, shed light on the interlinkages between modes of production (defined in the Marxian sense) and their corresponding modes of resource use (as defined here). In the natural economy of hunter-gatherers, of course, the mode of production is simultaneously the mode of resource use. In the peasant and industrial modes the links are more complex. In the former case, even where relations around land are accepted (however grudgingly) as asymmetrical, peasants insist that they have full access to the 'free gifts' of nature. And when enclosure-minded landlords and hunting monarchs violate this tacit agreement by restricting the exercise of common rights, peasants resist and the levels of conflict escalate. For the *stable* functioning of the mode of production, therefore, there must exist some discrepancy between the rights of the overlord to land (and a portion of its produce), and to living resources, respectively.

The stable functioning of the industrial-capitalist mode likewise requires a discrepancy between property relations in the field/factory, and in the forest. While private ownership predominates in the former case, forests are, to a much greater extent, owned and controlled by the state. However, the underlying rationale of government intervention is precisely to safeguard the stability of the industrial mode, by taking the long-term view available only to the state, thereby harmonizing conflicts between individual capitalists.

## TABLE 1.1

FEATURES OF TECHNOLOGY AND ECONOMY IN A SOCIETY
PREDOMINANTLY FOLLOWING ONE OF THE FOUR MAJOR
MODES OF RESOURCE USE

| | Gathering (including shifting cultivation) | Nomadic pastoralism | Settled cultivation | Industry (including fossil-fuel-based agriculture) |
|---|---|---|---|---|
| *Energy resources used* | Human muscle power, fuelwood | Human and animal muscle power, fuelwood | Human and animal muscle power, fuelwood; coal, water power to some extent | Fossil fuels, hydro-electricity, nuclear power; fuelwood, human and animal muscle power much less important |
| *Material resources utilized* | Stone | Plant and animal material | Stone, plant and animal material, some uses of metals | Extensive use of metals and synthetic materials |
| *Abilities to store resources* | Very rudimentary | Domestic animals serve as meat supply on hooves | Grain and domestic animals make long-term storage of food possible | Even highly perishable materials like fleshy fruit and meat can be stored over long periods |
| *Abilities to transport resources* | Very rudimentary | Domestic animals like horses make long distance transport possible | Domestic animals make long distance transport possible | Fossil-fuel based vehicles render transport over great distances easy |

| | *Gathering (including shifting cultivation)* | *Nomadic pastoralism* | *Settled cultivation* | *Industry (including fossil-fuel-based agriculture)* |
|---|---|---|---|---|
| *Abilities to transform resources* | Very rudimentary | Rudimentary | Low, including metal making, weaving | Very extensive |
| *Spatial scale of resource catchments* | Small, mostly of order of a few hundred or thousand km$^2$ | Could be quite extensive | Moderate | Global |
| *Quantities of resources consumed* | Very moderate | Moderate for most | Moderate for most, a small elite may consume large quantities of commodities | Large numbers consume enormous quantities |

## TABLE 1.2

FEATURES OF SOCIAL ORGANIZATION AND IDEOLOGY IN A
SOCIETY PREDOMINANTLY FOLLOWING ONE OF THE FOUR
MAJOR MODES OF RESOURCE USE

| | Gathering (including shifting cultivation) | Nomadic pastoralism | Settled cultivation | Industry (including fossil-fuel-based agriculture) |
|---|---|---|---|---|
| *Size of social groups* | Small, a few thousand people | Moderate, several thousand people | Moderate, several thousand people | Very large, in hundreds of thousands |
| *Extent of kinship within social groups* | Very strong | Strong | Strong, but interactions with non-kin increasing | Very weak |
| *Extent of attachment of social groups to particular localities* | Often strong | Weak | Often strong | Very weak |
| *Division of labour* | Rudimentary | Sex-age based | Rudimentary sex-age based | Considerable, based on specialized skills and knowledge |
| *Role of division of labour in formation of social groups* | Very weak | Very weak | Moderate to strong | Strong |

|  | Gathering (including shifting cultivation) | Nomadic pastoralism | Settled cultivation | Industry (including fossil-fuel-based agriculture) |
|---|---|---|---|---|
| *Idiom of social transactions* | Totally informal, based on face-to-face contact | Informal, primarily based on face-to-face contacts | Social conventions and codified transactions | Codified transactions with legal sanctions very important |
| *Within group differentiation of access of resources* | Weak | Weak | Considerable | Very extensive |
| *Mechanisms governing access to resources* | Community-based decisions | Community-based decisions | Private control of farmland, community and state control of non-cultivated lands | Private, state and corporate ownerships predominant; community ownership delegitimized |
| *Perception of working of nature* | Nature viewed as autonomous, capricious | Nature largely seen as capricious | Nature viewed as partially law-bound, controllable | Nature viewed largely as lawful |
| *Idiom of man-nature relationship* | Man part of community of beings | Man potentially conqueror of nature | Man a steward of nature | Man above and apart from nature, fully capable of controlling it |

### Recapitulation

The salient features of the four modes of resource use are reproduced in Tables 1.1 to 1.4. To recapitulate, as human societies move temporally from hunting and gathering through pastoralism, agriculture, and finally into industrialization, five distinct but closely interrelated processes occur. First, there is an increasing intensity of resource use and exploitation. Second, there is a secular increase in the level of resource flows across different geographical regions and across different levels of any economic/political system. Third, there is an integration of larger and larger areas into the domain of any given political/economic system. Fourth, there is, at the global level, a secular increase in population densities and in the extent of stratification and inequality with respect to the access, control and use of different natural resources. Finally, there is an intensification of rates of ecological change and ecological disturbance.

Of our four modes of resource use, in the Indian context nomadic pastoralism is best treated not as a separate mode but as being integrated with the peasant mode of resource use, within whose ecological zone it occupied a special niche. In Parts II and III we shall move from the ideal-typical to the historical, narrating the rise, decline and fall of different modes of resource use in the Indian subcontient.

TABLE 1.3

THE NATURE OF THE ECOLOGICAL IMPACT IN SOCIETIES
PREDOMINANTLY FOLLOWING ONE OF THE FOUR MAJOR MODES
OF RESOURCE USE

|  | Gathering (including shifting cultivation) | Nomadic pastoralism | Settled cultivation | Industry (including fossil-fuel-based agriculture) |
|---|---|---|---|---|
| Land transformation | Little, some regression of patches of forest to successional stages or grassland | Some extension of grasslands, deserts | Forests, grasslands extensively converted to fields | Large-scale deforestation, desertification, built-up habitats |
| Habitat diversity | Enhanced | Somewhat reduced | Reduced | Substantially reduced |
| Biodiversity | Little affected | Some effect | Moderate effect | Considerable impact |
| Resource populations | Occasionally overharvested | May overgraze some grasslands and overharvest some prey populations | May overgraze some grasslands and overharvest some prey populations | Many resource populations overharvested |
| Substances poisonous to life | Nil | Nil | Nil | A large range of synthetic chemicals |
| Modification of biogeochemical cycles | Very little | Very little | Little | Substantial |
| Modification of climate | Highly unlikely | Highly unlikely | Unlikely | Quite likely |

## TABLE 1.4

FEATURES OF SOCIAL ORGANIZATION, IDEOLOGY AND ECOLOGI-
CAL IMPACT IN SOCIETIES IN RELATION TO CHANGES IN THEIR
RESOURCE BASE

| | In equilibrium with resource base | Resource base expanding | Resource base shrinking |
|---|---|---|---|
| *Fluidity of social groups* | Low | High | Often considerable |
| *Extent to which group interests prevail over individual interests* | Group interests quite significant | Individual interests more important | Group interests may crumble |
| *Perception of man-nature relationship* | Man as steward of nature | Man as conqueror of nature | Man helpless |
| *Extent to which sustainable resource use prevails* | Quite often | Rarely | Rarely |
| *Level of ecological impact* | Low to moderate | High | High |

APPENDIX TO CHAPTER I

## Note on Population

In our analysis of modes of resource use, we have classified these modes according to technology, pattern of resource flows, social structure, dominant ideologies, and systems of conservation. In this appendix we deal briefly with another dimension which has an important bearing on ecological history—namely the density of human population. The reflections which follow are prompted by the question—under what conditions will human populations regulate rates of reproduction in individual or group interests?

In utilizing the resources of their environment at very low intensities, gatherers also maintain low population densities. As suggested above, in stable environments they are likely to maintain fixed territories, with high levels of conflict with neighbouring groups over territorial control. Under these conditions their interests lie in maintaining the population density at a level which ensures that the territorial group does not have to face occasional resource shortages—shortages that may weaken it in a conflict with neighbouring groups. It is possible that the variety of population-control mechanisms noted among gatherers may serve the function of maintaining their population densities in the group's interest. Such mechanisms may, of course, break down when gatherers lose control over their resource base—upon coming into contact with people practising more advanced modes of resource use.

Coming next to pastorals, their population densities tend to be low because they inhabit the more arid and unproductive regions. Very likely, these levels are maintained by natural checks; it is unlikely that the inhabitants of such variable environments, who have no fixed territories, would have evolved cultural practices to deliberately maintain their populations at low levels.

Characterized by a substantial intensification of resource

use, agricultural societies maintain population densities far higher than those of either gatherer or nomadic pastoral societies. However, most members make rather low-level demands on the resource base, though members of the non-agricultural elite may have high levels of demand for many non-essential commodities. Given these low levels of resource demand, and the possibilities of gradually improving yields from cultivated lands through technical change, agricultural communities are likely to be characterized by slow population growth. Further, the larger states of which they are a part generally ensure that there are no serious territorial conflicts within their limits—conflicts in which overpopulation may be a serious handicap. Yet the military strength of agricultural states may depend on their overall population; therefore the state apparatus would tend to encourage population growth. One therefore expects neither individual nor group-level pressures to deliberately check population growth in agricultural societies; *natural* checks and balances, such as diseases and disasters, may take a periodic and heavy toll of peasant populations.

The population history of the industrial mode is captured in the phrase 'demographic transition'. While in the initial phase of industrialization the population of European stock (both within Europe and in newly-colonized lands) expanded rapidly, over the last century it has grown much more slowly, stabilized, and in some places even declined. This transition seems related to the fact that the attempts of each individual in enhancing his/her resource consumption limits the quantity of resources available to raise offspring. Further, parents try to endow each offspring with a high level of ability to garner resources for itself. The need to invest in the *quality* of children again implies a severe limitation on the *quantity* of children produced. Industrial societies have thus generally stabilized their populations. At the same time, their *per capita* resource consumption remains high and on the increase.

PART TWO

# Towards a Cultural Ecology of Pre-modern India

# CHAPTER TWO

# Forest and Fire

## Geological History

When mammals made their appearance on earth 180 million years ago, the Indian subcontinent had recently broken off from the great southern land mass which earlier included South America, Antarctica and Africa. The land mass which is now India was crossing the equator when the age of dinosaurs came to an end, sixty-five million years ago. Fifty-four million years ago its northern tip bumped into the Asian plate; in another four million years it had established contact along the remaining boundary. The aftermath of this collision has been the Himalaya, rising at a rate of two centimetres a century to produce the tallest mountains in the world (Coward *et al.* 1988; Meher-Homji 1989). This barrier has been continually going up over the last fifty million years, so that most mammals have come to India through passes to the west, and some to the east of this mountain chain. Hominid fossils first appeared on the subcontinent in the Himalayan foothills thirteen million years ago; they continued till seven million years BP (before present) as a part of the mammalian communities of wooded habitats. Then they disappeared, presumably victims of habitat changes brought about by the continuing upward movement of the Himalaya (Badgeley *et al.* 1984).

Much later, members of our own genus, *Homo,* moved onto the subcontinent, probably from their place of origin in the savannas of East and South Africa. Tool-using hominids arose in Africa around two million years ago; by one million years BP they had reached Java. Most likely, they colonized India around this time; but firm evidence of human occupation appears in the form of artefacts such as hand-axes somewhat later, between 700,000 to 400,000 BP (Rendel and Dennell 1985). By this time the lifting up of the Himalaya was done and over, and a seasonal monsoon climate had become established (Rajaguru *et al.* 1984).

The hunter-gatherer populations of our own species, *Homo sapiens*, came to cover, in a thin way, much of the country through the remaining part of the late Pleistocene. At this time the climate fluctuated between periods of weak, moderate and strong monsoon. However, the wet and hilly tracts of the Western Ghats, the west coast, and the north-eastern hill regions as well as the Gangetic plain remained unoccupied until the terminal Pleistocene of 20,000 years BP, when the monsoons became distinctly weak at the height of glaciation in the northern latitudes (Rajaguru *et al.* 1984; Pant and Maliekel 1987; see table 2.1).

### Prudent Predators

We can only speculate on the ecological-niche relationships of these hunter-gatherer populations. In the productive, stable, tropical environments that they inhabited, most would have been organized in the form of bands with strong bonds to their territories, often in conflict over land- and water-use with neighbouring groups. Each endogamous tribe would have adapted itself culturally to its biological and physical environment, having learnt by trial and error what to eat and what to avoid how to look for food and how to keep away from predators. Each such tribe may then be thought of as occupying a distinctive, spatially disjunct, niche. As discussed in Chapter 1, such human communities would be expected to develop cultural traditions of sustainable utilization of their resource base (Gadgil 1987).

There is naturally no guarantee that such conservation practices would have led to the long-term persistence of all the elements of their biological environment. But so long as the total demand on resources remained limited, the human populations would tend to reach an equilibrium with their resource base after the elimination of elements that were over-utilized. Even at the hunting-gathering stage, the demand on some resources used as commodities, for instance ivory, could increase without limit and lead to their over-utilization. Drastic environmental changes could also result in a disturbance of the equilibrium that human populations may have reached. Climatic changes attendant on the withdrawal of Pleistocene glaciation 10,000 years ago seem to have resulted in the extinction of many species all over the world; this could in part have been due to the over-extension of human hunting after major changes had taken place in the prey populations. The baboon and hippopotamus became extinct in India at this time, perhaps as a result of such processes (Rajaguru *et al.* 1984; Badam 1978).

While more advanced agricultural societies have replaced hunter-gatherer societies over large parts of the moist tropical forest tracts of India, there are, even today, extensive areas where hunter-gathering shifting-cultivator societies persist. These include the humid forest tracts to the north-east of the Brahmaputra valley, and parts of central India where the eastern end of the Vindhya ranges join the north-eastern tracts of the Eastern Ghats. The difficulties of settled cultivation in this hilly terrain, its poor access from major centres of human populations, and low population densities resulting from malaria and inter-tribal wars—all these have contributed to the persistence of hunter-gatherer shifting-cultivator populations in these regions. Some of these populations have converted to Chris-

## TABLE 2.1
### CLIMATIC HISTORY OF THE INDIAN SUBCONTINENT

| Geological period | Years BP | Climate | Geomorphic data | Human population |
|---|---|---|---|---|
| Late Holocene | < 4,000 | Moderate monsoons | Saline lakes in western India | Agricultural settlements cover the subcontinent |
| Early Holocene | 10,000–4,000 | Strong monsoons | Fresh-water lakes, entrenched streams, stable dunes | Beginning of agriculture, denser populations |
| Terminal Pleistocene | 20,000–10,000 | Distinctly weak monsoons | Hypersaline lakes, choked rivers, active dunes | Human population spread throughout the subcontinent at low densities |
| Late Pleistocene | 70,000–20,000 | Weak to moderate monsoons | Entrenched streams, stable dunes | Hunter-gatherers in small groups, wetter tracts not colonized |
| Early-Late Pleistocene | 125,000–70,000 | Strong monsoons | Reddish soils (dating doubtful) | Hunter-gatherers in small groups, nomadic |
| Middle Pleistocene | 700,000–125,000 | Monsoonic seasonal climate | — | First evidence of human occupation |

| Geological period | Years BP | Climate | Geomorphic data | Human population |
|---|---|---|---|---|
| Lower Pleistocene | 2.0 million–0.7 million | Relatively dry seasonal climate | Volcanic ashes, streams aggrading | Hominids found in both Africa and Java; but no definite evidence of human occupation in India |
| Pliocene | 8 million–2.0 million | Tropical equatorial to strongly monsoonic(?) | Volcanic ashes | — |
| Miocene | 25 million–8 million | Tropical equatorial (?) | — | — |

SOURCE : S.N. Rajaguru (personal communication), R.K. Pant (personal communication) tianity, drastically changing traditional resource use practices in the process. In the non-Christian tracts, however, we see the persistence of social practices favouring prudent resource use. For example, in Meghalaya, Mizoram and the tribal belts of Orissa, we have evidence of sacred groves and quotas on wood extracted from fuel-woodlots, as well as practices with incidental conservation consequences such as protection given to totemic animals and plants (Fernandes and Menon 1987; Fernandes *et al.* 1988).

## Neolithic Revolution

Historians have argued that climatic change at the time of the withdrawal of glaciers 10,000 years BP led to a creeping back of the forest cover and, in time, to a food crisis. This crisis, very likely, prompted hunter-gatherer societies to domesticate animals and cultivate plants. The crisis was not equally acute everywhere; it was perhaps most serious in the Middle East, where the domestication of animals and the cultivation of plants began to gather momentum some 10,000 years ago (Hutchinson *et al.* 1977). Plants such as wheat, barley and lentils, and animals such as cattle, sheep and goats, were first domesticated here. This process undoubtedly provided the stimulus for the beginnings of agriculture and animal husbandry in the Indian subcontinent. The earliest evidence of this comes from Mehrgarh, in what is now the Pakistani state of Baluchistan, around 8000 years BP (Jarrige and Meadow 1980). There is a disputed claim for the origin of rice cultivation in the Gangetic valley of as early as 7000 BP (Sharma 1980); but it is more likely that rice was domesticated in India some 3000 years later, or diffused here from outside (Chaudhuri 1977). It is, however, certain that a number of pulses like horse gram, hyacinth bean, green gram and black gram were indigenously brought under cultivation in India around 4000 years ago. The humped cattle, zebu, is also likely to have been independently domesticated in the Indian subcontinent. These cattle and pulses give Indian agriculture and animal husbandry its special character (Kajale 1988).

Agricultural-pastoral people spread over the Indian subcontinent in many phases. Without metal tools they could not readily penetrate the moister forests, such as those of the Gangetic plains or the west coast. The habitat most favourable to them for cultivation was along the smaller water courses in the relatively drier tracts of north-western India, the Indus plains, and the Deccan peninsula. This is where agricultural settlements developed over the period 6000 to 1000 BC (Possehl 1982; Dhavalikar 1988; Allchin and Allchin 1968). There was also animal husbandry, including nomadic cattle-herding, and this mode of resource use has left traces—such as the ash mounds of the Deccan. It has been suggested that there was a gradual deforestation in parts of the Deccan over this period, with timber fences slowly giving way to stone walls surrounding pastoral camps (Allchin 1963). In the moister tracts over the rest of the country there would have been some shifting cultivation, but this has left no trace. Hunting-gathering, along with shifting cultivation, might then have continued to dominate all the moister tracts of the subcontinent (Misra 1973).

## River-valley Civilizations

The first urban civilization of the Indian subcontinent embraced a very wide region of the north-west. Archaeological evidence suggests that this culture was familiar with the use of the plough. They had also begun to add indigenous rainy-season crops, such as rice and pulses, to winter-season crops—wheat, barley and lentils—that were of West Asian origin (Mehra and Arora 1985). The agricultural surpluses thus produced permitted the establishment of many towns where the surplus served to promote further processing and exchange of materials—i.e. artisanal and trade activities. Exchange over long distances, as opposed to barter on a small scale, called for the maintenance of records, and the Indus Valley civilization offers the first evidence of literacy in Indian history (Possehl 1982). The gradual weakening and disappearance of the urban centres of this civilization have been attributed to a variety of possible causes. The explanation with the best documented evidence relates to the shifting of river courses, on account of geological changes associated with the continuing lifting up of the Himalaya. Satellite imagery clearly shows the palaeo-channels of the river Saraswati, which dried up when the Sutlej shifted its course westward to join the Indus, and the Yamuna eastward to join the Ganges (Fig. 2.1). There have also been suggestions of climatic change, as evident from palaeobotany, the flooding of the Indus, and of the salination of agricultural soils on account of irrigation (Agarwal and Sood 1982).

Iron was introduced to India by about 1000 BC, being associated with the Painted Grey Ware pottery culture of the north-west, the Black and Red Ware pottery culture of central India, and the megalithic cultures covering much of the peninsula. Iron, along with fire, made it possible to bring the middle Gangetic plains under intensive agricultural-pastoral colonization, with wet paddy cultivation as a key element (Kosambi 1970).

With the pattern of resource use becoming grounded in a continual march of agriculture and pastoralism over territory held by food gatherers, the belief system of the colonizers would naturally take a form very different from that appropriate to food gatherers, who had a great stake in the conservation of the resource base of their territories. Since the forest, with its wild animal populations, served as a resource base for the enemy, its destruction, rather than its conservation, would now have assumed priority. Supernatural power would now no longer reside in specific trees, groves or ponds, but would be the more abstract forces of nature: earth, fire, wind, water, and sky, whose assistance could be invoked in the task of subordinating hunter-gatherers and colonizing their resource base. Fire to clear the forest, and water to nourish crops in the fields, would be the most valuable of these forces; therefore Agni and

Varuna were the major deities. The main ritual was fire worship, the Yajna, a ritual in which huge quantities of wood and animal fat were consumed.

The burning of the Khandava forest, as depicted in the Mahabharata, beautifully illustrates the operation of this belief system. In this episode Krishna and Arjuna are at a picnic in the great Khandava forest which lies on the banks of the Yamuna, where the city of Delhi stands today. A poor Brahman appears begging for alms. On being granted his desire, the Brahman reveals himself as Agni, the fire god. He then asks that his hunger be satiated by the burning of the Khandava forest, along with every creature within it. Krishna and Arjuna agree to this, whereupon Agni gives them a fine chariot, and bows and arrows, to perform the task. The forest is set on fire, and Krishna and Arjuna patrol its perimeter, driving back all the creatures who attempt escape. This includes *nagas* (cobras)—probably the appellation for food-gathering tribes which venerated snakes.

Arjuna evidently wants to clear the Khandava forest to provide land for his agricultural/pastoral clan, and to build their capital city, Indraprashtha. The burning of the forest, and the killing of wild animals and tribal food gatherers is couched in the terminology of a great ritual sacrifice to please Agni. Agni's appearance as a Brahman begging alms is significant, because Brahmans, who presided over fire sacrifices (Yajnas) played an important role in the process of colonization. They served as pioneers, establishing their outposts in forests and initiating rituals which consumed large quantities of wood and animal fat. Thus provoked, the native food gatherers, termed demons or Rakshasas, would attempt to disrupt the holocaust and save their resource base in order to retain control over their territories. Specialist warriors, Kshatriyas, would then rush to the rescue of the Brahmans who had furnished them with appropriate provocation to invade these territories. This process is represented in the Mahabharata, in Dushyanta's visit to the abode of the Brahman sage Kanva. Dushyanta combs the

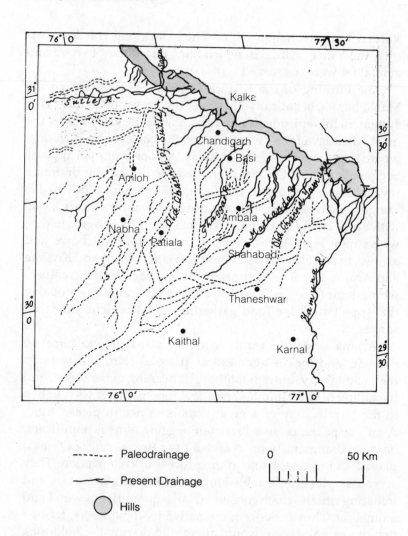

------- Paleodrainage

~~~ Present Drainage

◯ Hills

0       50 Km

**Figure 2.1**

The Indo-Gangetic divide showing paleo-river channels.

forest with the help of hundreds of assistants, killing wild animals with complete abandon. It appears reasonable to conclude that the purpose of this slaughter was to destroy the resource base of hunting and gathering tribals who lived in the forest (Karve 1974).

The archaeological evidence that becomes available from this period also makes it possible to make more definite statements about human population changes, and to relate these changes to the ecological setting. The land-to-man ratio would have been high in the phase of expansion—the phase analogous to that of r-strategists of ecology. For instance, Lal (1984) suggests a density of 0.75 persons per $km^2$ for Kanpur district in the Ganga-Yamuna divide for the Black and Red Ware period around 1350 BC; this would also have been accompanied by a high livestock-to-man ratio. Meanwhile, more forests would have to be burnt and colonized as older lands were over-used and exhausted. An ethic of exhaustive resource use, with the Yajna as its cornerstone, is likely to have been the ruling belief system in this period. If the human population did indeed quadruple over the next eight centuries of agricultural growth, as Lal (1984) has argued, then much of the fertile land would have come under the plough, resulting in a lower livestock-to-man ratio and a declining availability of forest land for further colonization. Now, with the society moving into a phase analogous to the k-strategists of ecology, people would have required a new belief system, stressing more careful and sustainable patterns of resource use. Such a belief system is likely to have appealed to the agricultural-pastoral component of the population, but to have been opposed by the Brahmans, votaries of the Yajna system. Buddhism and Jainism, sects often described as heterodox, i.e. in opposition to the Brahmanic religion, appear to have been responses to this need. Both these religions are known to have protested against the hegemony of Brahmans and the wasteful burning of endless quantities of clarified butter and wood, and the slaughter of animals in sacrificial rituals (Thapar 1984).

As we noted in Chapter 1, the belief system of food gatherers, with its conservation orientation, saw spirits that were respected in trees, groves, ponds, rivers, and mountain peaks. In contrast, in the belief system of food producers conquering new territories, worship focussed on elemental forces such as fire and water, and the great god of war, Indra. As the frontier receded and there were no longer large fertile tracts to move towards, there emerged the belief systems of Buddhism and Jainism, which rejected the supernatural and advocated a rational arrangement of human affairs. These belief systems included an important element of conservation: they advocated that resources should not be used wastefully.

## Social Organization

By the time of Gautam Buddha, some 2500 years BP, settled agriculture and pastoralism had covered wide tracts of the country. These continued into the drier tracts of western India, from where they made their way into the peninsula. The central parts of peninsular India are semi-arid, like much of western India, and here too small-scale cultivation and pastoralism took root along the tributaries of major rivers. The highest concentration of agricultural/pastoral populations were of course along the Gangetic plain. In all these tracts chiefdoms would have sprung up, their scale depending on the scale of cultivation and the extent of expropriable surplus. Food gatherers continued in the hilly tracts of the Himalaya and north-eastern India, on the Central Indian plateau, and on the Eastern and Western Ghats.

In the tracts brought under settled agriculture, the original territorial boundaries of food-gathering tribes would necessarily have broken down in this period. Food producers, whose populations must have undergone substantial expansion, moved into these lands, as might have other food gatherers displaced from their territories. An important question arises here—what would happen to the barriers of endogamy and culture which earlier existed in populations that were now thrown together within the larger territories of chiefdoms? One possibility is that these barriers of endogamy and cultural differentiation may have broken down, either partially or completely, producing a more or less homogeneous population. However, an agricultural/pastoral society, with its specialized crafts, trade, administration and fighting forces, is far more heterogeneous than a society of food gatherers because it involves a substantial division of labour and differentiation of status. A more likely possibility therefore is that, instead of merging, the different endogamous groups remain distinct and are assigned different tasks and status within the society. While there has undoubtedly been some merging of earlier endogamous groups and some redifferentiation in Indian society, barriers of endogamy among different tribal groups seem to have been largely retained, being converted into barriers of endogamy among caste groups (Karve 1961).

This transition from tribe to caste probably permitted the elites, which were involved in spreading food production and mopping up the surplus, to assign tedious and low-status tasks to various food-gathering tribes. The process itself was rationalized in the *varna* system, which divided the society into Brahmans or priests, Kshatriyas or warriors, Vaishyas or traders, Sudras or peasants and higher-status artisans, and Panchamas (Untouchables) or lower-status artisans and labourers. The conquered food gatherers were then assigned to the Sudra and Panchama categories, both to till the land and to perform lower-status artisanal and service tasks. However, these categories were by no means genetically homogeneous entities, and in any region a very large number of endogamous caste groups made up any one varna. Thus in western Maharashtra today the Sudra varna is made up of many endogamous groups of peasant cultivators such as Kunbis, of artisans such as Kumbhars, of pastorals such as Dhangars, and so on. Further, the potter Kumbhars or pastoral Dhangars are themselves a cluster of many distinct endogamous groups. Among the pastoral Dhangars of western Maharashtra, for example, we find the Gavli Dhangars who maintain buffaloes in hill forests, the Hatkar Dhangars who maintain sheep in the semi-arid tracts the Khatik Dhangars who are butchers, the Sangar Dhangars who are weavers of wool, and the Zende Dhangars who used to maintain ponies (Malhotra and Gadgil 1981). It is therefore appropriate to focus on individual endogamous groups as the primary unit of Indian society, and reserve for them the term 'caste' (*jati*) (Karve 1961). A set of castes with similar occupations and with some cultural affinity may then be characterized as a caste cluster. Endogamous castes that make up a caste cluster such as the Dhangars may often be genetically as well as culturally quite varied. Varna itself is a largely artificial construct, with member castes or caste clusters being extremely varied, both genetically and culturally. Even the two higher varnas, Brahman and Kshatriya, who have some cultural affinity among their member castes, have been shown to be genetically quite heterogeneous. For instance, in western Maharashtra the Rigvedic Deshastha Brahmans are genetically closer to the local Shudra Kunbi castes than to the Chitpavan Konkanastha Brahmans (Karve and Malhotra 1968).

This process of forcing the newly-assimilated groups into lower-status tasks was rationalized by the elite in two ways. It was justified first on the grounds that these groups were, at least in some ways, biologically distinctive, and second by attributing birth in lower-status groups to sins committed in a previous birth. A very interesting dialogue of the Buddha addresses this issue. When a questioner equates the different endogamous groups to different species of plants, the Buddha's rejoinder is to point out that the different biological species are separated by barriers to reproduction, with hybrids being either sterile or impossible to produce. Different human groups, on the other hand, are clearly inter-fertile, and, in keeping with his rationalist approach, the Buddha advocates their merger. Buddhism and Jainism, however, did not succeed in destroying the social hierarchy of Indian society.

## The Age of Empires

The eight centuries from 500 BC to AD 300, which followed the colonization of the fertile lands of northern India, seem to have been characterized by the availability of large surpluses of agricultural production for activities outside food production. The river valleys of peninsular India—for example of the Krishna, Godavari, Kaveri, and Vaigai—were also being brought under the plough at this time. The larger the surplus available, the larger the scale controlled by the elite which organized the usurpation of this surplus. Large surpluses would also have promoted large-scale resource exchange—i.e. trade. The chiefdoms of earlier times therefore gave way to larger states—to those of the Mauryas and the Kushanas in the north, and to the Chalukyas and Sangam Cholas in the peninsula (Thapar 1966). These states indulged in a vigorous trade, both internal and overseas.

Sustained by the surplus thus garnered, the primary interest of these states would be to generate ever-larger surpluses within their own territory, and to acquire as much surplus as possible from the territories of other states. The first aim can be achieved through an extension of the area under cultivation, and by enhancing the productivity of agriculture through the provision of irrigation facilities; the latter objective is carried out through external trade and warfare. These preoccupations are reflected in the activities of the Mauryan state, as recorded especially in Kautilya's *Arthasastra* (Kangle 1969). Based within the large tracts of fertile and cultivated lands in the Gangetic plain, the Mauryan state was keenly interested in pushing the frontiers of cultivated tracts further and further back, for instance by colonizing Kalinga (modern Orissa). This, of course, represented a continuation of the conflict between food gatherers and food producers which accompanied the colonization of the Gangetic plain itself. As observed earlier, that effort was pushed forward by smaller chiefdoms, with Brahmans and their Yajnas serving as forward probes. The continuation of such expansion under the Mauryan state was a more organized effort in the deliberate colonization of river valleys of the hilly tracts that now bounded the empire. The domesticated elephant probably played a significant role in the invasion of these tracts; a temple frieze in Orissa shows an elephant picking up in its trunk and flinging to death a man with the physical features characteristic of a tribal. At the same time, by providing irrigation works the Mauryan state attempted to boost the productivity of land already under cultivation.

By this time the elephant was established as an important component of the war machine, and Kautilya's *Arthasastra* discusses in some detail the quality of elephants, their capture and care, as well as the conservation of elephant forests. It remarks that every king should attempt to have in the army as large a number of these beasts as possible. Since elephants were never bred in captivity, the *Arthasastra* advocates setting aside, under strict protection, elephant forests on the borders of the state (Trautmann 1982). Such forests were undoubtedly inhabited by food-gathering tribals, and, since the killing of elephants attracted the death penalty, these tribals would be forced to give up the consumption of elephant meat. It is noteworthy that while elephants are hunted for meat in the north-eastern provinces of Mizoram, Nagaland and Arunachal Pradesh—areas that never came under the sway of any Indian state till British times—tribals in the rest of the country do not consume elephant meat.' Since these tribals readily consume Gaur, a wild relative of the sacred cow, it is likely that the taboo on elephant meat in peninsular India was a consequence of other taboos enforced by early states such as the Mauryan.

The Mauryan kings also maintained some forested areas as hunting preserves. Parks close to their capitals contained dangerous animals; these seem to have been shorn of their teeth and claws before being released for the hunting pleasures of the nobility. Further away there existed natural forest areas where no one except the nobles was allowed to hunt. This might, in part, have been a strategy to dominate food-gathering tribals by cutting off a major resource. The demarcation of these areas for exclusive use also provided practice in warfare and the pleasures of high life for the warrior classes.

Elephant forests and hunting preserves brought in a new form of territorial control over living resources—control by the state. As noted earlier, to the territorial control exercised by food-gathering tribes, peasants added control of individual fields by families, of forests and pastures by village communities, and of a much larger territorial entity by the chiefdom. Notwithstanding the large claims of the chiefdom, the actual control over cultivated and non-cultivated land vested with villagers. In contrast, elephant forests and hunting preserves were carved out of non-cultivated lands over which the state now claimed direct control, very likely by de-recognizing some of the rights of local food-gathering tribals and/or peasant communities. The latter were perhaps permitted to continue gathering plant material, and to hunt animals which were not explicitly protected. At the same time, the state attempted to regulate the clearing of forests to establish new agricultural settlements.

## Conservation from Above

This radical transformation led on the one hand to a considerable breakdown of local autonomy at the food-gathering stage, and on the other to the organized outflows of agricultural produce as well as commodities such as elephants, musk and sandal. Both processes must have profoundly affected man–nature interactions, chiefly through the breakdown of local traditions of resource conservation and a gradual over-harvesting and erosion of the resource base. As the frontiers closed, and as the resource crunch mounted, there would very likely have been an increasing social awareness of the need for readjustment through the more efficient and conservative use of resources. It was argued earlier that Buddhism and Jainism represented such a response, with the abandonment of Yajnas being another way of adapting to the changed circumstances. By loosening the hold of Brahmanism, these religions especially attracted the support of traders who flourished in the heyday of high surpluses. Buddhist and Jain monasteries then helped in the opening up of trade routes and the organization of trade, just as the camps of Brahman sages had earlier catalysed the opening up of forests for cultivation (Kosambi 1970).

Buddhism and Jainism then began to play a role in once again designing social conventions which promoted the prudent use of resources. In part, such conservation practices would have been founded on earlier ones, inherited from food-gathering societies. Apart from their appeal to traders, these religions perhaps appealed most to the lower social strata, composed largely out of food gatherers. In fact the leader of the erstwhile Untouchable castes, the late B.R. Ambedkar, believed that the bulk of the Untouchables were adherents of the Buddhist faith in this historical period (Ambedkar 1948). Buddhism and Jainism may thus have played a role for these people by bolstering traditions by which protection was given to various plants, animal species, and various elements of the landscape, such as groves and ponds. In fact the Buddha himself is said to have been born in a sacred grove full of stately sal trees dedicated to the goddess Lumbini.

The best-known ancient state-sponsored conservation campaign was undertaken by the Mauryan emperor Ashoka, following his conversion to Buddhism. The Ashokan edicts advocate both restraint in the killing of animals and the planting and protection of trees. One such edict, from the third century BC, in Dhauli (in present-day Orissa), goes in translation as follows:

> The king with charming appearance, the beloved of the gods, in his conquered territories and in the neighbouring countries, thus enjoins that: medical attendance should be made available to both man and animal; the medicinal herbs, the fruit trees, the roots and tubers, are to be transplanted in those places where they are not presently available, after being collected from those places where they usually grow; wells should be dug and

shadowy trees should be planted by the roadside for enjoyment both by man and animal.

In the heyday of Buddhism and Jainism, therefore, there appears to have been a widespread perception of the need to moderate the harvesting of plant and animal resources. But, as pointed out earlier, it is difficult to arrive at precise prescriptions that work in any given local situation, and even more difficult to arrive at prescriptions that work under a variety of conditions. Buddhism and Jainism did not attempt to prescribe practices applicable to each given local situation any more than did the religious beliefs of food-gathering tribes. These religions had perforce to suggest broad principles, such as compassion towards all living creatures, a ban on killing animals, and planting as well as protecting trees. Jainism, especially in its Digambara branch, carried this to its extreme logical conclusion. Digambara Jains are against the killing of any organism, plant or animal. They permit the consumption of biological products only so long as this does not involve 'killing': for instance grain and milk can be consumed, but meat is completely taboo. Digambara Jain monks wear no clothes: these may trap and kill insects and other small living organisms. The monks sweep the ground as they walk, to eliminate stepping on living things that normally escape notice.

This extreme ethic of non-violence has had a pervasive influence on Indian society. It has led towards the complete ban on the slaughter of cattle at ritual sacrifices, and to the taboo against beef by all the upper and most of the lower castes. Protection to cattle has undoubtedly been important in shaping the practices of mixed agriculture and animal husbandry so characteristic of India. A significant proportion of India's caste population has also come to assume vegetarianism. Notably, this proportion is relatively high in the drier, less productive tracts, such as Rajasthan, Gujarat and northern Karnataka, where animal power is more critical to agriculture and where deforestation is likely to have been always ahead of that in other, wetter regions. This ethic must have been important in the continuation of old, and in the development of new, traditions of protection *vis-à-vis* wild animals and plants. Traditions such as these have played a significant role in influencing the pattern of utilization of biological resources in India over the last several centuries, especially until the arrival of European colonial power.

CHAPTER THREE

# Caste and Conservation

## Resource Crunch

The 'ecological' implications of Buddhism and Jainism notwithstanding, the Indian subcontinent entered a period of decline in trade and urban centres in the fourth century of the
Christian era, the Gupta period (Sharma 1987). This decline
very likely related to a lowering of surpluses from agriculture,
which in turn could have been on account of three kinds of
changes, operating singly or in conjunction. First, there may
have been climatic changes that resulted in a decline in rainfall.
This is plausible, for we do know that there have been significant changes in patterns of rainfall over the historical and
geological time-scale. There is, however, no definite evidence
of rainfall decline over the Indian subcontinent in this specific
period. Second, there may have been a fall in agricultural
production due to a depletion of soil fertility, occasioned by the
failure to adequately replace soil nutrients removed by the
harvest of crops. These nutrients need to be replaced by a
natural deposition of silt during river floods, or by adding leaf
manure, cattle dung or fish to the fields. It is possible that such
replenishment was inadequate, perhaps because the sources of

organic manure were overharvested or because dung had to be increasingly used as fuel. Again, while this is a distinct possibility, there is no direct evidence to confirm that this was indeed the case. Last, the decline in available surplus may have been a direct consequence of growth in human population resulting in a lower land–man ratio. For instance, Lal (1984) estimates that the population in the Kanpur district grew from densities of 2.35 / $km^2$ in the Painted Grey Ware period to 12.82 / $km^2$ by the early historical period, i.e. over the course of about 1000 years. Demographic changes such as these could obviously make a major difference to the available surplus.

Whatever be the reason(s), there is little doubt that the Indian subcontinent experienced a major resource crunch over the fourth to tenth centuries of the Christian era. In this period, there existed few possibilities for the expansion of productive agriculture, except through the provision of irrigation facilities. Some expansion of such irrigation facilities did indeed occur in this period, for example in the Kaveri delta of southern India. However, there remained many regions with little cultivation, while even the cultivated tracts had large areas of non-cultivated land. The reasons for this were several: in north-western India and the centre of the Indian peninsula the rainfall was too scanty and variable; in the terai at the foot of the Himalaya the conditions were swampy and malarial; on the banks of the Ganges there was an excessive fury of annual floods; and in the Eastern and Western Ghats the topography was too steep. These non-cultivated tracts of land nevertheless provided resources of value to humans, as did lakes, rivers, estuaries and the sea. These resources could be tapped in many different ways, by hunting or fishing, by the maintenance of flocks of sheep grazed over a large area with regular seasonal movements, by grazing cattle that provided motive power and dung for agriculture, and by the harvest of leaf manure, fuelwood and timber. The need to ensure the sustainability of agriculture by providing adequate replenishment of soil nutrients, and the importance of sustainable harvests from non-cultivated tracts, would have become increasingly evident as the resource crunch progressed.

## Conservation from Below

In these conditions the problems of ensuring the prudent use of renewable resources would be faced by a society very different- ly organized from the territorial groups of hunter-gatherer- shifting cultivators. This society was an agglomeration of tens of thousands of endogamous groups, the different castes. These endogamous groups resembled tribal groups in being largely self-governing, with all intra-group regulation in the hands of the council of group leaders. Like tribal groups, caste groups were distributed over restricted geographical ranges. However, unlike tribal groups, which tended to occupy exclusive ter- ritories, caste groups overlapped with many others. Finally, unlike tribal groups, which were largely self-sufficient and carried out a broad range of activities, caste groups tended to pursue a relatively specialized and hereditary mode of subsis- tence. With their overlapping distributions and occupational specializations, the different caste groups were linked together in a web of mutually supportive relationships. This is not to claim that caste society was at all egalitarian. It was in fact a sharply stratified society, with the terms of exchange between different caste groups weighted strongly in favour of the higher status castes (Karve 1961; Srinivas 1962; Dumont 1970).

The subcontinent retained tribal hunter-gatherer-shifting cultivator groups in several pockets of hilly and often malarial country. Many of these tribes, too, interacted with caste society, exchanging forest produce such as ivory or honey for com- modities like metal tools and salt. Caste society, the Indian variant of the peasant mode of resource use, was further di- vided into sedentary, nomadic and semi-nomadic caste groups. Nomads were either pastorals who moved around with their animal herds, taking advantage of the seasonal availability of grazing grounds, or non-pastoral communities such as traders, entertainers, mendicants and specialist artisans. Sedentary villages were, then as now, populated by groups practising mixed agriculture and animal husbandry, and by service and artisan castes. Along rivers or on the coast, there were specialized fishing castes as well.

The basic unit of social organization, the village, most often included populations of several endogamous caste groups. These village communities were in many ways self-sufficient and dealt with the machinery of the state as a unit. Thus, taxes in the form of surplus production of grain were usually paid to the state by the village community as a whole, and not by individual householders. Governed by a council of leaders from the different caste groups, in many areas the community also retained the rights of reassigning cultivable land among different cultivator families, although the cultivated land was ordinarily never alienated from the group of tillers. Just as caste councils regulated most intra-caste affairs, village councils regulated most inter-caste matters, with recourse to any outside authority a most infrequent occurrence. Each sedentary village was regularly visited by certain nomadic lineages which had established a customary relationship with it.

The caste-based village society had developed a variety of institutions to regulate the use of living resources. Each settlement had a few families whose hereditary caste occupation was to serve as village guards. These guards were responsible for monitoring the visits of all aliens, as well as for maintaining information on the property rights of different families within the village. In Maharashtra, this function was generally served by members of the Mahar caste. In a fascinating record of pre-British Maharashtra, Aatre (1915) mentions that the Mahars also had the function of preventing any unauthorized wood-cutting in village common land. Additionally, they had to harvest and deliver all wood needed by village households. Since a Mahar family expected to be part of the village community and carry on its hereditary occupation, its interests would obviously lie in maintaining harvests from the village common lands at a sustainable level. Elsewhere, too, there are records of the harvests from village common lands being governed by a variety of regulations, notably quotas on the amount harvested by different families and in different seasons (Gadgil and Iyer, 1989; Guha, Ramachandra 1989a).

A resource from common lands such as fuelwood would of course be required by all village households. However, many other types of produce would not necessarily be required by all households. Here, caste society had developed an elaborate system of the diversified use of living resources that greatly reduced inter-caste competition, and very often ensured that a single caste group had a monopoly over the use of any specific resource from a given locale. We might exemplify this remarkable system of ecological adaptation with the help of two case studies, one of a relatively simple two-caste society, the second of a much more complex multi-caste village.

In the high rainfall tracts near the crest of the hill range of the Western Ghats of Maharashtra, the population density is low and the villages often made up of just two caste groups, the Kunbis and Gavlis. Here the Kunbis lived, and still do, in the lower valleys, while the Gavlis lived—they still do—on the upper hill terraces. The major occupation of the Gavlis was keeping large herds of buffaloes and cattle. They curdled the milk, consuming the buttermilk at home and bartering the butter for cereal grains (produced by the Kunbis) and for their other necessities. The protein requirements of the Gavlis were met from buttermilk, and they almost never hunted. They also practised a little shifting cultivation on the upper hill terraces.

The Kunbis, on the other hand, practised paddy cultivation in the river valleys and shifting cultivation on the lower hill slopes. They kept only a few cattle for draught purposes; these produced very little milk. To meet their protein requirements the Kunbis hunted a great deal. Thus cultivation in the valleys and lower hill slopes was restricted to the Kunbis, and that in the hill terraces to the Gavlis; the maintenance of livestock and the use of fodder and grazing resources were largely with the

Gavlis, while the Kunbis had the monopoly of hunting wild animals (Gadgil and Malhotra 1982).

Our second example comes from a village, Masur-Lukkeri, situated on an island in the estuary of the river Aghanashini, close to the town of Kumta in the state of Karnataka (14° 25′ N lat. and 24° E long.). Spurs of the hill ranges of the Western Ghats run all the way to the sea in this region, creating a rich mosaic of territorial, riparian and coastal habitats, supporting a great diversity of natural resources. Populations of as many as thirteen different endogamous groups live on this island of 4 km². There are, even today, hardly any marriages outside the endogamous groups, whose members continue largely to follow their traditional, hereditary occupations. Some of the hereditary occupations are no longer viable—e.g. toddy tapping or temple dancing—while a number of new ones have been introduced, such as rice milling, truck transport and white-collar jobs. Nevertheless, a fairly accurate reconstruction of the traditional subsistence patterns of the groups is possible on the basis of Gazetteers prepared in the nineteenth century, as well as the folk traditions collected over the period 1984-8 (Campbell 1883, Gadgil and Iyer 1989). The thirteen endogamous groups may be placed in seven categories based on their primary traditional occupations.

1. Fishing communities: This island has a population of one community of specialized river fishermen, the Ambigas.

2. Agriculturists: The traditional agricultural communities include the Halakkis, the Patgars and the Naiks. Of these, the Naiks were traditionally soldiers and petty chieftains as well.

3. Horticulturists: The Haviks, a priestly Brahman caste, are the only endogamous group of this region who specialize in horticulture. They are strict vegetarians but consume milk and milk products in substantial quantities.

4. Entertainers: Two endogamous groups: the Bhan-

daris and the Deshbhandaris, traditionally served as musicians and dancers, usually attached to the temples. The Bhandaris also used to tap toddy from coconut palms. Both the groups are now mainly dependent on agriculture.

5.  Service castes: There are two service castes, the Kodeyas (barbers) and the Madivals (washermen).

6.  Artisans: These include Shet (goldsmith), Achari (carpenter and blacksmith) and Mukhri (stone-workers).

7.  Traders: The Brahman caste of the Gowd Saraswats are the chief traders in the region. Unlike most other Brahman castes, they eat fish and seafood.

Apart from the Haviks, the other twelve castes relish fish and meat, a great deal of which are still acquired by hunting and fishing from natural populations. The Gowd Saraswat traders, the Shet goldsmiths and the Kodeya barbers do not hunt or fish themselves, but acquire animal food through barter or purchase. The other nine groups actively hunt or fish, the Ambiga males as their full-time activity, the others as a part-time activity. For this purpose they employ as many as thirty-three different methods, each of which tends to concentrate on a somewhat different habitat and a somewhat different mix of prey species. Interestingly enough, different endogamous caste groups prefer different methods of hunting and fishing, and thereby tap a different segment of the animal prey community. Thus deep water fishing, conducted from a craft and with more elaborate nets, is a monopoly of the Ambigas, the specialized fishing caste. Hunting larger mammals such as deer is a monopoly of the Halakkis, while trapping fruit bats is restricted to the Patgars and the Madivals. The Naiks are the only group to regularly trap bandicoot rats, while the Acharis, the smiths, are the only group to have taken to attracting fish with gaslights. Apart from hunting and fishing, the caste groups also exhibit an interesting diversification in the use of plant material for special purposes, such as mat weaving. Two of the caste groups engage in such activities; but while the Halakkis weave mats from the spiny leaves of *Pandanus*, the Patgars use *Cyperus* reeds.

Thus, sedentary endogamous groups living together, or sympatrically, to use a biological term, display a remarkable diversification of resource use. This pattern of adaptation is also exhibited by nomadic groups that largely tap resources away from village limits. Thus the Tirumal Nandiwallas, the Vaidus and the Phasepardhis are the three major nomadic castes hunting in the uncultivated tracts away from villages in the semi-arid region of Western Maharashtra (19° 5′ N lat. and 74° 45′ E long.). The Phasepardhis were primarily hunter-gatherers, bartering some of the game for other goods. The Tirumal Nandiwallas and the Vaidus had other primary occupations, such as displaying performing bulls, dispensing herbal medicines, selling trinkets, midwifery, etc., but hunted extensively for their own consumption.

We initiated an investigation of the hunting practices of these three castes with the presumption that they hunted much the same animals in the same tract.

However, our investigations showed that the three groups differed markedly in the hunting techniques used. The Tirumal Nandiwallas specialized in hunting with dogs, the average number of dogs per household being five. These dogs are used in locating, chasing and killing much of their prey, which includes the hyena, the leopard cat, the wild pig, the hare and the porcupine. The Vaidus kept a smaller number of dogs, an average of 1.5 per household in contemporary times. By contrast, they specialized in catching smaller carnivores such as the mongoose, the toddy cat, and the domestic cat, using traps often baited with squirrels. In the past they also specialized in catching fresh-water animals such as crabs, turtles and crocodiles (Avchat 1981). The Phasepardhis never used dogs, but instead used a trained cow to infiltrate a herd of blackbuck or deer, laying snares as they moved behind the cow. They also snared birds, particularly partridges, quails and peafowl, on a large scale.

## TABLE 3.1

RELATIVE DEPENDENCE IN TERMS OF PERCENTAGE OF REPORTED
BIOMASS CONSUMED OF THE DIFFERENT
PREY SPECIES BY THE THREE NOMADIC CASTES

|  | Hunted animal species | Tirumal Nandiwallas | Vaidus | Phase-pardhis |
|---|---|---|---|---|
| I | Small carnivorous mammals (e.g. toddy cat, mongoose) | 8.94 | 41.09 | 0.28 |
| II | Large carnivorous mammals (e.g. leopard cat, hyena, fox) | 24.99 | 7.86 | 0.07 |
| III | Small herbivorous mammals (e.g. hare, porcupine) | 15.53 | 5.14 | 0.01 |
| IV | Blackbuck | 1.19 | 0.00 | 70.43 |
| V | Wild pigs | 29.85 | 4.14 | 14.67 |
| VI | Birds (e.g. doves, quails, partridges, peafowl) | 2.09 | 3.59 | 14.21 |
| VII | Monitor lizards | 1.49 | 12.43 | 0.24 |
| VIII | Aquatic animals (e.g. fish, crab, turtle) | 15.92 | 25.75 | 0.00 |

SOURCE: Malhotra, Khomne and Gadgil (1983).

Table 3.1 provides present-day estimates of the relative importance of different prey species for the three castes (Malhotra, Khomne and Gadgil 1983). Admittedly, the exact quantitative estimates in this table are not to be taken literally, since our samples are small (2, 3 and 36 households for the three castes) and the abundance of prey has drastically declined in recent years. Nevertheless, the differences reflected in this table are very real, being a direct consequence of the employment of very distinctive hunting techniques. What is striking is that while the hunting techniques employed differ in this fashion, none of them are sophisticated enough to preclude their adoption by another caste. Thus the Phasepardhis could have easily added the Vaidus' baited traps to their own snares. The fact that they do not do so points to a genuine cultural adjustment to reduce competition with other castes hunting in the same region.

Notably enough, while sympatric caste groups exhibit resource-use diversification, groups that are identical in their resource-use patterns tend to have a non-overlapping geographical distribution in what may be thought of as an analogue of Gause's principle of competitive exclusion (Hardin 1968).

The operation of this principle is nicely illustrated by two castes of Nandiwallas which had, and largely continue to have, an identical mode of subsistence. These two castes, the Tirumal Nandiwallas and the Phulmali Nandiwallas, are both non-pastoral nomads, making a living by the display of the sacred bull, by selling trinkets and by hunting. They both originated from common ancestral stock in Andhra Pradesh. The Tirumal Nandiwallas migrated into Maharashtra about 800 years ago, while the Phulmali Nandiwallas did so only 300 years ago. While they have developed complete reproductive isolation, their way of making a living and their culture have remained essentially identical (Malhotra 1974; Malhotra and Khomne 1978). They thus exemplify castes with completely identical ecological niches.

It is notable, therefore, that the Tirumal Nandiwallas and the Phulmali Nandiwallas show no geographical overlap whatsoever, i.e. they are completely allopatric. The base village of the Tirumal Nandiwallas is Wadapuri in Pune district, while the thirty-nine base villages of the Phulmali Nandiwallas are distributed over the districts of Ahmednagar, Bhir, Aurangabad and Nasik (Malhotra, *et al.* 1983). Both these Nandiwalla castes traditionally spend the rainy season in their base camps, and then spread out over the dry-season territory to display bulls and sell trinkets in sedentary villages. There was, and there is even now, a complete absence of overlap in the dry-season migratory range of the two Nandiwalla castes (see Fig. 3.1).

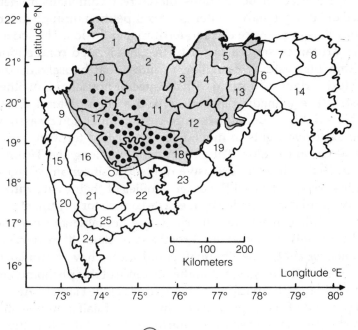

○ Area of operation of Phulmali Nandiwallas

○ Area of operation of Tirumal Nandiwallas

● Base village of Phulmali Nandiwallas

○ Base village of Tirumal Nandiwallas

Names of districts of Maharashtra: 1. Dhulia, 2. Jalgaon, 3. Buldhana,
4. Akola, 5. Amravati, 6. Wardha, 7.Nagpur, 8. Bhandara, 9. Thana,
10. Nasik, 11. Aurangabad, 12. Parbhani, 13. Yeotmal, 14. Chanda,
15. Kolaba, 16. Pune, 17. Ahmednagar, 18. Bhir, 19. Nanded,
20. Ratnagiri, 21. Satara, 22. Sholapur, 23. Osmanabad, 24. Kolhapur,
25. Sangli.

**Figure 3.1**
The administrative districts of Maharashtra and other localities referred to in
the text, base villages and dry season operating ranges of the two castes of
Nandiwallas.

Different caste populations, therefore, traditionally moderated or largely removed inter-caste competition through diver-
sifications in resource use and territorial exclusion. The sedentary artisan or service castes further moderated competition
within the caste by assigning to individual households the
exclusive rights of dealing with the specific households of other
castes. The nomadic pastoral as well as non-pastoral castes
achieved this moderation of intra-caste competition by assigning exclusive rights to move over a specified territory to individual households. Thus every family among the Tirumal
Nandiwallas has had exclusive rights to visit certain villages,
rights respected by all the other families of their caste, with a
heavy punishment levied by the caste council for any transgres-
sion of this convention. The rights are heritable and may be
sold, but only to another family of the same clan within the caste
(Malhotra 1974). The Phulmali Nandiwallas have a similar if
less well-defined system. Another instance of this phenomenon
is provided by the pastoral-nomadic caste of the Hatkars. About
18 per cent of the population of this caste of half a million still
practises nomadic sheep-keeping; the rest have taken to cultivation over the last few centuries. These shepherds spend the
rainy season in their base villages in the semi-arid tract and
move over a wide territory during the eight months of the dry
season to graze their flocks. The total caste population is
divided into a number of groups of families, each of which has
the exclusive privilege of grazing over a certain defined territory. This pattern is illustrated for the village of Dhawalpuri
in Ahmednagar district in Fig. 3.1. This village comprises four
different settlements within a kilometre of each other. While
setting out on the migration after the rains, each settlement
leaves as a single band, moving in a direction predetermined
by tradition. As the band moves, it continues to split along
kinship lines into progressively smaller groups, each moving
in its own specific direction, till the group of families constituting the ultimate unit of the flock reaches its own territory. This
composite territory of the small group of families is hereditarily

handed down from generation to generation, and may be encroached upon by other shepherd families only with special permission in times of serious distress (Malhotra 1982).

## An Eclectic Belief System

In this manner, diversification and territorial exclusion helped minimize inter- and intra-caste competition over living resources. We believe that this unique system of cultural adaptation to the natural environment was devised by Indian society in response to the resource crunch which it faced in the Gupta and post-Gupta periods, i.e. between the fourth and ninth centuries AD. It was then that social organization crystallized in the form of caste society, defined by its hereditarily prescribed modes of subsistence. Over this same period, Buddhism and Jainism lost the sway they held in earlier periods, with Buddhism being pushed out of India. The belief system that developed then was also distinct from the worship of fire and the rain gods that prevailed while the Indo-Gangetic plains were being colonized. The new system drew on beliefs in the sanctity of individual plants and animals, as well as of elements in the landscape, which the thousands of endogamous groups had inherited from their food-gathering days. The retention of these beliefs was probably aided by the continuity of culture facilitated by the barriers of endogamy. At the same time, these beliefs were, in part, turned into rituals meant to ensure the fertility of fields.

These distinctive local belief systems were now woven together into a composite fabric by identifying many of the spirits with a few key gods in the Hindu pantheon. Predominant among these were Shiva or Ishwara (male phallic worship), and the mother goddess, Parvati, (female fertility worship). A good proportion of local spirits was identified with these two. Others were then associated with them; thus, elephant worship became the worship of Ganesha, one of the sons of Shiva and Parvati. Shiva is also Pasupati—Lord of Beasts. He rides a male bull, Nandi, and around his neck are entwined cobras—both being creatures associated with fertility. The sacrifice of animals, and sometimes even of humans, continued, though the fire ritual now largely lost its central place. However, an important element added was devotion or *bhakti*. The communal pursuit of devotion to deities, synchronized on specific days, now played the role of cementing different endogamous groups together.

This belief system had a clear role in regulating and moderating the use of natural resources. It legitimized in a new framework the protection accorded to certain elements of the landscape—for instance groves or ponds near temples, and protection to certain species such as *Ficus religiosa* (peepal) or *Presbytis entellus* (the Hanuman langur)—as sacred to a variety of deities. Some of these prescriptions may have been functional in resource conservation, others neutral or even malfunctional. However, identifying them with deities from the Hindu pantheon was an effective way of continuing these practices.

Even more important than religious sanctions, in this regard, were social conventions. Thus the basket-weaving community of the Kaikadis in Maharashtra did not use the palmyra palm as a raw material simply because it violated *jatidharma*, the social duty enjoined on the endogamous group of the Kaikadis. And the Baigas, a group of slash-and-burn cultivators in Central India, believed that it was a sin to lacerate the breast of mother earth by plough cultivation (Elwin 1939). Similarly, the Agarias, whose hereditary occupation was charcoal iron-making, held all cultivation to be against their own jatidharma (Elwin 1942). Violations of caste duty were punished by the displeasure of other group members in the village, and much more effectively by social sanctions such as excommunication, enforced by the council of their own endogamous group.

Many of these social conventions helped resource conservation. We may cite here the case of the Phasepardhis, the hunter-gatherer group in Western Maharashtra discussed earlier. The Phaseparadhis had a monopoly on deer hunting in their territory, yet they let loose any pregnant does or young deer caught in their snares. The Phasepardhis are aware that this convention may facilitate better prey availability in the future. However, the practice was enjoined on all group members as jatidharma, as the right thing to do, and enforced through caste sanctions against any member who violated it.

The endogamous caste groups of Indian society might with profit be compared to biological species, although of course, as Gautam Buddha pointed out, they are all fully interfertile and in no way genetically differentiated from each other. Nevertheless, like biological species, they have been largely, though not absolutely, isolated reproductively through cultural barriers. Like the species within a biological community, the different caste groups also have characteristic modes of subsistence, and often tend to occupy distinct habitats. It is therefore appropriate to talk of the ecological niches of these various caste groups in terms of the habitats they occupy, the natural resources they utilize, and the relationship they bear to the other caste groups with whom they interact. Two interesting statements follow from this perspective, namely that the castes seem by and large to have narrow niches, and that the caste groups that live together have so diversified that they tend to have low niche overlaps.

We have already discussed the concept of a low niche overlap. This is a consequence of the low niche widths of individual caste groups. This narrowness of niches in a human community, developed in a tropical setting, offers very interesting parallels to the narrow niches of biological species in such an environment. Such narrow niches within biological species are considered a result of the high levels of productivity and low levels of variability with time, thus permitting high levels of specialization. The fact that there developed, and for centuries persisted, groups which are so specialized that their entire economy depends on tapping palms for toddy or making baskets from bamboo, suggests that there are genuine parallels. The fairly high levels of productivity, and the relative constancy from year to year, coupled to the low level of resources necessary to subsist in a tropical environment, have permitted high levels of specialization in the various modes of subsistence within Indian caste society. With such levels of specialization, the overlap of niches among caste groups living side by side has been kept at a low level. This has ensured that small numbers of people linked together by bonds of kinship, and by a common culture, have had a monopoly over specified resources in specified localities. As we argued in Chapter 1, this is expected to promote practices of prudent resource use. Indeed many such traditions—including sacred groves and sacred ponds as refugia, the protection to keystone species and to critical life-history stages, and the moderation of harvests from village wood-lots—have persisted in Indian society over the historical period, sometimes down to the present day.

## The Village and the State

The Gupta and post-Gupta pre-Muslim period in Indian history, from the fourth to the tenth centuries, was by and large a period of low trade, low levels of urbanization, and many small kingdoms. A change could be wrought in this situation in one of several ways. It could come about through an improvement in climate, or by the introduction of new technologies, such as irrigation tanks, which would lead to a rise in agricultural productivity and production. It could also be brought about by a great improvement in modes of transport, communication and coercion, enabling the accumulation of low levels of surplus, but from over broad areas, to support an extensive state. Again, there is no definite evidence for any climatic change. However, we know that new technologies and vastly improved modes of transport, communication and coercion did come into play in the period following the tenth century.

An example of improved technology is the tank irrigation system with its sluice gates, as well as the social organization of irrigation management, centred on the temple, that permitted the agricultural colonization of large areas of Tamilnadu under the Cholas, from AD 900 to 1200. Interestingly, the priestly castes of Hinduism again spearheaded the spread of this new technology, with land grants to Brahmans accompanying the establishment of new tanks.

Six thousand years after domesticated crops and animals revolutionized that subcontinent's economy, certain specific military techniques and technologies—skilled horseriding, the co-ordination of attack by cavalry, and the use of gunpowder—all came to India from the same north-western route and revolutionized political organization (Digby 1971). The period beyond AD 900–1000, indeed, shows the re-emergence of larger states, such as the Chola state, and the Vijayanagar and Mughal empires. Their emergence might in part have been made possible by an improvement in productivity, in part by an ability to better mobilize the surplus over larger areas. Questions on what surplus the state claimed, and what fraction of the production this constituted, are of some importance to our enquiry. They can be answered with a degree of certainty for Mughal India, for which excellent records exist.

In Mughal India the state essentially claimed only the surplus from grain production in agriculture. It also claimed some tax on the animals maintained by farmers above a threshold number. The Mughals did not tax horticulture, sheep-raising or fisheries. Nor did they tax forest holdings. In fact when the Mughal governor in Kashmir attempted to tax sheep-rearing and fishery, there were strong protests, leading to the removal of the governor. The emperor Shahjahan then promulgated an edict stating that the tax on sheep and fishing was cancelled as being against the custom. One can see, to this day, this edict carved upon a mosque in Srinagar. The proportion of surplus expropriated from the peasants might have been as large as 50 per cent; this would have been shared at several levels, beginning with the village headman. The peasants could not be dispossessed of land, and the state had no direct claims over lands other than hunting preserves. Fairly extensive tracts were, in fact, claimed as the emperor's property, wherein the general hunting of game was prohibited, the local population being only permitted to gather other materials (Moosvi 1987).

It has been claimed that outside Mughal India, for instance in Tamilnadu, the proportion of agricultural produce expropriated from the cultivator was much smaller (Dharampal 1986). Be that as it may, there is no doubt that the control and management of local resources did all this time vest in local communities who designed a variety of practices for effective resource conservation. There were, of course, some commodities that the state was interested in acquiring from non-cultivated land. Elephants were one such important element, and the ever-growing demand for elephants did indeed lead to their gradual elimination from one forest area after another. At the time of Kautilya's *Arthasastra*, elephants were noted from parts of Punjab and Saurashtra; by the time of the Mughals, some 1000 to 1500 years later, these areas no longer held elephants.

The demands for timber by the state were not very heavy, except perhaps where shipbuilding was concerned. But there is the memorable story of a sect, founded in the Rajasthan desert some five hundred years ago, that gave absolute protection to Khejadi (*Prosopis cinerarea*), a multi-purpose leguminous tree of great utility to the villagers. Followers of this sect, the Bishnois, never uproot or kill, nor suffer to be uprooted or killed, any Khejadi tree in their villages. Some 350 years ago the prince of Jodhpur needed wood to fuel his lime kilns, which were being deployed to build a new palace, and for this purpose he attempted to fell a grove of Khejadi trees in a nearby village. It is narrated that several Bishnois laid down their lives to prevent these trees being cut. It is quite clear that the growth of Khejadi trees in these desert areas is of much value to the local people. They harvest its pods, leaves and thorny branches as food, fodder, manure and fencing material. Such conservation practices, cast in a religious idiom, were continually arising and were not merely a carry-over from the food-gathering phase (Sankhala and Jackson, 1985).

## Conclusion

Our interpretation of the caste system as a form of ecological adaptation may also be used to illustrate the two different paths by which conflicts between different modes of resource use are resolved. The first path, which we call the *path of extermination*, explains both the victory of the neolithic revolution in Europe and the successful establishment of neo-Europes in the colonized territories of the New World. In this scenario, the earlier modes are more or less wiped out. While Eskimoes continued their hunting and trapping in the harsh environments of the Arctic, and fishermen were allowed to occupy niches in marine ecosystems, over the most part of Europe the victory of agriculture and pastoralism was complete. In more recent times the indigenous population of the Americas and Oceania have succumbed to the cultural and ecological expansion of European civilization. In parts of South America the last acts of this eco-drama are being played out now, with utterly predictable results. For the eventual fate of the indigenous population of the Amazonian rain forest is not likely to be any different from that of their counterparts in other areas of the New World.

The alternative pattern, which we call the *path of selective incorporation*, better fits the history of the Indian subcontinent prior to its colonization by the British. Insofar as the history of India exhibits the far greater *overlap* and *coexistence* of different modes of resource use, one can qualitatively distinguish the Indian experience from the European and New World paradigms of eco-cultural change. These differences in the forms of modal competition and co-operation are explicable only through a combination of ecological and cultural factors. The geographical diversity of the subcontinent, and the productivity of hilly and forested areas, enabled the continuance of hunting-gathering and shifting agriculture in large expanses where the plough could not penetrate. The persistence of earlier modes was also helped by patterns of disease which kept population densities at sufficiently low levels, allowing hunters sole possession of tracts of forest such as the malarial Terai. However, one must not underestimate the role of cultural factors in the maintenance of this diversity. As opposed to the exclusivist approach favoured by monotheistic religions such as Islam and Christianity, Hinduism has (at least until recently) relied more heavily on an *inclusive* framework that tries to incorporate rather than reject or convert apparently hostile sects and worldviews. The institutional mechanism for this process of incorporation is, of course, the caste system. And by accepting a distinctive if subordinate position within the caste hierarchy, hunters and gatherers could forestall extinction by continuing their traditional mode of resource use, though only at the cost of a larger subordination to the victorious peasant mode. These two complementary strategies, of leaving some ecological niches (hills, malarial forests) outside the purview of the peasant mode, and reserving certain niches within it for hunter-gatherers and pastorals, helped track a distinctive path of inter-modal co-operation and coexistence.

# PART THREE

# *Ecological Change and Social Conflict in Modern India*

# CHAPTER FOUR

# Conquest and Control

Despite the grave inequalities of caste and class, then, pre-colonial Indian society had a considerable degree of coherence and stability. This permitted a rapid turnover of ruling dynasties without major upheavals at the level of the village. On the one hand, cultural traditions of prudence ensured the long-term viability of the system of production, and of the institution of caste which was its central underpinning. On the other hand, remarkably strong communal institutions—existing at caste and supra-caste levels—oversaw the political, economic and juridical spheres of everyday existence. The agrarian system was well integrated with the highly sophisticated system of artisanal production, operating for local consumption and for trade. Relations between agriculture and industry, and between state and peasant, operated between loosely defined limits that defined the scope of each party.

Even the Mughals, whose religion was Islam, were unable, or perhaps unwilling, to radically alter the existing patterns of resource use and the social structures in which they were embedded. It was an entirely different story *vis-à-vis* India's contacts with Christian Europe. When this contact began in the sixteenth century, Europe was on the threshold of the momentous process of social change known as the Industrial Revolution. While the economic, political and social changes that came in the wake of the Industrial Revolution have been the staple of scholarly work for decades, it is only now that the ecological implications of this process are being unravelled.[1] The revolution in the mode of resource use brought about by industrialization enormously enlarged the possibilities of transforming resources from one form to another, and of transporting them over large distances. With these technological advances, a great range of objects became commodities, objects for which the demand could go on increasing indefinitely, almost limitlessly. Wood, for instance, would be consumed in a subsistence economy on a limited scale as domestic fuel, in the construction of implements, and for shelter. It could now be converted into paper, or burnt as fuel for the steam engines of trains and ships. While there can be a definite limit on the per capita demand for domestic fuel or agricultural implements, there can be no such limit on the consumption of paper, or on fuel for transport.

This conversion into commodities of a whole range of objects also radically transformed the flow of energy and materials. As outlined in Chapter 1, at the food-gathering stage such flows are largely confined to the territory of each group. With settled agriculture starts an outflow of grain from the countryside to the towns, where the non-food producers are concentrated. The advances in technology brought about by the industrial mode, however, prompt an outflow of a much greater range of resources from both cultivated and non-cultivated lands, and from water bodies.

These changes in the patterns of material flows had, of course, profound implications for patterns of resource use. Non-cultivated lands and waters were no longer dedicated

---

[1] Karl Polanyi's *The Great Transformation*, first published in 1944, has some remarkable passages on the ecological implications of the Industrial Revolution. Polanyi's insights have for the most part remained undeveloped.

exclusively to providing for flows within the local region. Communal control by food producers, the peasants, over these lands became a bone of contention, with the elite attempting to usurp rights over these lands and put them to other uses. Such territories were now capable of generating resources of high value, and their control by the elite became attractive because of technological advances in transport, transformation and coercion. The benefit : cost ratio thus changed, making possible the conversion of large tracts of non-cultivated lands and waters into private or state property. This process is well known in European history as the movement for the enclosure of commons by landlords and the state. Over a period of several centuries, forests, fens and waters controlled and used by local peasant communities were converted into the property of landlords and the state. At the ideological level too, private and state property was being upheld, and the validity of communal control questioned.

At the time the Europeans came to India, therefore, they were experiencing at home a far-reaching revolution in patterns of natural resource use. Three elements of that revolution are of particular importance to our story. First, it lowered the emphasis on resource gathering and food production for subsistence, focusing instead on the gathering, production, transport and transformation of resources for use as commodities. The proportion of the population engaged in subsistence gathering and the production of food declined; that of people engaged in manufacturing, transporting or using resources as commodities increased. Second, co-operation with neighbours of long standing, characteristic among people engaged in subsistence gathering and food production, became less and less important. With this came the breakdown of cohesive local communities. Human societies became atomized, with individuals acting largely on their own. This was relatively easy in Christian Europe, where the different human groups were loosely organized, with broad niches and large niche overlaps.

Finally, and perhaps most importantly, the changing 'hardware' of resource use was accompanied by equally dramatic changes in its 'software'. In the now atomized societies, with their possibilities of unlimited expansion in the consumption of resources, the capacity of individuals to command access to resources was at a premium. With manufacture and commerce the dominant activities, markets became the focal point for organizing access to resources. The new belief-system that developed therefore transferred to the institution of the market the veneration reserved for spirits resident in trees by food-gatherers, and in an abstract God by Christian food-producers. Success and status were now clearly measured in terms of money, the currency of the market.

These three characteristics of the industrial mode of resource use are central to a proper understanding of the ecological encounter between India and Britain. For the elevation of commercial over subsistence uses, the delegitimization of the community, and the abandoning of restraints on resource exploitation—all ran counter to the experience of the vast majority of the Indian population over which the British were to exercise their rule. This was a clash, in more ways than one, of cultures, of ways of life.

## Colonialism as an Ecological Watershed

The ecological history of British India is of special interest in view of the intimate connection that recent research has established between western imperialism and environmental degradation. World ecology has been profoundly altered by western capitalism, in whose dynamic expansion other ecosystems were disrupted, first through trade and later by colonialism. Not only did such interventions virtually reshape the social, ecological and demographic characteristics of the habitats they intruded upon, they also ensured that the ensuing changes would primarily benefit Europe. Colonialism's most tangible outcome (one whose effects persist to this day) related to its global control of resources. The conquest of new areas meant that the twenty-four acres of land available to each European at the time of Columbus' voyage soon increased to 120 acres per European. Large-scale settlement was only one way in which Europeans augmented their 'ghost acreage'. For the world-wide control they exercised over mineral, plant and animal resources also contributed to industrial growth in the metropolis. In turn, political hegemony enabled these manufactured products to capture non-European markets. Finally, colonialism introduced Old World diseases among vulnerable populations, often with fatal consequences (Webb 1964; Wallerstein 1974; Tucker and Richards 1983).

One of the finest studies of this process of 'ecological imperialism' is Alfred Crosby's monograph of that name (Crosby 1986). The firepower of the European vanguard, and the complex of weeds, animals and diseases it brought, devastated the flora, fauna and human societies of the New World. However, the extermination of native ecosystems and populations only paved the way for the creation of 'Neo-Europes', the extensive and enormously productive agricultural systems that dominate the New World today. In this process, suggests Crosby, the plants, animals and diseases that accompanied the migrating Europeans were as important in ensuring their success as superior weaponry.

In his fascinating yet chilling account of the biological expansion of Europe, Crosby pauses briefly to investigate the areas that he believes were 'within reach' but 'beyond grasp'—the complex Old World civilizations in the Middle East, China and India. He argues that population densities, resistance to disease, agricultural technology and sophisticated socio-political organizations—all made these areas more resistant to the ecological imperialism of Europe. Thus 'the rule (not the law) is that although Europeans may conquer the tropics, they do not Europeanize the tropics, not even countrysides with European temperatures' (Crosby 1986, p. 134).

As Crosby so skilfully demonstrates, their 'portmanteau biota' (his collective term for the organisms the colonizing whites brought with them) enabled the European powers to easily overrun the temperate regions of North and South America, as well as the continent of Oceania. They had to perforce adopt a different strategy in older—and more ecologically resistant—civilizations like India and China. Here they set up initially as traders and military advisers, later making the transition to rulership, either directly or by proxy. This does not mean, however, that European colonialism had an insignificant impact on these ecosystems, as Crosby's account seems to suggest. In India, the Europeans could not create neo-Europes by decimating the indigenous populations and their natural resource base; but they did intervene and radically alter existing food-production systems and their ecological basis. If in the neo-Europes, ecological imperialism paved the way for political consolidation, in India the causation ran the other way, their political victory equipping the British for an unprecedented intervention in the ecological and social fabric of Indian society. Moreover, by exposing their subjects to the seductions of the industrial economy and consumer society, the British ensured that the process of ecological change they initiated would continue, and indeed intensify, after they left India's shores.

This and the following chapter focus on possibly the most important aspect of the ecological encounter between Britain and India—the management and utilization of forest resources.

## The Early Onslaught on Forests

By around 1860, Britain had emerged as the world leader in deforestation, devastating its own woods and the forests of Ireland, South Africa and north-eastern United States to draw timber for shipbuilding, iron-smelting and farming. Upon occasion, the destruction of forests was used by the British to symbolize political victory. Thus in the early nineteenth century, and following its defeat of the Marathas, the East India Company razed to the ground teak plantations in Ratnagiri nurtured and grown by the legendary Maratha admiral Kanhoji

Angre (Campbell 1883). 'Of all European nations', one of the early conservators of forests in India, Henry Cleghorn, observed—

> the English have been most regardless of the value of the forests, partly owing to their climate, but chiefly because England has been so highly favoured by vast supplies of coal; and the emigrants to the United States have shown their indifference to this subject by needless destruction of forests in that country of which they now feel the want (Cleghorn 1860, p. ix).

Their early treatment of the Indian forests also reinforces the claim that 'the destructive energy of the British race all over the world' was rapidly converting forests into deserts (Webber 1902, p. 338). Until the later decades of the nineteenth century, the Raj carried out a 'fierce onslaught' on the subcontinent's forests (Smythies 1925, p. 6). With oak forests vanishing in England, a permanent supply of durable timber was required for the Royal Navy as 'the safety of the empire depended on its wooden walls' (Stebbing I, p. 63). In a period of fierce competition between the colonial powers, Indian teak, the most durable of shipbuilding timbers, saved England during the war with Napolean and the later maritime expansion. To tap the likely sources of supply, search parties were sent to the teak forests of India's west coast (Edye 1835). Ships were built in dockyards in Surat and on the Malabar coast, as well as from teak imported into England (Albion 1926). An indication of the escalating demand is provided by the increase in tonnage of British merchant ships (i.e. excluding the Royal Navy) from 1,278,000 tonnes in 1778 to 4,937,000 tonnes in 1860 (Leathart 1982). A large proportion of the wood required came from Britain's newly-acquired colonies. As late as the 1880s, the Indian forest department was entertaining repeated requests from the British admiralty for the supply of Madras and Burma teak.[2]

The revenue orientation of colonial land policy also worked

[2] See National Archives of India (hereafter NAI), Department of Rev. & Agrl. (Forests), (hereafter Forests).

towards the denudation of forests. As their removal added to the class of land assessed for revenue, forests were considered 'an obstruction to agriculture and consequently a bar to the prosperity of the Empire' (Ribbentrop 1900, p.60). The dominant thrust of agrarian policy was to extend cultivation and 'the watchword of the time was to destroy the forests with this end in view' (Stebbing I, pp. 61-2).

This process greatly intensified in the early years of the building of the railway network after about 1853. While great chunks of forest were destroyed to meet the demand for railway sleepers, no supervision was exercised over the felling operations; a large number of trees was felled and lay rotting on the ground (Stebbing I, pp. 298-9). The sub-Himalayan forests of Garhwal and Kumaon, for example, were all 'felled in even to desolation', and 'thousands of trees were felled which were never removed, nor was their removal possible' (Pearson, 1869, pp.132-3). Private contractors, both Indian and European, were chiefly responsible for this destruction, the forests of Indian chiefs not escaping their hands (Paul 1871). Before the coal mines of Raniganj became fully operative, the railway companies drew upon the forests for fuel as well. The fuelwood requirements of the railways in the North Western Provinces, at a high level into the 1880s, caused considerable deforestation in the Doab (Whitcombe 1971) Meanwhile, in Madras, as a member of the Indian Famine Commission noted, the demand for fuel of locomotives was 'large enough to cause a heavy drain, if not an utter exhaustion, of the particular forests, from which the supplies are drawn'. 'It appears certain', predicted a Madras official in 1876, 'that for a considerable time to come the main supply of railway fuel must come from the natural jungles'. In the districts of North Arcot and Chingleput the alternating cycles of flood and drought, which such destruction had caused, seriously affected irrigation, and thereby food production.[3]

[3] NAI, Forests, progs no. 134–35, March 1878, minute by Richard Temple, 18 April 1877; NAI, legislature dept, progs 43-142, March 1878, appendix KK,

One of the most vivid descriptions of the transformation in the ecological landscape wrought by the railways is found in Cleghorn's work, *The Forests and Gardens of South India*. The Melghat and North Arcot hills, formerly crowned with timber, were 'now to a considerable degree laid bare' by the insatiable demand of the railways. All around the tracks, where once there was forest, there now lay wide swathes of cleared land stripped bare of cover, and consequently of protection to wild animals. Thus the progress of the railway 'produced marvellous changes on the face of the country as regards tree vegetation'. In the Madras Presidency over 250,000 sleepers (or 35,000 trees) were required annually from indigenous sources. To meet this demand contractors resorted more and more to *sequential over-exploitation*. On the one hand they cleared jungles further and further away from the railway lines, while on the other they utilized more and more unsuitable species as those less favoured were rapidly exhausted. Although only half a dozen species were considered suitable for use as railway sleepers, more than fifty were tried out. Not surprisingly, sleepers expected to last five or six years only lasted a third of the time. In one consignment, out of 487 sleepers supplied, 458 (or 92 per cent) were found to be of unauthorized woods (Cleghorn 1860, pp. 2-3, 33, 63, 77-8, 253-4).

The pace of railway expansion—from 1349 kms of track in 1860 to 51,658 kms in 1910 (GOI 1964)—and the trail of destruction left in its wake brought home forcefully the fact that India's forests were not inexhaustible. Railway requirements were 'the first and by far the most formidable' of the forces thinning Indian forests (Cleghorn 1860, p. 60). Dubbing forest administration upto the 1857 rebellion a melancholy failure, the Governor-General had called in 1862 for the establishment of a department that could ensure the sustained availability of the enormous requirements of the different railway companies for

---

no. 2149, 23 December 1876, from acting secretary to government, revenue department, Madras, to secretary to GOI, legislative department.

sleepers, which 'has now made the subject of forest conservancy an important administrative question' (quoted in Trevor and Smythies 1923, p. 5). The magnificent forests of India and Burma, one official recalled, 'were being worked by private enterprise in a reckless and wasteful manner and were likely to become exhausted if supervision were not exercised' (Webber 1902, pp. xii–xiii).

The crisis had assumed major proportions as only three Indian timbers—teak, sal, and deodar—were strong enough in their natural state to be utilized as railway sleepers. Sal and teak, being available near railway lines in peninsular India, were very heavily worked in the early years, necessitating expeditions to the north-western Himalaya in search of deodar forests. The deodar of the Sutlej and Yamuna valleys was rapidly exhausted in the years following the inception of the forest department—over 6,500,000 deodar sleepers were supplied from the Yamuna forests alone between 1869 and 1885 (Paul 1871; Hearle, 1888).[4]

The imperial forest department was formed in 1864, with the help of experts from Germany, the country which was at that time the leading European nation in forest management. The first inspector-general of forests, Dietrich Brandis, had been a botanist at Bonn University before his assignment in India. The awesome task of checking the deforestation of past decades required, first and foremost, forging legal mechanisms to assert and safeguard state control over forests. It was in this dual sense that the railways constituted the crucial watershed with respect to forest management in India—the need was felt to start an appropriate department,[5] and for its effective functioning legis-

---

[4] Each mile of railway construction requires 860 sleepers, each sleeper lasting between 12 to 14 years. In the 1870s it was calculated that well over a million sleepers were required annually. While European sleepers were imported in some quantities, the emphasis was always on substituting them by Indian timbers. See D. Brandis, 'Memorandum on the Supply of Railway Sleepers of the Himalayan Pines Impregnated in India', *Indian Forester*, V, 1879.

[5] Although state control was not unknown in the early decades of colonial

lation was required to curtail the previously untouched access enjoyed by rural communities. This was an especially difficult task as, in many cases, the proprietary right of the state in forests had been 'deliberately alienated' in favour of peasant and tribal communities (Brandis 1897, p. 52; cf. also Ribbentrop 1900, and Dasgupta 1980). Before its late recognition of the strategic importance of forests, the policy of the colonial state had been to recognize forests and waste land as the property of the village communities within whose boundaries these fell (Stebbing II, pp. 464ff).

The first attempt at asserting state monopoly was through the Indian Forest Act of 1865. This was replaced thirteen years later by a far more comprehensive piece of legislation. Yet the latter act (to serve in time as a model for forest legislation in other British colonies) was passed only after a prolonged and bitter debate within the colonial bureaucracy. This controversy, enumerated in some detail below, has a particular resonance for us today: the protagonists of the earlier debate put forth arguments strikingly similar to those advanced by participants in the contemporary debate about the environment in India.

### An Early Environment Debate

Hurriedly drafted, the 1865 act was passed to facilitate the acquisition of those forest areas that were earmarked for railway supplies. It merely sought to establish the claims of the state to the forests it immediately required, subject to the proviso that existing rights not be abridged. Almost immediately, the search commenced for a more stringent and inclusive piece of legislation. A preliminary draft, prepared by Brandis in 1869, was circulated among the various presidencies. A conference of forest officers, convened in 1874, then went into the defects

---

rule, the setting up of a separate department marked a qualitative shift in colonial perceptions of the strategic value of forests. For early attempts to enforce state monopoly over trees such as teak and sandalwood, see Cleghorn (1860).

of the 1865 act and the details of a new one. The conference provided the basis for a memorandum on forest legislation, prepared by Brandis in 1875. The latter memorandum, further worked on by Brandis and a senior civil servant, B.H. Baden-Powell, culminated in the Indian Forest Act of 1878.

In these successive iterations the state was concerned above all with removing the existing ambiguity about the 'absolute proprietary right of the state'. A vocal advocate of state monopoly deplored the 'unfortunate but irrevocable action of government authorities in days past' which had taken many forest areas wholly out of the category of state property. Even forests where the state had in theory retained its 'absolute' proprietorship were 'everywhere used by all classes to get what they wanted'. Villagers had got accustomed to graze cattle and cut wood wherever they wished, the writer complained, because 'nobody cared whether [they] did or not' (Baden-Powell 1875).

Here colonial officials were, perhaps wilfully, confusing 'open access' with 'common property'—for the peasantry's customary use of forests was not random but governed and regulated by community sanctions. Clearly, a firm settlement between the state and its subjects over their respective rights in the forest represented the chief hurdle to be overcome. As Brandis put it, 'Act VIII of 1865 is incomplete in many respects, the most important omission being the absence of all provisions regarding the definition, regulation, commutation and extinction of customary rights . . . [by the state]'.[6] In the heated debate about how best to accomplish this separation of rights, three distinct positions emerged. The first, which we call *annexationist*, held out for nothing less than *total* state control over *all* forest areas. The second, which one may call *pragmatic*, argued in favour of state management of ecologically sensitive and strategically valuable forests, allowing other areas to remain under communal systems of management. The third position (a mirror image of the first) we call *populist*. This completely

[6] NAI, Forests, B progs, nos 37-47, December 1875, 'Explanatory Memorandum on Draft Forest Bill', by D. Brandis, IGF, 3 August 1869.

rejected state intervention, holding that tribals and peasants must exercise sovereign rights over woodland. These three perspectives on state control dovetailed with three distinct views on the sociology, history, politics and ecology of forest resource use. They deserve to be reconstructed in full, for the issues they raised and debated with such intensity a hundred years ago are very much with us today.

The bedrock of the *annexationist* position was the claim that all land not actually under cultivation belonged to the state. Of course it was not easy to wish away the access to forests—subject naturally to the norms of the community—that peasants so patently exercised down the centuries, right until the formation of the forest department. Officials argued that such customary use, however widespread and enduring, was exercised only at the mercy of the monarch. Here they used precedents, selectively citing Tipu's edict banning the cutting of sandalwood as proof that India's rulers had ultimately reserved to themselves the right of ownership over forests and forest produce. Baden-Powell claimed 'the state had not, it is true, exercised that full right: the forests were left open to any one who chose to use them; *but the right was there*' (Baden-Powell, 1875, pp. 4-5, emphasis in original). In this strictly legalistic interpretation only those rights of use which were *explicitly granted by the state* (presumably only in writing) were to be entertained. Thus Baden-Powell made a clever distinction between 'rights' defined as 'strict legal rights which unquestionably exist, and in some instances have been expressly recorded in land settlement records'—and 'privileges'—defined as 'concessions of the use of grazing, firewood, small wood, etc., which though not claimable as of legal right, are always granted by the policy of the government for the convenience of the people'.[7]

This tortuous distinction (in effect, a legal sleight-of-hand) was buttressed by an early version of the theory of 'oriental despotism' in which eastern chiefs were believed to have

[7] NAI, Forests, B progs, nos 37-47, Dec. 1875, 'Draft Forest Bill prepared by Mr. Baden-Powell'.

powers far more extensive than those enjoyed by their contemporaries in medieval Europe. Adducing no evidence, Baden-Powell nevertheless claimed that 'the right of the state to dispose of or retain for public use the waste and forest area, is among the most ancient and undisputed of features in oriental Sovereignty'.[8] 'In India', an official primer on forest law likewise affirmed, 'the government is by ancient law . . . the general owner of all unoccupied and waste lands' (Anon. 1906, p. 20). Whatever the historical evidence for this claim (scanty, at best), its purpose was clear: to pave the way for the formal assertion of ownership over forests and waste by the colonial state. The 'right' of oriental governments in the forest, insisted Baden-Powell, 'passed on to, and was accepted by, the British government' (Baden-Powell 1882, p. 9). Others were more blunt. 'The right of conquest', thundered one forest official, 'is the strongest of all rights—it is a right against which there is no appeal.' (Amery 1875, p. 27).

Counterposed to this claim of an age-old right was the total denial of the legitimacy of any state intervention in the forest. The Madras government, which emerged as the most articulate official spokesman for village interests, rejected Baden-Powell's tendentious distinction between legally proven 'rights' and 'privileges' exercised without written sanction. 'All instances of the use of the forest by the people', it argued, 'should be taken as presumptive evidence of property therein'.[9] Both 'private grantees and village and tribal communities', an early nationalist organization likewise pointed out, 'have cherished and maintained these rights [in the forest] with the same tenacity with which private property in land is maintained elsewhere'.[10]

---

[8] B.H. Baden-Powell, 'Concessions', *Indian Forester*, XXV, 1899, p. 358.

[9] NAI, legislative department, A progs, nos 43-142, March 1878, 'Remarks by the Board of Revenue, Madras', 5 August 1871. This file, containing the major documents pertaining to the 1878 act, is hereafter referred to as LD file of 1878.

[10] Memorial, 3 March 1878, from Puna Sarvajanik Sabha, and the inhabitants of the city and camp of Pune, in LD file of 1878.

If this view was to be allowed, then it seemed the claim of the state was virtually non-existent. For—

> There is scarcely a forest in the whole of the Presidency of Madras which is not within the limits of some village and there is not one in which, so far as the Board can ascertain, the state asserted any rights of property unless royalties in teak, sandalwood, cardomoms and the like, can be considered as such, until very recently. All of them, without exception are subject to tribal or communal rights which have existed from time immemorial and which are as difficult to define as they are necessary to the rural population . . . . Nor can it be said that these rights are susceptible of compensation, for in innumerable cases, the right to fuel, manure and pasturage, will be as much a necessity of life to unborn generations as it is to the present . . . . (In Madras) the forests are, and always have been, common property, no restriction except that of taxes, like the Moturpha and Pulari, was ever imposed on the people till the Forest Department was created, and such taxes no more indicate that the forests belong to the state than the collection of assessment shows that the private holdings in Malabar, Canara and the Ryotwari districts belong to it.[11]

Intermediate between those two extreme positions was the moderate voice of the inspector-general of forests, Dietrich Brandis, the exemplar of what we have termed the 'pragmatic' approach. Brandis allowed that in certain cases the state had indisputable rights; however, he was with the Madras board of revenue in disputing Baden-Powell's contention that rights had to be 'proved' in writing before they could be said to exist. In most forest areas, he believed, villagers were accustomed to freely graze their cattle, cut wood, etc., subject only to some restrictions which rulers imposed from time to time. Drawing on a cross-cultural comparison, he pointed out that

> the growth of forest rights in India has been analogous to the growth of similar rights of user in Europe. There are many

[11] Source cited in footnote 9.

well-known cases in which forest rights in Europe have arisen out of a specific growth and in such cases the extent of the right is construed by the terms of the grant and is not necessarily restricted by the limitations adverted to. In most instances however, they have grown up out of the use by the surrounding villages of the common waste and forest. Forest rights in India have had a similar origin and development as in Europe, with that important difference that the arbitrary dealings of the Native Rulers have interfered with the growth of these rights and have in many cases restricted or extinguished them.

Not that he was approving of such arbitrary action. Contesting Baden-Powell's invocation of the case of the Amirs of Sindh (chieftains who had enclosed forests for hunting), Brandis said:

the fact that the former Rulers in many cases have extinguished such customary usage of the forest in a summary manner and without compensation is hardly an argument in point, for these were cases of might versus right. As against other individuals and communities the customary rights to wood and pasture have as a rule been strenuously maintained (Brandis, 1875, pp. 13-14).[12]

Other officials used the European analogy to quite different ends. If for Brandis the forest history of Europe called for a similar treatment of village rights in India, for others it merely served as a warning *not* to grant these rights. Thus a committee set up by the government of Bombay in 1863 went so far as to claim that in *both* India and England the governing powers could assert a total state monopoly whenever they wished. Claiming the right of the sovereign in England, since ancient times, to 'make forests of any extent over the lands of his subjects', they urged the application of a similar principle in India.[13] For

[12] One does not know whether Brandis was unaware of this, but in Europe too there were many instances of the monarch abruptly extinguishing customary rights in the forest. See, for example, Thompson (1975) for England, Agulhon (1982) for France, and Linebaugh (1976) for Germany.

[13] See NAI, legislative department, A progs, nos 32-38, Feb. 1865.

Baden-Powell the lesson of European forest history was not to allow, in India, the building up of rights which could encumber timber production by the state (Baden-Powell 1892, p. 7). Moreover, even if the development of common rights was analogous, the rights permissible to free-born citizens in England were not feasible in a colonial territory. Thus—

> in England by the Common Law, all proprietors of land own everything up to the sky and down to the centre of the earth except gold and silver mines, which by prerogative, belong to the Crown. But in a ceded and conquered country like India, this English Common Law and Crown prerogative does not apply at any rate beyond the limits of the Presidency towns (Baden-Powell 1882, p. 50).

Brandis, a comparative newcomer to colonial administration—and from a country which was at the time very much a beginner in the colonial game—took a different stance. The settlement of rights, he insisted, must be done in a 'just and equitable manner' (Brandis 1875, p. 12). For him, identical acts could call upon the same legal justification, even if one was committed in colonial India, the other in 'free' England. 'There has been much thoughtless talk', he told a meeting in Brighton in 1872,

> as if the natives of India, in burning the forests and destroying them by their erratic clearings were committing some grave offence. If the matter is carefully analysed they will be found to have the same sort of prescription which justifies the commoner in the New Forest to exercise his right of pasture, mast and turbary (Brandis 1897, p. 53).

If Baden-Powell concluded from European history that the colonial state must be more emphatic in asserting its claims, and Brandis that it must grant to its subjects the rights prevalent in Europe, the Madras government sharply challenged the principle of state forestry in Europe itself. The board of revenue compared Brandis' first draft of a forest act to the French Code

Forestier, that 'most stringent of forest rules' and itself the consequence of 'a long feudal tyranny, followed by the storms of a revolution, and the despotism of an empire'; indeed 'no system could be more opposed either in its history or its provisions to the corresponding circumstances in India'. Faced with the Indian Forest Act of 1878, the governor of Madras had to go even further back to find an appropriate analogy from European history. Condemning the new legislation as a 'Bill for confiscation instead of protection', he said 'it was probably the same process which the Norman Kings adopted in England for their forest extension'.[14]

Very different proposals flowed from these different readings of Indian and European forest history. The 'annexationists' urged the constitution of all wooded areas as state forests, following a settlement of rights on the interpretation advanced by Baden-Powell. Thus the agricultural secretary, Allan Octavian Hume, sharply rebuked the vocal Madras official W. Robinson for his opposition to state encroachment on customary rights. Commenting on his decision to recognize rights of ownership in forests cultivated in rotation by swidden agriculturists in South Kanara, Hume cautioned 'the government of India to watch carefully and satisfy itself that [Robinson's] kindly and warm-hearted sympathy for the welfare of the semi-savage denizens of the Kanara forests does not lead him into a too lavish dissipation of the capital of the state'.[15] Sentiments such as Hume's informed the proposals for a generalized takeover of forest and waste by the state. As Brandis found to his dismay, the majority of the participants at the crucial Forest Conference of 1874 disagreed with his more modest claims for state control, holding out for the constitution of all forests as state reserves controlled by the forest department.[16]

[14] 'Remarks by the Board of Revenue, Madras', 5 August 1871, and minute by the governor of Madras, 9 February 1878, both in appendix SS, LD file of 1878.

[15] NAI, Forests, B progs, no 10, Sept. 1876, note by A.O. Hume, 24 June 1876.

[16] NAI, Forests, B progs, no 3-8, July 1874, 'Memorandum by D. Brandis,

In sharp contrast, the Madras government believed that state intervention should be minimal, fully respecting existing rights. The collector of the Nilgiris, for example, insisted that forest officials actively seek out evidence of communal rights, rather than wait for headmen to put forward their claims. Objecting to Brandis' draft bill of 1869, he remarked that the procedure it envisaged would pit two unequal antagonists. While its provisions were 'too arbitrary, setting the laws of property at defiance', the bill left 'the determination of the forest rights of the people to a Department which, in this Presidency at all events, has always shown itself eager to destroy all forest rights but those of Government'. Showing a prescient awareness of the territorial aspirations of the forest department, the board pleaded for the involvement of civil courts in the arbitration of forest rights. As it stood, the bill had stacked the dice heavily against the interests of villagers, for the forest department, 'which acquires new importance by every forest right which it strangles will be the arbitrator'.[17]

Typically straddling these two positions, Brandis advocated the *restricted* takeover of forests by the state. He justified this middle course both on the grounds of equity—respect for age-old rights—and efficiency—as the only feasible course. Brandis urged the administration to 'demarcate as state forests as large and compact areas of valuable forests as can be obtained free of forest rights of persons', while leaving the residual areas—smaller in extent but more conveniently located for their supply—under the control of village communities. He hoped for the creation of three great classes of forest property based on the European model: state forests, forests of villages and other communities, and private forests. State ownership had to be restricted on account of the 'small number of ex-

---

IGF, on several matters discussed at the Forest Conference, dated the 1st June 1874'.

[17] 'Remarks by the Board of Revenue', 5 August 1871, and appendix A to the above ('Abstract of Collectors' Reports on the Forest Bill'), both in the LD file of 1878.

perienced and really useful officers' in the colonial forestry service, *and* out of deference to the wishes of the local population. Thus:

> the trouble of effecting the settlement of forest rights and privileges on limited well-defined areas is temporary and will soon pass away, whereas the annoyance to the inhabitants by the maintenance of restrictions over the whole area of large forest tracts will be permanent, and will increase with the growth of population.[18]

Walking a tightrope between the imperatives of colonial administration and the claims of social justice, Brandis, quite remarkably for his time and milieu, placed considerable trust in the ability of village communities to manage their own affairs. He wrote appreciatively of the extensive network of sacred groves in the subcontinent, which he termed 'the traditional form of forest preservation' (Brandis 1897, pp. 14-15). Displaying an early 'ethnobotanical' interest in indigenous systems of tree and plant classification, he circulated a list of local names, urging 'younger officers, with more leisure and more extensive opportunities, to take up the study of the names of trees and shrubs used by the [tribes] of Central India' (Brandis 1876, appendix). He also praised Indian rulers for their forest sense, singling out the Rajasthan chiefs who, in strenuously preserving brushwood in a dry climate, 'have set a good example, which the forest officers of the British government will do well to emulate' (Brandis 1873, pp. 24-5). He was especially keen on reviving and strengthening village communal institutions. When his requests for official initiative in the formation of village forests were repeatedly turned down, he argued the fruits of well-managed communal forests, recognized by law, as existing in several European countries. On a visit to Mysore he expressed the hope that at some future date he would 'find

[18] This paragraph draws on three memoranda by Brandis, namely Brandis (1875), and the unpublished memoranda of 1869 and 1874, cited in footnotes 6 and 16 above.

the advantages of true communal forests recognized not in Mysore only but in all parts of India where a village organization exists', for it seemed to him 'particularly desirable to strengthen the old village organization by consolidating and ameliorating the grazing grounds, forest and waste land of the village community'.[19]

Brandis' task was an uphill one. His sentiments may have been noble but they were not shared by his peers and masters in the British colonial system. Rapping him on the knuckles, Brandis' boss, the agricultural secretary, said the inspector-general's 'views as to rights of aboriginal tribes, forest villages etc., are to my mind clearly in advance of my own, and *a fortiori* of those of the government of India'.[20] Meanwhile, the Madras government was chastised by the secretary of state for its 'laxity with respect to the forest rights of Government'.[21] Under pressure from both London and Fort William, it finally capitulated, and in 1882 agreed to an act closely modelled on the 1878 Indian Forest Act.[22]

And so the internal resistance to the 1878 act crumbled. As the next chapter documents, it prefigured the far more widespread *popular* resistance to colonial forest management. Within official circles, however, the question was firmly resolved in favour of the 'annexationists', and a policy of state annexation was embarked upon. The concrete proposals were embodied in Brandis' memorandum of 1875 which, with Baden-Powell's paper in the Forest Conference of the previous year, formed the basis of the 1878 act.

Based on Baden-Powell's distinction between 'rights' and

[19] See Brandis' memoranda of 1869, cited in footnote 6, and 'Memorandum by D. Brandis, IGF, on the district forest scheme of Mysore', 1 June 1874, in NAI, Forests, B progs, nos 12-15, June 1874.

[20] NAI, Forests, progs nos 43-55, March 1875, note by A.O. Hume, 19 August 1874.

[21] NAI, Forests, B progs nos 9-10,1879, no. 13 (revenue forests), London, 28 August 1879, from secretary of state to governor, Madras.

[22] See NAI, Forests, progs nos 18-27, July 1882, and progs nos 9-31, January 1883.

'privileges', the act was a comprehensive piece of legislation which, by one stroke of the executive pen, attempted to obliterate centuries of customary use by rural populations all over India. It provided for three classes of forest. 'Reserved forests' consisted of compact and valuable areas, well connected to towns, which would lend themselves to sustained exploitation. In reserved forests a legal separation of rights was aimed for, it being thought advisable to safeguard total state control by a permanent settlement that either extinguished private rights, transferred them elsewhere, or in exceptional cases allowed their limited exercise. In the second category, the so-called 'protected forests' (also controlled by the state), rights were recorded but not settled. However, control was firmly maintained by outlining detailed provisions for the reservation of particular tree species as and when they became commercially valuable, and for closing the forest whenever required to grazing and fuelwood collection. Given increased commercial demand and their relatively precarious position from the government's point of view, protected areas were gradually converted into reserved forests where the state could exercise fuller control. Thus the 14,000 square miles of state forest in 1878 (the year the act was passed) had increased to 56,000 square miles of reserved forests and 20,000 square miles of protected forests in 1890—the corresponding figures a decade later being 81,400 and 3300 square miles respectively (Stebbing, I, pp. 468ff). The act also provided for the constitution of a third class of forests—village forests—although the option was not exercised by the government over the most part of the subcontinent. Finally, the new legislation greatly enlarged the punitive sanctions available to the forest administration, closely regulating the extraction and transit of forest produce and prescribing a detailed set of penalties for transgressions of the act.

## Forest Policy Upto 1947

As advocates of state monopoly well understood, the strict

regulation (and preferably extinction) of traditionally exercised rights was a *sine qua non* for the activity of commercial timber production. Under the provisions of the 1878 act, each family of 'rightholders' was allowed a specific quantum of timber and fuel, while the sale or barter of forest produce was strictly prohibited. This exclusion from forest management was, therefore, both *physical*—it denied or restricted access to forests and pasture—as well as *social*—it allowed 'rightholders' only a marginal and inflexible claim on the produce of the forests. The principle of state monopoly also formed the cornerstone of the important forest policy statement of 1894. Influenced by a devastating official indictment of the commercial orientation of forest management (Voelcker 1893), the policy was also a response to the 'serious discontent among the agricultural classes' caused by strict forest administration. While apparently more favourably disposed to village needs, the policy cautioned that these should be met only 'to the utmost point that is consistent with imperial interests'.[23]

Superbly equipped to maintain strict state control over forest utilization, the 1878 act provided the underpinnings for the 'scientific' management of forests, enabling the working of compact blocks of forest for commercial timber production. Consistent with this legal and institutional structure, the administration of the forests reserved by the state—some 99,000 square miles in 1947—was contingent on the imperial interests it served—first, during the era of railway expansion, and later, during the two world wars. At the same time, the forest department had to generate an adequate revenue, in keeping with a cardinal principal of imperial policy, namely that the administrative machinery had to be self-supporting. As such, a constant endeavour was to find markets for the multiple species of

[23] Regional Archives Dehradun, list no. 22, file no. 244, circular no. 22F, 19 October 1894, revenue & agricultural (forests). One must distinguish between policy statements and legislative enactments. Whereas it is always possible to make conciliatory gestures in the former, it is the latter which will actually be in operation.

India's tropical forests, of which only a few, often comprising less than 10 per cent of the canopy, were readily saleable. The inaccessibility of many forest areas and the stagnant nature of industrial development further inhibited the full commercial utilization of forest produce (Smythies 1925, pp. 57ff).

TABLE 4.1

REVENUE AND SURPLUS OF FOREST DEPARTMENT 1869–1925

| Yearly average for the period | Revenue (Rs million) | Surplus (Rs million) | Per cent of column 3 to column 2 |
|---|---|---|---|
| 1869-70 to 1873-74 | 5.6 | 1.7 | 30 |
| 1874-75 to 1878-79 | 6.7 | 2.1 | 31 |
| 1879-80 to 1883-84 | 8.8 | 3.2 | 36 |
| 1884-85 to 1888-89 | 11.7 | 4.2 | 36 |
| 1889-90 to 1893-94 | 15.9 | 7.3 | 46 |
| 1894-95 to 1898-99 | 17.7 | 7.9 | 45 |
| 1899-1900 to 1903-4 | 19.7 | 8.4 | 43 |
| 1904-1905 to 1908-9 | 25.7 | 11.6 | 45 |
| 1909-1910 to 1913-14 | 29.6 | 13.2 | 45 |
| 1914-1915 to 1918-19 | 37.1 | 16.0 | 43 |
| 1919-1920 to 1923-4 | 55.2 | 18.5 | 34 |
| 1924 to 1925 | 56.7 | 21.3 | 38 |

SOURCE: Stebbing III, p. 620.

Nevertheless, as Table 4.1 indicates, the department consistently showed a handsome surplus on account. This was made possible by the requirements of urban centres for fuelwood, furniture, building timber, etc., while supply was facilitated by the improved communications which the railway network brought about (Tucker 1979). Thus the Himalayan forests provided bamboo, sal and several species of conifer for the urban markets of Punjab and the United Provinces, and for the military cantonments and hill stations that were a creation of colonial rule (Walton 1910; Hearle 1889). The railways, by now wholly government owned, continued to be an important customer of the forest. Urging closer co-operation, one committee pointed out that 'in many cases the interests of the two departments are identical' (GOI 1929, p. 30).

After years of research, the treatment of several woods for use as railway sleepers was made possible on a commercial scale in 1912. These timbers included the chir and blue pines. In the years that followed, the extensive pine forests of Garhwal and Kumaun were reserved—largely obviating any need to import wood or to search for metal and concrete substitutes (Guha, Ramachandra, 1989a). The teak export trade continued to pay, with well over one million sterling worth of teak wood being imported annually into Britain (Anon. 1920). The development of 'Minor Forest Produce' (hereafter MFP)—i.e. excluding timber—was also taken up in the twentieth century. This MFP was found to have a variety of industrial uses. And as India was the only source in the empire for several of the more valuable MFPs, e.g. resin and turpentine, tanning materials such as kath and myrabolans, and essential oils, foreign trade in these products showed a rapid rise. For example, the export trade in shellac was valued at more than two and a half million pounds in 1917-18 (Smythies 1925, pp. 82-91).

During this period the large areas of forest under India's princes were also drawn in, though indirectly, into the orbit of colonial capitalist expansion. The exploitation of these forests, either by the agency of the colonial state or directly by the princes, introduced qualitative changes in the relationship between ruler and subject. Over time, native rulers became 'generally very much alive to the value of their forest properties' (Smythies 1925, p. 27). The magnificent deodar forests of the Himalayan state of Tehri Garhwal had early attracted the attention of the colonial state, and in 1865 it successfully negotiated a lease on the important deodar and chir forests in Tehri Garhwal. These forests were divested of the existing rights of user of the surrounding population and commercially managed by the Imperial Forest Department from 1865 to 1925. Their enormous value being evidenced by the average annual profit of over Rs 1.6 lakhs (for 1910-25), the king did not renew the lease, henceforth working the forests—with professional help from the forest department and with legislation modelled on the 1878 act—himself (Guha, Ramachandra, 1989a). The deodar forests of the adjoining hill states of present-day Himachal Pradesh were also leased in by the colonial government, with deodar being 'valued daily more and more for sleepers'. The British similarly cast a covetous eye on the magnificent sal forests in the Himalayan principality of Sikkim.[24] Meanwhile, several states in Central India, such as Rewa, became important suppliers of MFP such as lac and myrabolans (Smythies 1925).

The strategic value of India's forests, first made evident in the building of the railway network, was forcefully highlighted during the world wars. In the war of 1914-18 timber and bamboos were supplied for the construction of bridges, piers, wharves, buildings, huts and ships. In little over a year (April 1917 to October 1918) 228,076 tonnes of timber (excluding railway sleepers) were supplied by the specially created 'timber branch' of the munitions board, and 50,000 tonnes of fodder grass exported to help military operations in Egypt and Iraq. Approximately 1.7 million cubic feet of timber (mostly teak) were exported annually between 1914 and 1919, and the indigenous resin industry proved to be a great boon at a time when American and French supplies were unavailable (Stebbing I, p. 36; III, ch. XIX; Smythies 1925, pp. 13, 82, 84, etc.).

The impact of the Second World War was more severely felt on the forests of the subcontinent. Early in 1940, a timber directorate was set up in Delhi to channel supplies of forest produce from the provinces. India was the sole supplier of timber to the Middle East theatre, and later to the Allied forces in Iraq and the Persian Gulf. The war became an 'over-riding

---

[24] NAI, Forests, progs nos 30-35 for October 1878, and B progs nos 83-5 for December 1876. See also *Indian Forester*, XLVII, 1921, pp. 80-2.

objective' as timbers had to be found to replace the few that hitherto ruled the market and had become unobtainable. The cessation of the import of structural steel too brought an urgent demand for wood substitutes (GOI 1948; Champion and Osmaston 1962, ch. 4).

To meet the exigencies of war, 'fellings and sawings were pushed to the remotest corners of the Himalayas and the densest forests of the Western Ghats'. In the United Provinces the demand for chir and sal timber was 'virtually unlimited', and the forest department's instructions were 'to produce the maximum outturn possible'. By the end of 1944 it had supplied, to the defence department alone, 909,000 tonnes of timber. Not surprisingly, working-plan prescriptions in reserved forests were 'considerably upset' during the war. In Bombay too the yield prescriptions were widely departed from, the margin in some cases being upto 400 per cent (Champion and Osmaston 1962).

Table 4.2 gives some indication of the war demand and efforts to meet it. The figures relate only to recorded fellings; the available evidence points to considerable over-fellings; in forests, both of private owners and native states, as well as unrecorded felling in government forests. The fellings recorded show an increase, over pre-war outturn, of 65 per cent. However, it was admitted that due to the varying circumstances in which felling took place, an accurate estimate of the damage done to forest capital was not possible. For the government forests, these excess fellings were estimated to be on average as much as six annual yields (GOI 1948, pp. 107-8). As can be seen from Table 4.2 the accelerated fellings in the last years of the war coincided with a sharp drop in the area covered by working plans, implying that much felling was carried out without the supervision required to ensure proper regeneration. In general, the damage done in areas not covered by working plans would have been greater, and at the same time undetectable.

TABLE 4.2

INDIA'S FORESTS AND THE SECOND WORLD WAR

| Year | Outturn of timber and fuel (m. cuft) | Outturn of MFP (Rs m) | Revenue of FD (Rs. m.) | Surplus of FD (Rs. m.) | Area sanctioned under working plans (sq. miles) |
|---|---|---|---|---|---|
| | | | (at current prices) | | |
| 1937-38 | 270 | 11.9 | — | — | 62,532 |
| 1938-39 | 299 | 12.3 | 29.4* | 7.2* | 64,789 |
| 1939-40 | 294 | 12.1 | 32.0 | 7.5 | 64,976 |
| 1940-41 | 386 | 12.5 | 37.1 | 13.3 | 66,407 |
| 1941-42 | 310 | 12.7 | 46.2 | 19.4 | 66,583 |
| 1942-43 | 336 | 12.9 | 65.0 | 26.7 | 51,364 |
| 1943-44 | 374 | 15.5 | 101.5 | 44.4 | 50,474 |
| 1944-45 | 439 | 16.5 | 124.4 | 48.9 | 50,440 |

NOTE :     * Average for the period 1934-5 to 1938-9.

SOURCE :   Compiled from *Indian Forest Statistics, 1939-40 to 1944-45*
(Delhi, 1949).

## The Balance Sheet of Colonial Forestry

In so far as the main aim of the new department was the
production of large commercial timber and the generation of
revenue, it worked willingly or unwillingly to enforce a separation between agriculture and forests. This exclusion of the
agrarian population from the benefits of forest management,
anticipated by opponents of the 1878 act, continued to draw
sharp criticism from within the ranks of the colonial intelligentsia. In the words of an agricultural chemist, the forest
department's objects

> were in no sense agricultural, and its success was gauged
> mainly by fiscal considerations; the Department was to be a
> revenue paying one. Indeed, we may go so far as to say that its
> interests were opposed to agriculture, and its intent was rather

to exclude agriculture from than to admit it to participation in
its benefits (Voelcker 1893, pp. 135-6).

While advocating the creation of fuel and fodder reserves
to more directly serve the interests of the rural population, Dr
Voelcker used the characteristic justification that the increased
revenue from land tax—which such a reorientation would
bring—would more than compensate any loss of revenue from
a decline in commercial timber operations. Faced with such
criticism, colonial foresters were unrepentant. 'It is the province
of the forest department to grow *timber*', stated one, 'pasturage
is the province of the agricultural department, and should be
taken up by that department'.[25] Grazing and shifting cultivation, the life-blood of tens of millions of Indians, were singled
out by foresters as activities to be totally banned in areas under
their control. Such hostility bears the mark of its origins, for,
in other countries dominated by German forestry techniques,
agriculture and forestry were likewise considered separate and
often opposed activities (cf. Raumolin, 1986; Peluso 1989).

The priorities of colonial forestry were essentially commercial in nature. This trend was reinforced by the financial crunch
faced by the government following the revolt of 1857, which
kept every department on its toes. At periodic intervals government committees asked the forest department to generate even
more revenue, and from Brandis onwards senior officials had
to justify the activities of the department on commercial
grounds (Stebbing I, pp. 463-4). Testifying to the Royal Commission on Agriculture, the inspector-general of forests explained that as 'the Forest Department has always been considered a commercial department', and revenue only came
from large timber forests, it was forced to neglect shrub forest
and pasture under its control (Anon. 1927, p. 256). Indeed, a
commercial orientation was built into the education of forest
officials. Thus the analogy with business enterprise, one wholly

[25] GKB, 'Some notes on the connection existing between forestry and
agriculture in India', *Indian Forester* XV, 1889, p. 331.

inappropriate for an activity with such a long gestation period (and from which societies derive inestimable and unquantifiable benefits in the form of environmental stability) was nonetheless widely used in training manuals for junior officials (cf. D'Arcy 1910).

To fulfil these twin demands of commercial timber and revenue, the forest department intensified its exploitation through two major devices. First, it made remote forests accessible by improving transportation networks. As one United Provinces forester proclaimed, 'communications are the veins and arteries along which forest revenue flows and from its earliest days the Department has been active in "opening up the jungles" with a proper system of cart roads and paths' (Ford Robertson 1936, pp. 15-16). Simultaneously, foresters made the accessible areas more profitable by increasing the proportion of commercially valued species in the growing stock. Silvicultural techniques of ringing, girdling and fire manipulation successfully transformed many mixed oak-conifer forests in the Himalaya into pure coniferous strands (Guha, Ramachandra, 1989a). From Brandis onwards, foresters have looked down upon oaks, the climax vegetation of the area, as having 'much too slow a rate of growth to justify their maintenance as component parts of the high forest' (Brandis 1882, p. 124). Likewise, working plans in South India prescribed the conversion of evergreen forests into single-species teak forests (Dhareshwar 1941).

We may briefly mention two related processes of ecological change under colonialism. First, there is the intimate connection between the Raj and shikar. From the middle of the last century, a large-scale slaughter of animals commenced in which white hunters at all levels—from the viceroy down to the lower echelons of the British Indian army—participated. Much of this shooting was motivated by the desire for large 'bags'. While one British planter in the Nilgiris killed 400 elephants in the eighteen sixties, successive viceroys were invited to shoots in which several thousand birds were shot in a single day in attempts to claim the 'world record'. Many Indian princes sought to emulate the shikar exploits of the British. The Maha·raja of Gwalior, for example, shot over 700 tigers early in this century. Although it is difficult to estimate the impact of such unregulated hunting on faunal ecology, some of the consequences were clear by the time India gained independence, as reflected in the steadily declining populations of wild species such as the tiger and the elephant (Bennet 1984; Elliot 1973; Sukumar 1989).

An equally major transformation in forest ecology concerned the sale, at extremely low prices of large expanses of woodlands to Europeans for the development of tea, coffee and rubber plantations. Although many areas had been taken over before 1864, in its early years the forest department was besieged with requests for land by coffee and tea planters who enjoyed considerable influence in the colonial administration.[26] In fact the state's desire to commercialize the forests went hand in hand with the allotment of vast areas to planters. The development of a road and railway network to facilitate the export of tea, coffee and rubber also served to hasten the pace of timber exploitation. The plantation economy itself had a high level of timber demand for fuel and packaging—the tea industry in eastern India alone requiring wood for over a million chests annually. In this manner the expansion of plantations bore a direct relation to the shrinking areas under forest cover (Pandian 1985; Sinha 1986; Tucker 1988).

However, perhaps the most serious consequence of colonial forestry was the decline in traditional conservation and management systems around the forest. Disregarding the proposals of Brandis and the Madras government, the state was quite lukewarm about the constitution of community forests. This was in line with overall colonial policy, for, as Voelcker pointed out,'the tendency of our system of government has, to a considerable extent been to *break up* village communities, and

---

[26] See evidence of A. Rodger, offg IGF, in Anon (1927), p. 256.

now for the most part they are heterogeneous bodies rather than communities' (Voelcker 1893, p. 16, emphasis in original). In the one presidency where *panchayat* (village) forests were promoted to any significant extent—Madras—it was a case of too little too late. While waiting thirty years after the 1878 act before creating village forests, the government had imposed a set of rules that quickly bureaucratized the panchayats and impaired their functioning (Pressler 1987). Elsewhere the provision in the act for constituting village forests remained a 'dead letter', and the stipulation that these forests be first constituted as state reserves aroused the suspicion of peasants (Ribbentrop 1900). The consequence of the alienation of peasants from forests they had earlier protected was anticipated by the Poona Sarvajanik Sabha in 1878. Contesting the new act's excessive reliance on state control, it argued that the maintenance of forest cover could more easily be brought about by

> taking the Indian villagers into the confidence of the Indian Government. If the villagers be rewarded and commended for conserving their patches of forest lands, or for making plantations on the same, instead of ejecting them from the forest land which they possess, or in which they are interested, emulation might be evoked between neighbouring villages. Thus more effective conservation and development of forests in India might be secured and when the villagers have their own patches of forests to attend to, Government forests might not be molested. Thus the interests of the villagers as well as the Government can be secured without causing any unnecessary irritation in the minds of the masses of the Indian population.[27]

But it was not to be. The state's takeover of forests and the subsequent working of these on commercial lines had, as one corollary, the diminution of customary rights, as well as a second—a slow but significant process of ecological decline. Attracted to India by among other things 'the seemingly immeasurable extent of its natural resources' (Whitcombe 1971,

[27] See, for example, NAI, Forests, progs nos 7-11, July 1875.

p. ix), the British had by the turn of the nineteenth century acquired firm control over these resources. However, the loss of forests and pastures, earlier communally owned and managed, severely undermined the subsistence economy of the peasant. Simultaneously, British land policy worked towards the increasing differentiation of the peasantry and the decline of communal institutions. Losing his autonomy, the peasant was forced further out of production for use and into the vortex of the market economy (cf. Scott 1976, for a similar process in colonial Indochina).

Forest management was easily the most significant element in the state takeover of natural resources which had earlier acted as a buffer for the peasant household. Yet colonial policy had other, often unforeseen, ecological consequences. Large-scale irrigation works in UP and Bihar resulted in waterlogging and soil salinity in some areas. According to Whitcombe this created an unprecedented strain in the economy of the Doab. In Bihar the breakdown of traditional small-scale community irrigation systems was believed to be a factor in rural impoverishment and the rise of *kisan sabhas* in the 1930s (Whitcombe 1971; Sengupta 1980). While there is limited research yet on the specific changes in different geographical regions, clearly the century and a half of colonial rule had introduced manifold social and ecological changes whose interdependence has rarely been appreciated.

# CHAPTER FIVE

# The Fight for the Forest

In the olden days small landholders who could not
subsist on cultivation alone used to eat wild fruits like
figs and jamun and sell the leaves and flowers of the
flame of the forest and the mahua tree. They could also
depend on the village grazing ground to maintain one
or two cows and two or four goats, thereby living
happily in their own ancestral villages. However, the
cunning European employees of our motherly government have used their foreign brains to erect a great
superstructure called the forest department. With all
the hills and undulating lands as also the fallow lands
and grazing grounds brought under the control of the
forest department, the livestock of the poor farmers
does not even have place to breathe anywhere on the
surface of the earth.
—Jotirau Phule, 1881, in Keer and Malshe 1969

From the perspective of the Indian villager, the ecological and
social changes that came in the wake of commercial forestry
were not simply an intensification of earlier processes of change
and conflict. Clearly, many of the forest communities (especial-

ly hunter-gatherers and shifting cultivators) had for several centuries been subject to the pressures of the agrarian civilizations of the plains. However, while these pressures themselves ebbed and flowed with the rise and fall of the grain-based kingdoms of peninsular India, they scarcely matched in their range or scope the magnitude of the changes that were a consequence of the state takeover of the forests in the late nineteenth century. Before this the commercial exploitation of forest produce was largely restricted to commodities such as pepper, cardamom and ivory, whose extraction did not seriously affect either the ecology of the forest or customary use. It was the emergence of timber as an important commodity that led to a qualitative change in the patterns of harvesting and the utilization of forests. Thus when the colonial state asserted control over woodland earlier controlled by local communities, and proceeded to work these forests for commercial timber production, it represented an intervention in the day-to-day life of the Indian villager which was unprecedented in its scope. Second, the colonial state radically redefined property rights, imposing on the forest a system of management and control whose priorities sharply conflicted with earlier systems of local use and control. Finally, one must not underestimate the changes in forest ecology that resulted from this shift in management systems. Significantly, the species promoted by colonial foresters—teak, pine and deodar in different ecological zones—were invariably of very little use to rural populations, while the species they replaced (e.g. oak, terminalia) were intensively used for fuel, fodder, leaf manure and small timber.

In these varied ways, colonial forestry marked an ecological, economic and political watershed in Indian forest history. The intensification of conflict over forest produce was a major consequence of changes in the patterns of resource use it initiated. This chapter analyses some of the evidences on conflicts over forest and pasture in colonial India. Based on both primary and secondary sources, it outlines the major dimensions of such conflicts by focussing on the genesis, the geographical spread, and the different forms in which protest manifested itself.

### Hunter-Gatherers: The Decline Towards Extinction

Until the early decades of this century, almost a dozen communities in the Indian subcontinent depended on the original mode of sustenance of human populations, namely hunting and gathering. Their distribution encompassed nearly the entire length of India, with the Rajis of Kumaun in the north to the Kadars of Cochin in the south. The abundant rainfall and rich vegetation of their tropical habitats facilitated the reproduction of subsistence almost exclusively through the collection of roots, fruit, and the hunting of small game. While cultivation was largely foreign to these communities, they did engage in some trade with the surrounding agricultural population, exchanging forest produce such as herbs and honey for metal implements, salt, clothes, and very occasionally grain. With minimal social differentiation, and restraints on the over-exploitation of resources through the partitioning of territories between endogamous bands, these hunter-gatherers, if not quite the 'original affluent society' (cf. Sahlins 1971) were, as long as there existed sufficient areas under their control, able to subsist quite easily on the bounties of nature.

Predictably, state reservation of forests sharply affected the subsistence activities of these communities, each numbering a few hundred—and with population densities calculated at square miles per person rather than persons per square mile. The forest and game laws affected the Chenchus of Hyderabad, for example, by making their hunting activities illegal and by questioning or even denying their existing monopoly over forest produce other than timber. The cumulative impact of commercial forestry, and the more frequent contacts with outsiders that the opening out of such areas brought about, virtually crippled the Chenchus. As suspicious of mobile populations as most modern states, in some parts the colonial government forcibly gathered the tribals into large settlements. Rapidly losing their autonomy, most Chenchus were forced into a relationship of agrestic serfdom with the more powerful cultivating castes. Further south, the Chenchus of Kurnool, almost in desperation, turned to banditry, frequently holding up pilgrims to the major Hindu temple of Srisailam (von Fürer Haimendorf 1943a; Aiyappan 1948).

As with the Chenchus, other hunter-gatherer communities were not numerous enough to actively resist the social and economic changes that followed state forest management. Forced sedentarization and the loss of their habitat induced a feeling of helplessness, as outsiders made greater and greater inroads into what was once their undisputed domain. Thus the Kadars succumbed to what one writer called a 'proletarian dependence' on the forest administration, whose commercial transactions and territorial control now determined their daily routine and mode of existence. In this manner the intimate knowledge of his surroundings that the Kadar possessed was now utilized for the collection of forest produce marketed by the state. In the thickly wooded plateau of Chotanagpur, meanwhile, the commercialization of the forest and restrictions on local use led to a precipitous fall in the population of the Birhor tribe—from 2340 in 1911 to 1610 in 1921 (Ehrenfeld 1952 ; Roy 1925).

While the new laws restricted small-scale hunting by tribals, they facilitated more organized shikar expeditions by the British. The disjunction between the favours shown to the white shikari and the clampdown on subsistence hunting had serious consequences. While there were few formal restrictions on the British hunter until well into the twentieth century, hunter-gatherers as well as cultivators for whom wild game was a valuable source of protein found their hunting activities threatened by the new forest laws.

The Baigas of Central India, for example, were famed for their hunting skills. They were 'expert in all appliances of the chase', and early British shikaris relied heavily on the 'marvellous skill and knowledge of the wild creatures' that they possessed. Yet the stricter forest administration, dating from the turn of the century, induced a dramatic decline. Writing in the

1930s, Verrier Elwin noted that while their love for hunting and meat persisted, the old skills had largely perished. There remained, however, a defiant streak, and as one Baiga said, 'even if Government passes a hundred laws we will do it. One of us will keep the official talking; the rest will go out and shoot the deer' (Ward 1870; Best 1935; Elwin 1939). In the Himalayan foothills, too, where there was an abundance of game, villagers continued to hunt despite government restrictions, taking care to be one step ahead of the forest staff—a task not difficult to accomplish, given their familiarity with the terrain.[1]

Among shifting cultivators, there was often a ritual association of hunting with the agricultural cycle. Despite game laws, the Hill Reddis of Hyderabad clung to their ritual hunt—called Bhumi Devata Panduga or the hunt of the earth god—which involved the entire male population and preceded the monsoon sowing. The reservation of forests also interfered with the movement of hunting parties across state boundaries. In 1929 a police contingent had to be called in to stop a party of Bison Marias from Bastar state, armed with bows and spears, from crossing into the British-administered Central Provinces. This, of course, constituted an unnatural intervention, as the ritual hunt was no respector of political boundaries. Nevertheless, in later years the authorities were successful in confining the Maria ritual hunt to Bastar, the game caught steadily declining in consequence (von Fürer Haimendorf 1943b; Grigson 1938).

### The 'Problem' of Shifting Cultivation

Shifting or jhum cultivation was the characteristic form of agriculture over large parts of India, especially in the hilly and forested tracts where plough agriculture was not always feasible. Jhum typically involves the clearing and cultivation of patches of forest in rotation. The individual plots are burnt and cultivated for a few years and then left fallow for an extended

[1] See 'Gamekeeper', 'Destruction of Game in Government Reserves During the Rains', *Indian Forester*, XIII (1887), pp. 188-90. Cf. also Corbett (1952).

period (ideally, a dozen years or longer), allowing the vegetation and soil to recoup and recover lost nutrients. Cultivators then move on to the next plot, abandoning it in turn when its productivity starts declining. Although in parts of the Western Ghats Hindu castes did depend on this mode of resource use, it was usually practised by 'tribal' groups for whom jhum was a way of life encompassing, beyond the narrowly economic, the social and cultural spheres as well. The corporeal character of these communities was evident in the pattern of cultivation, where communal labour predominated, with different families adhering to boundaries established and respected by tradition. The overwhelming importance of jhum in structuring social life was strikingly manifest, too, in the many myths and legends constructed around it in tribal cosmology.

As in many areas of social life, major changes awaited the advent of British rule. For, almost without exception, colonial administrators viewed jhum with disfavour as a primitive and unremunerative form of agriculture in comparison with plough cultivation. Occasionally the British, out of political compulsions, were constrained to allow the continuance of jhum, as in the hill tracts of the Western Ghats formerly controlled by the Maratha kingdom. The Marathas were among the last to fall to the East India Company, and shifting cultivators in their territories continued to rise periodically against the British, right until the insurrection led by Vasudev Balwant Phadke in the 1880s. Elsewhere, however, the British strove strenuously to curb or destroy shifting agriculture. Influenced both by the agricultural revolution in Europe and the revenue-generating possibilities of intensive (as opposed to extensive) forms of cultivation, official hostility to jhum gained an added impetus with the commercialization of the forest. Like their counterparts in other parts of the globe, British foresters held jhum to be 'the most destructive of all practices for the forest'.[2] There was a very good reason for this animosity—if 'axe cultivation

[2] Cf. Muhafiz-i-Jangal (pseud), 'Jhooming in Russia', *Indian Forester*, II (1876), pp. 418-19.

was the despair of every forest officer' (Elwin 1943, p. 8), it was largely because timber operations competed with jhum for territorial control of the forest. This negative attitude was nevertheless tempered by the realization that any abrupt attempt to curtail its practice would provoke a sharp response from jhum cultivations. Yet the areas cultivated under jhum often contained the most valued timber species.[3] In the circumstances, the curbing of jhum was an intractable problem for which the colonial state had no easy solution.

A vivid account of the various attempts to combat jhum can be found in Elwin's classic monograph on the Baiga (Elwin, 1939, esp. chapter 2), a small tribe that inhabited the Mandla, Balaghat and Bilaspur districts of present-day Madhya Pradesh. The first serious attempt to stop shifting cultivation in the 1860s had as its impetus the civilizational zeal of the chief commissioner of the province, Richard Temple. In later years, though, it was the fact that the marketable value of forest produce 'rose in something like geometrical proportion' which accounted for the 'shifting of emphasis from Sir Richard Temple's policy of benevolent improvement for their own sake to a frank and simple desire to better the Provincial budget'. A vigorous campaign to induce the Baiga to take to the plough culminated in the destruction of standing jhum crops by an over-enthusiastic deputy commissioner. When many tribals fled to neighbouring princely states, the government advised a policy of slow weaning from axe cultivation.

In fact, such difficulties had been anticipated by the settlement officer in 1870; he observed that 'it has been found quite impracticable, as well as hard and impolitic, to force the Baigas to give up their dhya (jhum) cultivation and take to the plough'. He advised limits on jhum rather than a total ban. A more cautious policy was dictated, too, by the dependence of the

[3] As the chief commissioner of the Central Provinces put it, 'the best ground for this peculiar cultivation is precisely that where the finest timber trees like to grow'. Sir Richard Temple, quoted in J.F. Dyer, 'Forestry in the Central Provinces and Berar', *Indian Forester*, LI (1925), p. 349.

forest department on the labour of the Baigas, for they were most proficient at wood-cutting and the collection of forest produce. As a consequence, the government established the Baiga *chak* (reserve) in 1890, covering 23,920 acres of forest, where they planned to confine all jhum cultivators. The area chosen was described as 'perfectly inaccessible [and] therefore useless as a timber producing area'. While permitting jhum within the reserve, the administration stressed an overall policy of discouraging it elsewhere. In this they were partially success- ful, as Baiga villages outside the chak, faced with the prospect of leaving home, accepted the terms of plough cultivation. While many Baigas continued to migrate into neighbouring princely states, within the chak itself the population of jhum cultivators steadily dwindled.

Baiga opposition took the form of 'voting with their feet', and other forms of resistance that stopped short of open con- frontation—such as the non-payment of taxes and the con- tinuance of jhum in forbidden areas. The new restrictions incul- cated an acute sense of cultural loss, captured in a petition submitted to the British government in 1892. After jhum has been stopped, it said—

> We daily starve, having had no foodgrain in our possession. The only wealth we possess is our axe. We have no clothes to cover our body with, but we pass cold nights by the fireside. We are now dying for want of food. We cannot go elsewhere. What fault have we done that the government does not take care of us? Prisoners are supplied with ample food in jail. A cultivator of the grass is not deprived of his holding, but the government does not give us our right who have lived here for generations past (Elwin 1939; Ward 1870).

In some areas tribal resistance to the state's attempt to curb jhum often took a violent and confrontationist form. This was especially so where commercialization of the forest was accom- panied by the penetration of non-tribal landlords and money- lenders who came to exercise a dominant influence on the indigenous population. Elwin himself, talking of the periodic disturbances among the Saora tribals of the Ganjam Agency, identified them as emanating from two sources: the exactions of plainsmen and the state's attempts to check axe cultivation. Thus, Saoras were prone to invade reserved forests and clear land for cultivation. In the late 1930s, several villages endeavoured to fell large areas of reserved forests in preparation for sowing. The Saoras were ready for any penalty—when the men were arrested and put in jail, the women continued the cultivation. After returning from jail the men cleared the jungle again for the next year's crop. As repeated arrests were unsuccessful in stopping Saoras from trying to establish their right, the forest department forcibly uprooted crops on land formally vested in the state (Elwin 1945).

Perhaps the most sustained resistance, extending over nearly a century, occurred in the Gudem and Rampa hill tracts of present-day Andhra Pradesh. Inhabited by Koya and Konda Dora tribesmen (predominantly jhum cultivators), under British rule the hills were subject to a steady penetration of the market economy and the influx of plainsmen eager to exploit its natural wealth. Road construction led to rapid development in the marketable trade of tamarind, fruit, honey and other forest products that were exported to urban centres and even to Europe. Traders, from the powerful Telugu caste of Komatis, also took on lease (from local chiefs) tracts of forest as well as the trade in palm liquor. As in other parts of India, they were actively helped by the colonial government, which had banned the domestic brewing of liquor (an important source of nutrition in the lean season) and farmed out liquor contracts in a bid to raise revenue. Simultaneously, commercial forest operations were begun on a fairly large scale, and, as elsewhere, the creation of forest reserves conflicted with the practice of jhum. Slowly losing control over their lands and their means of subsistence, many tribals were forced into relations of dependence with the more powerful plainsmen, either working as tenants and sharecroppers in the new system of market agriculture or as forest labour in the felling and hauling of timber.

Among the many small risings or *fituris* documented by David Arnold, several were directly or indirectly related to forest grievances. The Rampa rebellion of 1879-80 arose in response to the new restrictions concerning liquor and forest regulations. Complaining bitterly against various exactions, the tribals said that 'as they could not live they might as well kill the constables and die'. The rebellion broke out in March 1879 and spread rapidly to neighbouring areas. The rebels, led by a minor tribal chieftain, Tammam Dora, attacked and burnt several police stations, executing a constable as an act of ritual sacrifice. While Tammam Dora was shot by the police in June 1880, the revolt spread to the Golconda Hills of Vishakhapatnam and the Rekepalle country in Bhadrachalam. The latter territory had earlier been part of the Central Provinces, and its transfer to Madras led to greater restrictions on the practice of jhum. Here, protest emanated directly from forest grievances and, as in other fituris, police stations—a highly visible symbol of state authority—were frequent targets. It took several hundred policemen and ten army companies to suppress the revolt, a task not finally accomplished till November 1880.

The last recorded fituri was, like its predecessors, closely linked to restrictions on tribal access to the forest. This occurred in 1922-3 and was led by a high-caste Hindu from the plains called Alluri Sita Rama Raju, who was able to transform a local rising into a minor guerilla war. Including dispossessed landholders and men convicted for forest offences, Rama Raju's men were actively helped by villagers who gave them food and shelter. After raids on police outposts had netted a haul of arms and ammunitions, Raju's band was able to evade the police by its superior knowledge of the hilly and wooded terrain. Unsuccessful in his attempts to spread the rebellion into the plains, Rama Raju was finally captured and shot in May 1924 (Arnold 1982; von Fürer Haimendorf 1945a).

Interestingly, as the Indian princes sought to emulate their

British counterparts in realizing the commercial value of their forests, they too came in conflict with shifting cultivators. Regarding the state takeover as a forfeiture of their hereditary rights, in several chiefdoms tribals rose in revolt against attempts to curb jhum. A major rebellion took place in Bastar state in 1910, directed against the new prohibitions concerning jhum, restrictions on access to forests and their produce, and the *begar* (unpaid labour) exacted by state officials. The formation of reserved forests had resulted in the destruction of many villages and the eviction of their inhabitants. In order to draw attention to their grievances, some tribals went on hunger strike outside the king's palace at Jagdalpur. Mostly Marias and Murias, the rebels, affirming that it was an internal affair between them and their ruler, cut telegraph wires and blocked roads. Simultaneously, police stations and forest outposts were burnt, stacked wood looted, and a campaign mounted against *pardeshis* (outsiders), mostly low-caste Hindu cultivators settled in Bastar. Led by their headmen, the rebels looted several markets and attacked and killed both state officials and merchants. In a matter of days, the rebellion engulfed nearly half the state, or an area exceeding 6000 square miles. Unnerved, the king called in a battalion of the 22nd Punjabis (led by a British officer) and detachments of the Madras and Central Province police. Armed with bows, arrows and spears, the rebels unsuccessfully engaged the troops in battle. In a decisive encounter near Jagdalpur, over 900 tribals, of all ages from 16 upwards, were captured.[4]

In 1940 a similar revolt broke out in Adilabad district of Hyderabad state. Here, Gonds and Kolams, the principal cultivating tribes, were subjected to an invasion of Telugu and Maratha cultivators who flooded the district following the improvement of communications. Whole Gond villages fell to immigrant castes. In the uplands, meanwhile, forest conservan-

[4] Based on National Archives of India, New Delhi, foreign department, Secret—I progs, nos. 34-40, for August 1911, and nos. 16-17 for September 1910; Grigson (1939); Clement Smith (1945).

cy restricted jhum, with cultivated land lying fallow under rotation being taken into forest reserves. Following the forcible disbandment of Gond and Kolam settlements in the Dhanora forest, the tribals, led by Kumra Bhimu, made repeated but unsuccessful attempts to contact state officials. After petitions for resettlements were ignored, the tribals established a settlement on their own and began to clear forests for cultivation. An armed party which came to burn the new village was resisted by Bhimu's Gonds, who then took refuge within a mountain fastness. When the police asked them to surrender, they were met with the counter demand that Gonds and Kolams should be given possession of the land they had begun to cultivate. The police thereupon opened fire, killing Bhimu and several of his associates (von Fürer Haimendorf 1945b).

Elsewhere in Hyderabad state, the Hill Reddis of the Godavari valley were at the receiving end of the new forest laws. The restriction of jhum to small and demarcated areas forced the Reddis to shorten fallow cycles, or to prolong cultivation on a designated patch until deterioration set in. Interestingly enough, as forest laws were not quite as stringent across the Godavari in British territory, tribals moved across in response to a ban on jhum—returning to Hyderabad when the ban was lifted. While not resorting to open protest, the Reddis thus made evident their dislike of the ban (von Fürer Haimendorf 1943b; 1945b). Likewise the population of shifting cultivators in one taluk of Nasik declined by 24 per cent in a single year (1874): fed up with restrictions on jhum, they fled to neighbouring princely states.[5]

These repeated protests had a significant impact on government policy. In some parts of Madras Presidency, certain patches were set aside for tribals to continue jhum. For although 'the forest department would welcome the complete stoppage of podu [jhum] it is not done for fear of *fituris* [tribal uprisings]'

---

[5] National Archives of India (NAI), rev. agl. (forests) progs nos. 1-3, January 1877.

(Aiyappan 1948, pp. 16-17). Elsewhere the state found a novel way of pursuing commercial forestry without further alienating tribal cultivators. This was the 'taungya' method of agrosilviculture—developed in Burma in the nineteenth century—where jhum cultivators were allowed to grow food crops in the forest provided they grew timber trees alongside. Thus, after a few years, when the cultivator moved on to clear the next patch, a forest crop had been established on the vacated ground. Taungya, which rendered possible at a 'comparatively low cost' the establishment of the labour force necessary for forest works, is still widely in operation. It helped forestall the very real possibility of revolt among tribals who were to be displaced by a prohibition upon their characteristic forms of cultivation. (Sometimes, though, even taungya cultivators thwarted the state, for example by planting only upon those areas likely to be inspected by touring officials). Ironically enough, its success has even led to the reintroduction of jhum in tracts where it had died out or been put down at an earlier stage (Blanford 1925; Baden-Powell 1874; Champion and Seth 1968).

More commonly, the cumulative impact of market forces and state intervention forced the abandonment of jhum in favour of the plough or wage labour. Even where the practice continued, the disruption of the delicate balance between humans and forests—initially through the usurpation of forests by the state, and later through rises in population—has led to a sharp fall in the jhum cycle. A form of agriculture practised for several millenia has become unsustainable in the face of external forces over which it had very little control.

## Settled Cultivators and the State

Notwithstanding the spatial separation between field and forest, over the most part of India plough agriculturists (mostly caste Hindus) were scarcely less affected by forest reservation than cultivators. For they too depended on their natural habitat in a variety of ways. An adequate forest cover was ecologically necessary to sustain cultivation, especially in mountainous tracts where terrace farming predominated. And with animal husbandry a valuable appendage to cultivation, the forest was a prime source of fodder in the form of grass and leaves. Forests also provided such essential inputs as fuel, leaf manure, and timber for construction and agricultural implements.

Here too state reservation enforced changes in the traditional pattern of resource utilization, even if these changes were not quite as radical as in the case of shifting cultivators. Under the provisions of the 1878 act, the takeover of a tract of forest involved settling the claims of surrounding villages. Under the new 'legal' (i.e. codified) arrangements, the previously unlimited rights of user were severely circumscribed. These restrictions affected two distinct classes of agriculturists, and in somewhat different ways. In areas dominated by cultivating proprietors, and where differentiation was not too marked, those affected by state forestry primarily consisted of middle to rich peasants, many of whom were graziers rather than agriculturists. On the other hand, in tracts which exhibited more advanced forms of class differentiation, it was a different social stratum that was at the receiving end. These were adivasi (tribal) and low-caste communities which supplemented their meagre earnings as tenants and sharecroppers with the extraction and sale of fuel, grass and other minor forest produce.

An example of the first form of deprivation comes from the Madras Presidency. There, several decades after forest reservation, villagers had vivid memories of their traditional rights over the forest, continuing to adhere to informal boundaries demarcating tracts of woodland claimed and controlled by neighbouring villagers. The tenacity with which they clung to their rights was visibly manifest, too, in the escalation of forest offences (averaging 30,000 per annum)—with the killing of forest personnel a not infrequent occurrence. A committee formed to investigate forest grievances was puzzled to find that villagers interpreted the term 'free grazing' quite differently from the committee itself. While quite prepared to pay a small fee, peasants understood 'free grazing' to mean 'the right to graze all over the forests', i.e. a continuation of the territorial control over the forest that they formerly enjoyed. Thus the demand for grazing was accompanied by the demand for free fuel and small timber—in effect 'for the abolition of all control and for the right to use or destroy the forest property of the state without any restriction whatever'. Commenting on the widespread hostility towards state forest management, the committee observed that 'the one department which appears at one time to have rivalled the forest department in unpopularity is the salt department, which, like the forest department, is concerned with a commodity of comparatively small value in itself but an article for daily use and consumption' (Anon 1913; Baker, C.J. 1984).

Not surprisingly, the opposition to state forestry was far more intense among the lower castes and tribals. In the Thane district of coastal Maharashtra, an important source of income for tribal households was the sale of firewood to Koli fishermen. This trade was severely affected by the stricter control exercised over the forest since the later decades of the nineteenth century. Typically, the early manifestations of discontent were peaceful, e.g. petitioning the local administration. When this had no impact, collective protest turned violent. Surrounding the camp of a deputy collector, a group of villagers demanded that 'the forests be thrown open, palm tax be abolished, country liquor [be sold] at one anna a seer, salt at one anna a paili, rice at Re 1/1/4 per maund and that the Government should redeem their mortgaged land and restore it to them'. In another incident, a large number of tribals carrying firewood to the market were intercepted by the police. In protest, the adivasis stacked wood on a nearby railway line and refused to allow a train to pass. Sensing the prevailing mood of defiance, the officer in charge of the force allowed them to proceed to the market (Singh 1983).

A similar turn of events is reported from the Midnapur district of Bengal Presidency. In one area called the Jungle

Mahals, land owned by the Midnapur Zamindari Company (MZC)—an associate of the important British managing agency firm of Andrew Yuile—and other large landlords was cultivated by Santhal tribal tenants. While early lease deeds clearly specified that all land was to be handed over to the lessee, the coming of the railway and consequently of a thriving timber trade influenced the zamindars to impose sharp restrictions on the Santhals. Again, the tribals first tried the courts and other means of legal redress. However, the conditions of economic distress prevailing in the aftermath of World War I provoked a more militant response. Thus in 1918 the forest-dwelling Santhals proceeded on a campaign of *haat* (market) looting, their principal targets being upcountry cloth traders who were moneylenders as well.

Some years later, and after the intervention of Congress nationalists, the Jungle Mahals witnessed a movement more sharply focussed on the question of forest rights. In early 1922 Santhals working as forest labour went on strike. Following a scuffle between employees of the MZC and the strikers, the Congress directed the Santhals to plunder the forests. Further incidents of haat looting (including the burning of foreign cloth) and attempts to restrict the export of paddy were also reported. In one subdivision, Silda, Santhals began to plunder jungles leased to timber merchants. When a police party tried to confiscate the newly-cut wood, they were beaten up (Dasgupta 1980).

Another form of the assertion of traditional rights was manifest when Santhals began to loot fish from ponds controlled by individual zamindars. In April 1923 there was a wave of fish pond looting and breaches of the forest law over an area of 200 square miles, from Jhargram in Midnapur to Ghatshila in the Singhbhum district of Bihar. While recognizing this to be an 'illegal' act, the tribals argued that tank-raiding would force the zamindars to concede their customary rights over forests. The Santhals, the district magistrate commented, 'will let you know how in his father's time all jungles were free, and *bandhs* (ponds)

open to the public. Sometimes he is right . . . '. When the protests were supported by a dispossessed local chieftain, even the
belief that their acts were illegal was abandoned. Indeed, as
alarmed officials reported, 90 per cent of the crowd believed
that through their acts they were merely bringing back a golden
age when all jungles were free (Sarkar 1984).

The defiance of forest regulations also formed part of the
countrywide campaigns led by the Indian National Congress
in 1920-2 and 1930-2. Gandhi's visit to Cudappah in south-east-
ern India in September 1921 was widely hailed as an opportunity to get the forest laws abolished. In nearby Guntur, peasants actually invaded the forests in the belief that 'Gandhi Raj'
had been established and the forests were now open. Ten years
later, during the Civil Disobedience movement, the violation of
forest laws was far more widespread. In Maharashtra, where
women played a significant part, nearly 60,000 villagers in
Akola district marched into government forests with their cattle. In Satara district peasants, arguing that grazing restrictions
deprived the sacred cow of its daily food, resolved not to pay
the grazing fee. Encroachment on reserved forests was followed
by the felling of teak trees and the hoisting of the national
tricolour on a teak pole, in front of a temple dedicated to Shiva.
Women also played a key role in a similar campaign in the
coastal district of North Kanara (in present-day Karnataka),
garlanding and smearing ritual paste on men who went off to
the forest to cut the valued sandal tree. There too the timber was
loaded onto carts and stacked in front of a local temple. The
arrests of men inspired the women, who invoked Sri Krishna
(the deity who had gone to the forest), when symbolically
breaching the rules themselves. In the Central Provinces, meanwhile, tribals came forth in great numbers to participate in the
organized violation of forest laws. While formally conducted
under the rubric of the Congress, these movements actually
enjoyed a considerable degree of autonomy from that organization; moreover, the many violent incidents were clearly in
defiance of nationalist leaders, wedded as the latter were to an

ideology of non-violence (Sarkar 1980; Halappa 1964; Baker, D.E.U. 1984; Brahme and Upadhya 1979).

Perhaps the most sustained opposition to state forest management was to be found in the Himalayan districts of present-day Uttar Pradesh. Dominated by magnificent stands of coniferous species, these hill forests have been—as the only source of softwoods—one of the most valuable forest properties in the subcontinent. At the same time, the forests here have also played a crucial role in sustaining agriculture in the mountainous terrain, this role being strikingly reflected in the traditional systems of resource conservation evolved to inhibit the over-exploitation of village forests.

In the period of colonial rule this region was divided into two distinct socio-political structures—the princely state of Tehri Garhwal and the British administered Kumaun division. However, as the forests of Tehri Garhwal came under commercial management even earlier (*c.* 1865), in both areas peasant resistance to this encroachment on customary rights was remarkably sustained and uniform. In Tehri, important if localized movements occurred in 1904, 1906, 1930 and 1944-8, in all of which forest grievances played an important and sometimes determining role. Through the collective violation of the new laws and attacks on forest officials, the peasantry underscored its claim to a full and exclusive control over forests and pasture. As in other pre-capitalist societies where the ruler relied on a traditional idiom of legitimacy, protest was aimed at forest management and its back-up officials, not at the monarch himself. In Kumaun division, on the other hand, social protest was aimed directly at the colonial state itself, and at the most visible signs of its rule, namely pine forests under intensive commercial management, and government buildings and offices. It reached its zenith in the summer of 1921, when a wide-ranging campaign to burn forests controlled by the forest department virtually paralyzed the administration, forcing it to abolish the much disliked system of forced labour and to abandon effective control over areas of woodland. Largely autonomous of organized nationalist activity (as represented by the Congress), the movements of 1916, 1921, 1930 and 1942 in Kumaun division brought to the fore the central importance of forests in peasant economy and society. Notwithstanding differences in the social idiom of protest—not unexpected in view of the somewhat different socio-political structures and styles of rule —in both Tehri Garhwal and Kumaun division forest restrictions were the source of bitter conflicts, unprecedented in their intensity and spread, between the peasantry and the state (Guha, Ramachandra 1989a).

### Everyday Forms of Resistance: The Case of Jaunsar Bawar

In a penetrating study of rural Malaysia, the political scientist James Scott (1986) has observed that while most students of rural politics have focused on agrarian revolt and revolution, these are by no means the characteristic forms of peasant resistance. Far more frequently, peasants resort to methods of resisting the demands of non-cultivating elites that minimize the element of open confrontation—for example non-co-operation with imposed rules and regulations, the giving of false and misleading information to tax collectors and other officials, and migration. In colonial India, too, the peasantry often resorted to violent protest only after quasi-legal channels—such as petitions and peaceful strikes—had been tried and found wanting. Whereas the historical record is heavily biased towards episodes of violent revolt—the times when the peasant imposes himself rather more emphatically on the processes of state—it is important not to neglect other forms of protest which were not overtly confrontationist in form.

Often, these other forms of resistance preceded or ran concurrently with open conflict. Thus, in many areas the breach of forest laws was the most tangible evidence of the unpopularity of state management: the available evidence showing that typically the incidence of forest 'crime' followed a steadily escalating trend. While this would be true of regions where sustained protest did occur (such as those described above), the absence of an organized movement quite evidently did not signify an approval of state forestry.

That the conflict between villagers and colonial forest management did not always manifest itself in open revolt is clearly shown by the experience of Jaunsar Bawar, the hilly segment of Dehra Dun district that bordered Tehri Garhwal on the west. From the early 1860s the forests of Jaunsar Bawar had attracted the attention of the state. These forests were important for three reasons—as a source of wood for the railway, as 'inspection' forests for training students at the Forest School in nearby Dehra Dun, and for supplying fuel and timber to the military cantonment of Chakrata. In the ensuing settlement of 1868 the state divided the forests into three classes. While forests of the first class were wholly closed for protection, villagers had certain rights of pasturage and timber collection in the second class. The third class was to be kept for the exclusive use of peasants, with the caveat that they were not allowed to barter or sell any of the produce.

Early protests were directed at this government monopoly. The confused legal status of Class III forests, village leaders argued (it was not clear who held actual proprietary right, the state or the village), was compounded by the refusal to allow rightholders to dispose of their timber as they pleased. If peasants believed that they could not dispose of the produce of Class III forests as they liked, their control was only a formal one, the government on its part was loath to give up its monopoly over the timber trade. Extending over three decades, and conducted through a series of petitions and representations, this was in essence a dispute over the proprietary claims of the two parties. As the superintendent of the district observed, villagers were concerned more with the legal status of Class III forests than its extent—indeed 'they would be contented to take much less than they have now, if they felt it was their own'.[6] Nor were

---

[6] Uttar Pradesh Regional Archives, Dehra Dun (hereafter UPRA); Post-Mutiny Records (hereafter PMR); file no. 71, department XI, list no 2

they tempted by money for forests they claimed as theirs, with one frustrated official complaining: 'the Jaunsaris are a peculiar set and no reasonable amount of compensation will satisfy them'.

The unsettled state of forest boundaries had made the peasantry suspicious that the government would slowly take over Class III forests and put them under commercial management. On a tour of the district, the lieutenant-governor of the province encountered repeated complaints on the 'severity of the forest rules, dwelling chiefly on the fact that no forest or wasteland was made over to them in absolute proprietary right, and so they were afraid that at some future period government might resume the whole of it and leave them destitute'. As one hillman succinctly put it, 'the forests have belonged to us from time immemorial, our ancestors planted them and have protected them; now that they have become of value, government steps in and robs us of them'.[7] The official urged a revision of forest boundaries and the confirming of village proprietorship in Class III forests, as 'nothing would tend to allay the irritation and discontent in the breasts of the people so much as giving them a full proprietary title to all lands not required by government'.[8]

---

(hereafter, L2); Memorandum by H.G. Ross, superintendent of Dehra Dun, on verbal complaints made to the lieutenant-governor by Syanas (headmen) of Jaunsar and Bawar, n.d. (prob. 1871 or 1872); NAI, department of rev. agl. (forests), progs. nos. 9-10, July 1876, no. 24, 21 April 1876, from CF, NWP, to government NWP, PWD.

[7] A comparatively wealthy *inamdar* in Thana district of the Bombay Presidency, fighting to protect his forest rights, made a similar complaint about the avarice of the forest department: 'In the earlier days when the plantation was comparatively young, the forest department left it to your memorialist to foster and nurture the growth of the trees; as soon as they are come to be of any value, the department steps in with its prohibitory and threatening orders'. See memorial of Naro Ramachandra Parchurey, inamdar of Asnouli, Tanna collectorate, to Marquis of Salisbury, secretary of state, 6 July 1874, in NAI, rev. and agl. (forests), progs no. 34, April 1875.

[8] UPRA, PMR, file no. 2 department I, L2, superintendent, Dehra Dun, to

At the level of everyday existence, the restrictions on customary use under the Forest Act were regarded as unnecessarily irksome. Thus the government tried, not always with success, to restrict the use of deodar (cedar, the chief commercial species) by villagers, arguing that while the peasants were 'clearly entitled to wood according to their wants, nothing is said about its being *deodar*'. This legal sleight of hand did not always succeed, as villagers insisted on claiming deodar as part of their allotted grant—the wood being extensively used in the construction of houses. Again, the takeover of village grazing lands and oak forests to supply the fuel and grass requirements of Chakrata cantonment were grievances district officials acknowledged as legitimate, even if they could do little about it within the overall structure of colonial administration. Particularly contentious were proposals to regulate and ban the traditional practice of burning the forest floor before the monsoons for a fresh crop of grass. While this closure was regarded by the forest department as essential for the reproduction of timber trees, it led to the drying up of grass, and consequently a shortage of green fodder, as well as a proliferation of ticks.[9] Pointing to deodar forests where numerous young seedlings had sprung up despite constant grazing and even occasional fires, villagers were openly sceptical of the department's claim that closure was 'scientific'.[10] An additional reason for the persistent hostility towards grazing restrictions was the liberal allowance extended to nomadic cattle herders from the plains. These herdsmen were important as suppliers of milk to the cantonment and to lumbermen working in the forest. They were of the Muslim community of Gujars and, it was pointed out, they were allowed access to forest pasture even in areas

---

commissioner, Meerut div., no. 340, 15 September 1873.

[9] UPRA, PMR, L2, file no. 244, note by C. Streadfield, superintendent, Dehra Dun, 1 November 1898.

[10] See E.C. McMoir, 'Cattle Grazing in Deodar Forests', *Indian Forester*, VIII (1882), pp. 276-7.

where sheep and cattle belonging to the local peasantry were banned.[11]

The forest department also banned the use of the axe by peasants claiming their allotment of timber. Villagers demurred, arguing that the saw was too expensive, that they were not familiar with its use, that split wood lasted longer than sawn, and finally that since their forefathers had always used the axe, so would they. As a consequence, attempts to insert a clause in the land settlement of 1873 prohibiting the use of the axe came to naught. Although the settlement had considerably enhanced the land revenue, the main grievance expressed continued to be the infringement of village rights over forest. Village headmen first asked for a postponement of the settlement, and then drove a hard bargain, agreeing to the new revenue rates and the continuance of forest restrictions only on condition they were allowed the use of the axe in obtaining their grants of timber from forest land.[12]

If such petitions represented an appeal to the 'traditional' obligations of the state, the peasants of Jaunsar Bawar also resorted to extra-legal forms of protest that defied the government's control over forest extraction and utilization. Before an era of motorized transport, commercial forestry depended on fast-flowing hill rivers to carry felled logs to the plains, where they were collected by timber merchants and sold as railway sleepers. The floating logs were considered the property of the forest department; nearly 2 million sleepers were floated annually down the Yamuna and its chief tributary, the

[11] B.B. Osmaston, deputy conservator of forests, Jaunsar div., to assistant superintendent, Jaunsar Bawar, no. 483, 19 March 1899, source cited in fn. 6.

[12] UPRA, PMR, L2, file no. 244. Report on Forest Administration in Jaunsar Bawar, submitted by superintendent, Dehra Dun, to comm., Meerut Div., no. 520/244, 10 December 1900; L2, UPRA, PMR, L2, dept. XXI, file no., 244, E.C. Buck, offg secretary to board of revenue, NWP, to C.A. Elliot, secretary to government of NWP; UPRA, PMR, L2, file no. 2, no. 47, 17 February 1872, from settlement officer, Jaunsar Bawar, to comm., Meerut div.

Tons. Although villagers dwelling on the river banks had been 'repeatedly warned that government property is sacred', thefts were endemic. As 'every Jaunsari knows well all about the working of the Government forests and the floating of timber', officials tried to stop pilfering by levying heavier sentences than those sanctioned by the Forest Act. Thus, while each sleeper was worth only Rs 6, it was not unknown for villagers caught in possession of a sleeper to be sentenced to two months' rigorous imprisonment or a fine of Rs 30. Stiff sentences needed to be enforced, magistrates argued, as 'river thieves are pests and a deterrent fine is necessary'. Such measures failed to have the anticipated effect, and as late as 1930, a full sixty years after the state takeover of woodland, the superintendent of the district was constrained to admit that 'pilfering, misappropriating and stealing government and State timber' was 'a chronic form of crime in Jaunsar Bawar'.[13]

As in eighteenth-century England (Thompson 1975), while the infringement of forest laws was viewed as 'crime' by the state, it represented here too, as an assertion of customary rights, an incipient form of social protest. In Jaunsar Bawar the theft of floating timber and the defacement of government marks were accompanied by other forms of forest 'crime' wherein the peasantry risked a direct confrontation with the authorities—notably the infringement of laws preventing forest fires. In a fascinating incident, the head priest of the major temple of the area, which was dedicated to the god Mahashu Devta at Hanol, organized a firing of the pine forest to get rid of both the dry grass and the insects it harboured. The tall grass also attracted deer, who were a hazard to the adjoining croplands. Under the direction of the priest, Ram Singh, several villagers set fire to the forest on the night of 13 July 1915. Under section 78 of the Forest Act, villagers were liable to inform the forest staff of any fire in their vicinity. This they proceeded to

[13] See Trials, nos. 98 of 1925, 36 of 1927, 53 of 1930, and unnumbered trials dated May 1922, 15 June 1922, 7 April 1923, all in basta (box) for 1927-30 for Chakrata tehsil, in the Criminal Record Room, Dehra Dun collectorate.

do, but only several hours after the fire had been started. Ram Singh then advised a low-caste labourer, Dumon Kolta, to call the forest guard, but to go slowly.

While early enquiries clearly revealed that the fire was not accidental, its occurrence near the Mahashu Devta temple and the involvement of its priest made it difficult for the state to convict those accused.[14] Indeed, several prosecution witnesses, after a meeting with village headmen at the Hanol temple, suddenly retracted their confessions in court. Though the headmen were looked upon by the state to act as a bulwark of the administration, they underlined their partisan stance by appearing en masse for the defence. One elder, Ranjit Singh (whose fields were closest to the forest that was fired), expressed his disavowal of the *wajib-ul-arz* (record of rights) whereby headmen were personally required to put out fires and collect other villagers for the same purpose. As he defiantly told the divisional forest officer, 'such a wajib-ul-arz should be burnt and that his ancestors were ill-advised to have agreed to such a wajib-ul-arz with the Government'.[15]

Such organized and collective violations were hardly as frequent, of course, as the numerous acts of individual 'crime'. What differentiates Jaunsar from other forest areas where protest took a more open and militant form is the reliance on individual and largely 'hidden' forms of resistance. What is worthy of note is that this was an equally effective strategy in thwarting the aims of colonial forest administration. As an official reflecting on the history of state forestry in Jaunsar Bawar remarked, 'prosecutions for forest offences, meant as deterrents, only led to incendiarism, which was followed by more prosecutions and the vicious circle was complete'.[16]

[14] The oath in the court of Jaunsar Bawar was taken in the name of Mahashu Devta.

[15] See criminal case 98 of 1915, in source cited in fn. 13. Ram Singh and five others were sentenced to terms of imprisonment ranging from three months to a year.

[16] M.D. Chaturvedi, 'The Progress of Forestry in the United Provinces', *Indian Forester*, LI (1925), p. 365.

Clearly, these ostensibly individual acts of violation relied on a network, however informal, of consensus and support within the wider community. With all strata of village society uniformly affected by commercial forestry, every violation of the forest act could draw sustenance from a more general distrust of state control. And as individuals could quite easily be subject to the due processes of colonial justice, this resistance could hardly 'hope to achieve its purpose except through a generalized, often unspoken complicity' (Scott 1987). This complicity is strikingly evident in the refusal to testify, or alternatively the giving of false and misleading information to officials—as in the case of Ram Singh and the Hanol temple.

## The Decline of the Artisanal Industry

Apart from its all too visible impact on the cultivating classes, state forest management, by restricting access to traditional sources of raw material, also contributed to the decline of various forms of artisanal industry. Chief among these was bamboo, a resource vital to many aspects of rural life. Extensively used in house construction, basketweaving, for the manufacture of furniture and musical instruments, and even as food and fodder, this plant of enormous local utility was initially treated as a weed by colonial foresters—and early management plans advocated its eradication from timber producing areas. With the discovery in the early decades of this century that bamboo was a highly suitable raw material for papermaking, there was a radical shift: foresters now encouraged industrial exploitation while maintaining restrictions on village use. Many weavers were now forced to buy bamboo from government-run depots or the open market. Limited availability also led to new forms of social conflict within the agrarian population. Thus the Baigas, who had earlier supplemented slash-and-burn agriculture with bamboo weaving, lost this subsidiary source of income when the Basors, an artisanal caste specializing in basket work, asserted their 'trade union' rights to a monopoly of bamboo supplied by the forest department (Prasad and Gadgil 1981; Elwin 1939).

While bamboo, whether obtained surreptitiously from the forest or bought in the market, continues to play an important role in present-day village society, one form of indigenous industry that collapsed under colonial rule was the manufacture of charcoal-based iron. Again, we are indebted to Verrier Elwin for a sensitive study of the industry in its declining years. In his book on the Agaria, an iron-smelting tribe of the Central Provinces, Elwin describes in chilling detail how high taxes on furnaces and diminished supplies of charcoal led to a sharp fall in the number of operating furnaces—from 510 to 136 between 1909 and 1938. Although peasants preferred the soft, malleable metal of village smelters, the changing circumstances had virtually forced the Agaria out of business, especially as improved communications had made local iron uncompetitive when compared to imported British metal. Deeply attached to their craft, the Agarias resisted as best they could, by defying forest laws concerning charcoal burning or, alternatively, migrating to nearby chiefdoms where they were accorded more liberal treatment. In an extensive survey of Madras Presidency, the first inspector general of forests, Dietrich Brandis, provided confirmatory evidence of this decay, on account of limited fuel supplies and foreign competition, of an industry that was formerly very widespread (Elwin 1942; Bhattacharya 1972; Brandis 1882).

Significantly, proposals to set up iron works controlled by European capital did briefly evoke an interest in the conservation of trees for charcoal. Pointing out that the metallic content of Indian ores was nearly twice that of European ores, several administrators urged the reservation of large tracts of forest for the benefit of European-owned and managed works using the latest technological processes. Here, the expansion of charcoal-based iron production was based on the assumption that 'iron-making by hand in India will soon be counted among the things of the past'.[17] Brandis, while acknowledging that the abundance of wood in presently inaccessible areas made the promotion of charcoal iron a potential source of forest income, advocated a different form of utilization. Articulating an early version of 'intermediate' or 'appropriate' technology, he believed that any such attempt must build upon, rather than supplant, traditional forms of manufacture. In the event both proposals came to naught, and the industry died an inevitable if slow death.[18]

Other forms of artisanal industry, too, declined under these twin pressures—the withdrawal of existing sources of raw material and the competition from machine-made, largely foreign, goods. Thus the tassar silk industry, which depended on the collection of wild cocoons from the forest, experienced a uniform decline through most of India from the 1870s onwards. Here too decay could be attributed to the new forest laws: specifically to the enhanced duties levied on weavers collecting cocoons from the forest. Although, much later, the tassar industry experienced a revival under official patronage (chiefly in response to a growing export market), the household industry was in no position to compete with the newly-formed centres of production operating from towns. A parallel case concerns the decline of village tanners and dyers, likewise denied access to essential raw materials found in the forest.[19]

[17] Anon, 'Iron-making in India', Indian Forester, VI (1880), pp. 203, 208.

[18] See Diétrich Brandis, 'The Utilization of the Less Valuable Woods in the Fire Protected Forests of the Central Provinces, by Iron-Making', Indian Forester V (1879).

[19] This paragraph is based on information kindly supplied to the authors by Dr Tirthankar Roy of the Centre for Development Studies, Trivandrum, who is doing research on handicraft production during the colonial period. Fishing communities were also affected by forest laws, being forced to use inferior wood for canoes owing to the heavy duties levied on teak by the forest department. See Grigson (1938), pp. 163-4. Among other artisanal castes, evidence from Khandesh in Western India suggests that bangle-makers were almost ruined by the fee imposed on fuelwood. See Maha-

One group on whom we have comparatively little information, but who would definitely have been affected adversely by forest laws, are pastoral nomads. The Gaddis of the northwestern Himalaya (present-day Himachal Pradesh) were denied grazing for their sheep by the constitution of reserved forests. Turning to the uncultivated waste and protected forests, they found that the communities of settled cultivators who controlled these areas were reluctant to allow their sheep to graze. Forest reservation thus fostered conflict between nomads and peasants, with officials worried that the Gaddis would 'eventually go to the wall in the struggle with the village communities'. In fact the abrupt changes in the pastoral cycle pitted the Gaddis not merely against the peasants but against the forest administration as well. Ironically, even as restrictions on grazing increased, the simultaneous expansion of the lowland market for meat and wool propelled Gaddis to increase their flocks.[20]

## Conclusion

### The Social Idiom of Protest

In this chapter, through a synthesis of the available evidence from both primary and secondary sources, we have tried to indicate the quite astonishing range of conflicts over access to nature in British India. These conflicts were entirely consistent with the wide variety of ecological regimes, and correspondingly of social forms of resource use, which prevailed in the Indian subcontinent. Our survey reveals some interesting regularities in the forms within which protest characteristically expressed itself, notably against the state's attempts to abrogate

---

rashtra state archives, revenue department, file 73 of 1884 (personal communication from Dr Sumit Guha, St Stephen's College, Delhi).

[20] See NAI, rev. and agl. (forests), B progs no. 76, May 1875, and Tucker (1986).

traditional rights over the forest. In essence, state monopoly and its commercial exploitation of the forest ran contrary to the subsistence ethic of the peasant. To adapt a contrast first developed by E.P. Thompson in his study of the eighteenth-century food riot, if the customary use of the forest rested on a moral economy of provision, scientific forestry rested squarely on a political economy of profit (Thompson 1971).

If state monopoly severely undermined village autonomy, then what is striking about social protest is that it was aimed precisely at this monopoly. In many areas, peasants first tried petitioning the government to rescind the new regulations. When this had no visible impact they issued a direct challenge to state control, in the form of attacks on areas controlled by the forest department and worked for profit. Whether expressed covertly, through the medium of arson, or openly, through the collective violation of forest laws, protest focussed on commercially valuable species—pine, sal, teak, and deodar in different geographical regions. Quite often these species were being promoted at the expense of tree varieties less valuable commercially but of greater use to the village economy. While challenging the proprietary right of the state, peasant actions were remarkably discerning. Thus in the Kumaun movement cited above, the 'incendiary' fires of the summer of 1921 covered 320 square miles of exclusively pine forests. In other words, by design rather than accident, the equally vast areas of broadleaved forests *also controlled by the state* were spared as being of greater use to hill agriculture (cf. Guha, Ramachandra 1989a). As in peasant movements in other parts of the world, arson as a technique of social protest had both a symbolic and an utilitarian significance—the latter by contesting the claim of the state over key resources, the former by selectively choosing targets where the state was most vulnerable.

Historical parallels with other peasant movements far removed in time and space are evident, too, in the close association of protest with popular religion. A religious idiom also reflected the sense of cultural deprivation consequent on the loss of control over resources crucial to subsistence. In many areas the customary use of nature was governed by traditional systems of resource use and conservation that involved a mix of religion, folklore, and tradition in regulating both the quantum and form of exploitation (cf. Chapters 1, 2 and 3). The suppression and occasionally even obliteration of these indigenous systems of resource management under colonial auspices was acutely felt by different communities, albeit in somewhat different ways. The Baigas, for example, resisted attempts to convert them into plough agriculturists by invoking their myth of origin, in which they had been told specifically not to lacerate the breasts of Mother Earth with the plough. As Elwin observes, 'every Baiga who has yielded to the plough knows himself to be standing on *papidharti*, or sinful earth'. Even if not entirely a willing one, this conversion was not without divine retribution. And as one Baiga put it, 'when the *bewar* [slash and burn] was stopped, and we first touched the plough . . . a man died in every village' (Elwin 1939, pp. 106-7).

The Gonds, aboriginal plough cultivators, were similarly afflicted by a melancholia, or what Elwin has elsewhere called a 'loss of nerve' (Elwin 1943). Gonds were convinced that the loss of their forests signalled the coming of Kaliyug, an Age of Darkness, in which their extensive medical tradition was rendered completely ineffective. So insidious and seductive was the power of modern civilization that even their deities had gone over to the camp of the powerful. Unable to resist the changes wrought by that ubiquitous feature of industrial society, the railway, 'all the gods took the train, and left the forest for the big cities', where with their help the urban dweller prospered (Elwin 1958, p. 58, Elwin 1960, p. 80; Hivale and Elwin 1935, pp. 16, 17).

The belief that traditional occupations were sanctioned by religion was evident, too, in the obvious reluctance of the Agaria to abandon iron smelting. According to *their* myth of origin, both slash-and-burn and plough cultivation were sinful. In the old days, when they were faithful to iron, the Agaria believed they had enjoyed better health; now that government taxes and scarcity of charcoal had forced many iron workers to take to cultivation, their gods no longer provided immunity from diseases. The real point of conflict with authority concerned charcoal burning—vividly reflected in the numerous dreams that hinged on surreptitious visits to the jungle, and which often culminated in the Agaria being intercepted and beaten up by forest officials (Elwin 1942, pp. 264-8).

## The Mechanisms of Protest

Researches over the past two decades have quite convincingly demonstrated that while the peasant operates in a world largely composed of 'illiterates', and consequently many peasant movements lack a written manifesto, his actions are imbued with a certain rationality and internally consistent system of values. It is the task of the scholar, therefore, to reconstruct this ideology, an ideology that informs the peasant's everyday life as much as episodes of revolt, even where it has not been formally articulated. In this chapter too, we can discern, from a reconstruction of different episodes of social protest, a definite ideological content to peasant actions. Protest against enforced social and ecological changes clearly articulated a sophisticated theory of resource use that had both political and cultural overtones.

Of special significance is the wide variety of strategies used by different categories of resource users to oppose state intervention. Hunter-gatherers and artisans—small and dispersed communities lacking an institutional network of organization —were unable to directly challenge state forest policies. They did, however, try and continue their activities by breaking the new regulations, resorting chiefly to what one writer has called 'avoidance protest', i.e. protest that minimized the element of confrontation with the state (for example, migration and petty crime: cf. Adas 1981). In the long term, though, these groups were forced to abandon their traditional occupation and eke out

a precarious living by accepting a subordinate role in the dominant system of agricultural production. Both slash-and-burn
and plough agriculturists were able to mount a more sustained
opposition. Their forms of resistance ranged from individual to
collective defiance, from passive or 'hidden' protest to open and
often violent confrontation with instituted authority. Tightly
knit in cohesive 'tribal' communities, the characteristic response of jhum cultivators to forest laws was militant resistance, one that was almost wholly outside the stream of organized nationalism. The fate of this protracted resistance varied
greatly across different regions. Occasionally, the colonial state
capitulated, allowing the continuance of traditional forms of
cultivation. More frequently the state reached an accommoda-
tion with these communities, restricting but not eliminating
jhum cultivation. The resultant shrinkage of forest area available for swidden plots, when coupled with rising population,
led gradually to a reduction in fallow cycles and declining
yields. A large proportion of jhum cultivators, therefore, have
also had no alternative to becoming landless labourers.

Settled cultivators have perhaps been more successful in
retaining some degree of control over forest resources. The new
laws, while sharply limiting access, did not, unlike in the cases
given above, seriously threaten the livelihood of agriculturists
and graziers. With subordinate forest officials often hailing
from the same castes, the peasantry was often able to obtain
forest produce through bribing rangers and guards. In such
cases, while the *cost of access* may have increased significantly,
the deprivation of forest resources was very rarely total. Moreover, Hindu peasants protesting forest restrictions were more
successful in using the resources and strategies of modern
nationalism (petitions, litigation) in advancing their own interests.

Whatever the specific modalities of protest in different time
periods, and across different regions and forms of resource use,
it was in its essence 'social'—reflecting a general dissatisfaction
with state forest management, and resting heavily on tradition-

al networks of communication and co-operation. It is note-worthy that, almost uniformly, traditional leaders of agrarian society—for example clan and village headmen—played a key role in social mobilization and action. Since the colonial state looked upon them as local bulwarks of power and authority, such leaders were subjected to conflicting pressures; however, they usually decided to throw in their lot with their kinsmen. Second, one may indicate the tenuous hold exercised by the premier nationalist organization, the Congress, over most of the movements described in this chapter. Although individuals like Gandhi may have recognized the importance of natural resources such as salt and forest produce in the agrarian econ-omy, even protest formally conducted under the rubric of Congress often enjoyed considerable autonomy from its leader-ship. Social protest over forests and pasture pre-dated the involvement of Congress: even when the two streams ran con-current, they were not always in tune with each other. Finally, these conflicts strikingly presaged similar conflicts in the post-colonial period. Thus contemporary movements asserting local claims over forest resources have, if unconsciously, replicated earlier movements in terms of their geographical spread, in the nature of their participation, and in the strategies and ideology of protest (chapter 7).

The bitter and endemic opposition to colonial forestry, spanning as it did the entire spectrum of agrarian society, testifies to the foresight of those on the losing side in the debate over the 1878 act. Hostile to village control and sceptical of the peasant's ability to manage his own resources, forest officials claimed that the 'ryot is a short-sighted individual and probab-ly will not see that his cherished popular privileges must dis-appear in any case before railroads and increased cultivation and that his best chance, and in fact, only chance, is in a well conducted forest establishment'.[21] Such rationalizations were

[21] NAI legislature department, progs 43-152, February 1878, appendix KK, no. 1931, 24 February 1870, from offg. CF, Madras, to acting secretary to govt. of Madras.

rejected by early nationalists and sympathetic officials, both being aware that forests were much more than an economic resource for the majority of the Indian population. The act would leave 'a deep feeling of injustice and resentment amongst our agricultural communities', the Madras board of revenue pointed out with uncanny prescience; indeed the act would 'place in antagonism to Government every class whose support is desired and essential to the object in view, from the Zamindar to the Hill Toda or Korambar'.[22] While deploring the severance of the intimate connections between plough cultivation and the forest, peasants also pointed to the adverse effects the act would have on the forest tribes which earned their livelihood through hunting, gathering and the sale of forest produce. The new restrictions, they pointed out, would force these tribes to seek other modes of subsistence—perhaps even into plundering from the peasants themselves. For no fault of their own, therefore, these forest tribes would be converted by the colonial state into criminals, with thousands being 'thrown into the jails at the cost of the general public'.[23] While such opposition may have been in vain, we can retrospectively see its justification in the fulfilment of its direct predictions. And, as the following chapters document, the colonial forestry debate has more than a fleeting relevance to contemporary developments.

The movements described in this chapter were both short-lived and unsuccessful, yet their legacy is very much with us today:

[22] See Appendix SSS in ibid.

[23] NAI, dept. of rev. & agl. (forests), B progs, no. 54, March 1878, memorial to Baron Lytton, viceroy of India, from inhabitants of Kolaba collectorate, 21 December 1877 ('signed by about 5000 people'). In fact many of the tribes later designated by the colonial administration 'criminal'—for example the Chenchus of Hyderabad and the Lodhas of Bengal—were hunter-gatherers deprived of their livelihood by the new forest laws.

# CHAPTER SIX

# Biomass for Business

## Two Versions of Progress: Gandhi and the Modernizers

A detailed treatment of the transition from colonial rule to independence would take us too far afield (see Sarkar 1983). To properly appreciate the evolution of forest policy since 1947, however, we must set it in the context of the overall development strategy adopted by the government of independent India. The national movement had thrown up two alternative scenarios for the reconstruction of Indian economy and society. One vision was associated with Mahatma Gandhi, under whose tutelage the Indian National Congress emerged as the chief vehicle of mass nationalism. The kernel of Gandhi's social philosophy is contained in a slim tract, *Hind Swaraj*, written in 1909, which contains a massive indictment of modern western civilization. Rejecting modern industry in its totality, the pioneer organizer of indentured labour in South Africa pleaded for a revival of the organic village communities of the pre-colonial and pre-industrial past. Gandhi's idealization of the village community and the integration within it of craft production was central to the transformation from elite to mass nationalism. For it was his peculiar genius that enabled the Congress to reach out into the countryside and take into its orbit the broad masses of the people. In this political 'appropriation' of the peasantry, Gandhi's ideology, and more particularly the moral idiom in which it was couched, was to strike a sympathetic chord among social groups peripheral to the urban and Westernized sections of Indian society. Weighed down by the pressures of the state and intermediary classes, the peasantry saw in Gandhi's challenge an opportunity to recover its lost autonomy. Gandhi's appeal was strengthened by his distinctive personal style—ascetic, charismatic and all-inclusive in nature, it enabled him to fit easily into the saintly tradition of Hinduism. Another contributory factor was his reliance on a religious idiom and his invocation of symbols like the prayer meeting and hunger fast, though it has been suggested that his affinities lay more with the folk, as opposed to the classical tradition of Hinduism (cf. Chatterjee 1984; Nandy 1980).

In an apparent paradox, the Gandhian era of Indian politics saw the juxtaposition of a peasant-based politics with the increasing influence of Indian capitalists over the Congress organization. It is a central, largely unresolved question of Indian historiography that while Gandhi always gave 'theoretical primacy' to the peasant (Sarkar 1983, pp. 208-9), it was the Indian industrial class that was able to use nationalism as a vehicle to wrest concessions from the British. The isolation of the economy during World War I and the protective measures reluctantly offered to Indian industry in the years immediately following it had given a considerable filip to indigenous manufacture, notably in new areas like sugar, cement, and paper (the pioneering steel plant having already been set up by Jamsetji Tata in 1907). Simultaneously, the development of the Congress as a mass organization afforded entrepreneurs like G.D. Birla, growing in strength and confidence, an opportunity to serve as its major source of funds. While initially reluctant to support the Congress, they quickly perceived that their aims were in congruence with Gandhi's doctrinal insistence on non-violence and his desire to avoid a chaotic upheaval by mobilizing all social groups, including landlords and capitalists, against the British. Gandhi's theory of the capitalist as 'trustee' of the workers was sufficiently ambiguous, too, for the camouflaging of the shrewdly tactical support extended by industrialists to the Congress by protestations of good faith.

Whatever the anomalies in Gandhi's own thought and practice, it is clear that in the path of economic development eventually charted by the Indian nation the Mahatma's ideals were made redundant with a quite alarming rapidity. Most Indian nationalists drew a wholly different conclusion from the colonial experience, arguing that India's subjugation was a consequence of its intellectual and economic backwardness. In this perspective, as contrasted with the dynamic and progressive West, India was a once-great civilization that had stagnated and atrophied under the dead weight of tradition. Its revitalization could only come about through an emulation of the West, intellectually through the infusion of modern science, and materially through the adoption of large-scale industrialization. The person most closely associated with this view is of course Jawaharlal Nehru, Prime Minister for nearly two decades after independence.

As a nationalist leader who had spent many years in British prisons, Nehru's attitude towards the West was an ambivalent one—admiration for its industrial progress coupled with a bitterness resulting from the colonial experience. Other leaders more openly praised the West for awakening India out of its slumber. An early and influential statement may be found in the work of the Mysore engineer and statesman, Sir M. Visveswaraya. As Visveswaraya saw it, the choice before the Indian people was stark:

> They have to choose whether they will be educated or remain ignorant; whether they will come into closer contact with the outer world and become responsive to its influences, or remain secluded and indifferent; whether they will be organized or disunited, bold or timid, enterprising or passive; an industrial or an agricultural nation, rich or poor; strong and respected or

weak and dominated by forward nations. The future is in their own hands. Action, not sentiment, will be the determining factor (Visveswaraya 1920, pp. 273-4).

Operative here are the standard assumptions of modernization theory, even if Visveswaraya was writing several decades before that body of literature made its first appearance. Through rapid industrialization and urbanization, and the creation of a strong nation state, India could 'catch up' with the West. As exemplars of a radically compressed process of state-induced industrialization, Meiji Japan, Stalinist Russia, and Bismarckian Germany were variously held aloft as the beacon. Not surprisingly, this vision coincided with that of the rising Indian capitalist class. In the Bombay Plan of 1944, leading industrialists had agreed upon the importance of a strong and centralized state. This plan also stressed the need for government investment and control in heavy industry and public utilities, areas in which private investment would not easily be forthcoming (Thakurdas, *et al.* 1944; Chattopadhyay 1985).

What Visveswaraya termed the 'industrialize or perish' model of economic development was formally institutionalized in the Second Five Year Plan, for which underdevelopment was 'essentially a consequence of insufficient technological progress' (GOI 1956, p. 6; cf. also Mahalanobis 1963). Underlying a strategy of imitative industrialization was the adoption of the most 'modern' technologies, with little regard for their social or ecological consequences. In theory there were, of course, many options available to the Indian state. The technologies adopted could be capital or labour intensive; they could be oriented towards satisfying the demand for luxury goods or fulfilling the basic needs of the masses; they could degrade the environment or be non-polluting; they could use energy intensively or sparingly; and they could use the country's endowment of natural resources in a sustainable fashion or liquidate them; and so on.

In a sharply stratified society like India, these choices were critically affected by three interest groups: capitalist merchants and industrialists, the technical and administrative bureaucracy, and rich farmers. The influence of the capitalists was reflected in the massive state investments in industrial infrastructure—e.g. power, minerals and metals, and communications, all provided at highly subsidized rates—and in the virtually free access to crucial raw materials such as forests and water. Large landowners, for their part, ensured that they had an adequate and cheap supply of water, power, and fertilizer for commercial agriculture. Finally, the bureaucrat–politician nexus constructed an elaborate web of rules and regulations in order to maintain control over resource extraction and utilization. In this manner, the coalescence of economic interests and the seductive ideology of modernization worked to consolidate dominant social classes. This strategy willingly or unwillingly sacrificed the interests of the bulk of the rural population—landless labour, small and marginal farmers, artisans, nomads and various aboriginal communities—whose dependence on nature was a far more direct one.

## Forests and Industrialization: Four Stages

The continuity between the forest policies of colonial and independent India is exemplified by the national forest policy of 1952. Upholding the 'fundamental concepts' of its predecessor—the forest policy of 1894—it reinforces the right of the state to exclusive control over forest protection, production, and management. With the integration of the princely states into the Indian union, the forest department in fact considerably enlarged its domain in the early years of independence. While inheriting the institutional framework of colonial forestry, however, the new government put it to slightly different uses. The one major difference in the post-1947 situation has been the rapid expansion of forest-based industry. As we argue below, the demands of the commercial–industrial sector have replaced strategic imperial needs as the cornerstone of forest policy and management. Here, one can distinguish four stages in the industrial orientation of Indian forestry.

In the first stage, foresters relied exclusively on traditional 'sustained yield' selection methods to meet the growing commercial demand. Under this regime a proportion of the more mature trees is selectively extracted at some fixed time interval, such as thirty years, with the expectation that the stock will replenish on its own between two episodes of extraction. A powerful incentive to industrial expansion was provided in the shape of handsome state subsidies in the supply of forest raw materials. These subsidies cut across conventional political alignments. In Kerala, one of the first acts of the first elected communist government in the world was to commit the administration to supply bamboo and other raw materials at Re 1 per tonne for a rayon factory set up by the Birlas, the largest industrial house in the country. Meanwhile, existing forest operations were also intensified. In the Himalayan coniferous forests, for example, rotation periods were progressively reduced (from 160 to 100 years for the chir pine) as consumers were willing to settle for trees of smaller and smaller girth. Simultaneously, research helped find new uses for hitherto unmarketable trees. Thus the large areas of twisted chir pine in Kumaun—unsuitable both for resin tapping and railway sleepers—were sold to paper mills once their suitability for pulping had been established (Guha, Ramachandra 1989a, chapter 6).

While these concessions were consistent with the strategy of state subsidies to private industry, conventional management practices were unable to keep step with the escalating commercial demand. Due to a variety of reasons—including an inadequate data base, the failure to take account of rural demand, excessive grazing and fire, and the violation of standing prescriptions—the selection system did not generate the expected 'sustained' yields. A new strategy of intensive forestry was called for. One of its early and most influential proponents was a visiting expert of the Food and Agricultural Organization, who called upon the forest department to adopt a more dynamic approach. An 'expanding economy on the eve of modern industrialization', he observed, 'requires the *highest* tonnage of production of organic raw material within the *shortest* possible period, and at the *lowest* possible cost' (Von Monroy 1960, p. 8, emphasis added). Forest officials took up the challenge, arguing that the extremely low productivity was due to the 'uneconomic' and 'conservation oriented' approach that had hitherto characterized Indian forestry. The emphasis of the new strategy would be on the production of 'economically attractive resources', i.e. plantations of quick-growing, high-yielding tree species, to replace 'inferior' slow growing ones. Through their emphasis on rapid industrialization, foresters hoped the five year plans would 'compel' the government to 'revise' its policy from one of satisfying the demands of the rural population towards more directly serving the developing forest-industries sector (Sagreya 1979, pp. 15ff, 169; Nair 1967; Chandra and Srivastava 1968). Then in the early seventies the National Commission on Agriculture retrospectively set the official seal on industrial forestry by recommending a strategy whose first element 'would have to be production forestry for industrial wood production' (NCA 1976, p. 39).

From the early sixties the central government provided financial incentives, in the form of central plan funds, to encourage state governments to take up industrial plantations. 'Production' forestry has had as its mainspring the clear-felling of existing forest stands and their replacement by fast-growing commercial species. As dwindling forest stocks and traditional methods of 'sustained yield' harvesting were found inadequate to meet the growing commercial and industrial demand, foresters set aside working plans for programmes of clear-felling and the plantation of quick-growing, chiefly exotic, species. These plantations, the official manual of silviculture admits, were necessitated by the shortage of timber caused by the 'too rapid exploitation' of the mature stock of natural forests (Champion and Seth 1968, p. 444).

The attempts to increase the economic productivity of forests relied on two distinct kinds of monoculture. The first and more widespread strategy involved the raising of eucalyptus and tropical pine plantations for industrial raw material. Additionally, foresters were encouraged to grow species like teak and rosewood which could be converted into high-quality furniture and generate valuable foreign exchange. Indeed, on account of its revenue-generating potential, foresters were advised to plant teak, *wherever possible*, in and outside of its forest habitat (Champion and Seth 1968). Encouraging such trends, foreign aid agencies believed 'it would be highly advantageous for the Indian economy to replace a significant percentage of the mixed tropical hardwood species with man-made forests of desirable [*sic*] species such as eucalyptus, tropical pine and teak' (USAID 1970, p. viii).

The salient question, which the votaries of industrial forestry never stopped to ask, was 'desirable' for whom? By hastening the conversion of mixed forests into single species stands, monoculture was putting the final touches to the separation of forestry from subsistence agriculture. Indeed, the local populations adversely affected by the new plantations have reacted sharply to its imposition (see chapter 7). However, more proximate ecological causes underlie the failure of plantation forestry to realize projected yields. The expected yields from eucalyptus plantations varied from 10 tonnes/ha/year on poor sites, to 30 tonnes/ha/year on favourable sites. However, available data from 25 forest ranges in Karnataka show actual yield varying from 10 to 43 per cent of projected yield. In Haliyal division, a region with abundant rainfall, yields were as low as 0.86 to 2.22 tonnes/ha/year, with an average of 1.4 tonnes/ha/year, i.e. 10 per cent of the lower limit of the projected yields. In Bangalore, a dry zone, yields were 20 per cent of those projected (Gadgil, Prasad and Ali 1983). Similar results have been obtained from Kerala, where a study of 70 eucalyptus plantations revealed that annual increments were between 15 per cent and 30 per cent of those estimated by the forest department (Krishnakutty and Chundamannil 1986).

In some areas, in fact, productivity was close to zero. In many high-rainfall tracts of the Western Ghats, eucalyptus plantations were taken up on a large scale by clear-felling excellent rain forests. These were everywhere attacked by a fungus known as pink disease. In effect, tropical rain forests were converted into man-made deserts.

Even when judged on its own narrow terms, therefore, plantation forestry has been a miserable failure. Faced with a continuing crisis of raw-material supply, the wood-based industry has, since the middle of the last decade, adopted yet a third strategy to supplement harvesting from natural forests and the creation of man-made plantations. It has turned increasingly to farmers for the supply of fibrous raw material. Although the market price of wood is far higher than the subsidized rates at which it was being supplied produce from government forests, companies have preferred to deal directly with private farmers.

This third stage in the industrial orientation of forest policy has witnessed a veritable revolution in cash-crop agriculture. In the last decade, millions of farmers have taken to planting tree crops, which are hypothecated and sold on maturity to rayon and paper manufacturers. Ironically, the propagation of eucalyptus has been closely tied to a programme of 'social forestry' whose stated aims are quite different, namely supplying fuel, fodder, and small timber to agriculturists. In theory, social forestry programmes have three components: plantation on individual holdings ('farm forestry'), fuel woodlots on government land and along roads and canals, and community forests planted and managed by villages on common land. In principle, the programme emphasizes the last component—community forestry. However, its original aims were quickly subverted. A mid-term review by the World Bank, one of the programme's chief sponsors, conclusively showed that of its three components only farm forestry was at all successful.

In Uttar Pradesh, for example, community self-help woodlots achieved only 11 per cent of their targets, whereas farm forestry overshot its target by 3430 per cent, an unprecedented rate of success for a government-sponsored programme. Observable trends in other states were very similar. Within the farm-for-estry programme, the percentage of seedlings reaching big farmers (i.e. those whose holdings exceed two hectares) was over 60 per cent (CSE 1985; Blair 1986).

And what do these farmers plant? Overwhelmingly, com-mercial species such as eucalyptus. Indeed, nearly 80 per cent of all seedlings distributed by the forest department to farmers were of eucalyptus. There has been a veritable eucalyptus rev-olution in states such as Karnataka, Gujarat and Haryana, with perhaps a million hectares of farmland being brought under different varieties of the tree. Eucalyptus has been praised by foresters as the panacea for all ills and bitterly attacked by environmentalists as being totally unsuited to Indian condi-tions. Its critics invoke social as well as environmental argu-ments. Although the programme of clear-felling rain forests to raise plantations of exotic species was quite ill-advised, the ecological impact of eucalyptus on farmland (as compared to other cash crops such as sugarcane) is less clear. Perhaps the social implications are more serious. In some areas, it is replac-ing food crops (ragi), in other areas cash crops (cotton). With an assured market and handsome advances offered by industry, many farmers now find it profitable to switch from traditional agricultural crops to eucalyptus even on irrigated holdings. However, as it is much less labour intensive than the crops it replaces, this substitution is in effect displacing agricultural labour from existing sources of employment in conditions of high rural unemployment. Where eucalyptus displaces food crops, it may increase the cost of grains for the poor, in so far as they earlier obtained a portion of their wages as grain and are now forced to turn at a greater real cost to the market. Finally, it has been observed that eucalyptus, as it involves little super-vision and is not browsed by animals, is an attractive option for absentee landlords, especially the urban gentry. In these several ways, farm forestry is very likely intensifying the already grave inequalities within the agrarian sector (Shiva *et al.* 1982; Chandrashekar *et al.* 1987).

The explanation for the eucalyptus epidemic is a relatively straightforward one. As the state, upon whom the private sector had traditionally relied, exhausted the forests under its command, industrialists have turned to private farmers.

According to one study (Prasad, personal communication), whereas the average yield and cost of production for government plantations were 2 tonnes/ha/year and Rs 2000 respectively, the corresponding figures for farmers were greater than 10 tonnes/ha/year and between Rs 500–1000 respectively (one must note here that the price at which government supplied wood to industry from its plantations was vastly lower than the actual cost of production). So farmers are more productive than the industry-oriented plantations of the forest department. While indirectly availing of state subsidies in the form of free seedlings and technical help, private farmers have emerged as a far more reliable ally for an industry plagued with chronic raw-material shortages.

Indeed the hiatus between precept and practice is striking even for a government programme. Social forestry, claims a leading forester, 'involves the people at all levels with raising forests as their own assets for their own use'. It aims 'to provide forest goods and services in rural areas where these are needed the most' (Tewari 1981). The programme, a high-level committee likewise affirms, 'is the practice of forestry for meeting the felt needs of rural areas, as against that for meeting the (needs) of the commercial and industrial interests' (GOI 1980, p. 5).

In official rhetoric, therefore, 'social forestry' revives a plea made by Voelcker nearly a century earlier (Voelcker 1893) for the creation of 'fuel and fodder reserves' to overcome the separation between forestry and agriculture. Yet within a decade it stands exposed as oriented almost exclusively towards commercial farmers and industrial units. In so far as it has bypassed more pressing questions of social equity and environmental stability while pursuing commercial ends, farm forestry is akin to the earlier policy of clear-felling natural forests to raise industry-oriented plantations. In this sense, as a leading environmentalist has observed, the recent afforestation of parts of India has been as negligent of the needs of the vast bulk of the rural population as the continuing deforestation in other parts of the country (Agarwal 1986).

While abandoning the forest department for the commercial farmer, the paper and rayon industries have also been lobbying intensely for the allotment of government land as captive plantations. Along with a massive import of wood and paper pulp, the move for captive plantations can be said to constitute an emerging fourth stage in the industrial orientation of forest policy. In one very important case, a company of the giant Birla group, Harihar Polyfibres Ltd (KPC), has formed a joint-sector company, Karnataka Pulpwoods Ltd, with the government of the state of Karnataka. KPL has been allotted 75,000 acres of land, spread over six districts, for growing eucalyptus as raw material for the exclusive use of Harihar Polyfibres.

The formation of KPL has been challenged in the supreme court as being harmful to the interests of peasants, who were fulfilling their own biomass requirements from the land handed over to the new company. The court judgment (awaited at the time of writing) will be of considerable significance, for it may determine the principles on which public land can or cannot be used by private industry. The controversy raises with particular cogency the question of 'biomass for whom and for what?' If the judgment is in favour of KPL, it could inaugurate a fourth stage in the industrial orientation of forest policy, in which large industry is allowed to lease land at highly favourable terms. While the ministry of environment and forests in New Delhi has indicated its opposition to captive plantations (whether undertaken directly by the concerned industry or in the joint sector), many state governments are eager to go ahead with KPL-style schemes. Industries are eagerly queuing up in half a dozen states, including Orissa, Madhya Pradesh, and Maharashtra.

TABLE 6.1

FOUR STAGES OF INDUSTRIAL FORESTRY

| Period | Method | Species | Agency | Prime beneficiary |
|--------|--------|---------|--------|-------------------|
| 1947 | Selection felling | Indigenous commercial species | Forest department | Industry |
| 1960–85 | Clearfelling and mono-cultural plantations | Chiefly exotics | Forest department | Industry |
| 1975 | Farm forestry | Chiefly exotics | Commercial farmers | Commercial farmers and industry |
| 1985 | Import and captive plantations | Exotics | Joint sector | Industry, importers |

## The Balance Sheet of Industrial Forestry

Whereas between 1860 and 1947 forests were a strategic raw material crucial for imperial interests—such as railway expansion and the world wars—since independence the commercial–industrial sector has been the prime beneficiary of state forest management. The growth in wood-based industries, and the expansion in demand for processed wood products, have been the determining influence on forest policy in this period. While the ends may be different, the means to achieve these ends have been very similar in the two periods. Both the 'hardware' and 'software' of forest administration in independent India have been closely modelled on colonial forestry.

The continuity between colonial and post-colonial forestry

regimes is most clearly manifest in the system of ownership. The state has contrived to uphold its monopoly over forest ownership. In some ways it has even strengthened its hold, as for example through the Forest Conservation Act, 1980, which actively discourages the participation of individuals and communities in forest plantation and protection. Until very recently it has also inhibited tree plantations by individuals and communities, reserving to itself the right of disposal of commercially valuable species *wherever planted*. State monopoly is reinforced by the preference for macroplanning, wherein national priorities take precedence over local priorities. As enshrined in the National Forest Policy of 1952, the exclusion of local communities from the benefits of forest management is legitimized as being in the 'national interest', namely that the 'country as a whole' is not deprived of a 'national asset' by the mere 'accident [*sic*] of a village being situated close to a forest' (GOI 1952, p. 29).

The rationale for government ownership is the belief that private individuals and groups will not invest in tree crops whose gestation period often exceeds a lifetime. However, as Part I of this book shows, many castes and communities have had informal systems of natural resource management. Moreover, these decentralized and small-scale systems were far better adapted to the rhythms and demands of subsistence agriculture, husbandry, and crafts production. State management has substituted a set of *external* constraints for a set of *internal* constraints which operated in a local, communal context. Meanwhile as the 'national interest' has virtually been equated with the industrial sector, local populations have been further alienated from the activity and practice of forest management.

Another element of continuity is in the monopoly of technically trained experts on decisions concerning forest management. Foresters draw a very clear line between 'professionals' and 'non-professionals', recognizing neither the local user's knowledge about forest conditions nor his/her definition of a resource and priorities in its management. Resource use is thus separated from resource management. Again, the ideology of scientific forestry is clearly at variance with the facts. For peasants have, in using the forests for centuries, drawn upon an intimate knowledge of local ecological conditions accumulated and passed on through the generations (see Fortmann and Fairfax 1989).

TABLE 6.2

FOREST REVENUE AND SURPLUS, SELECTED YEARS
(AT CURRENT PRICES)

| Average for period | Revenue (Rs million) | Surplus (Rs million) |
|---|---|---|
| 1951-52 to 1954-54 | 240-1 | 133.9 |
| 1956-57 to 1958-59 | 418.4 | 247.2 |
| 1961-62 to 1963-64 | 693.8 | 371.5 |
| 1966-67 to 1968-69 | 1075.0 | 508.7 |
| 1969-70 to 1970-71 | 1358.7 | 623.8 |
| 1975-76 | 2927.0 | 1275.0 |
| 1976-77 | 3355.0 | 1466.0 |
| 1980-81 | 4725.5 | 1547.2 |

SOURCE : NCA (1976); PUDR (1982).

Thirdly, the forest department, assigned the role of a revenue generating organ by the colonial state, continued to be a money-spinner for the government of independent India (Table 6.2). The emphasis on revenue generation, encouraged by many state governments, has contributed to the overexploitation of growing stock. A narrow commercial orientation is also reflected in research priorities. A recent attempt at consolidating forest research produced individual bibliographies for commercially valuable species such as teak, sal and chir pine, whereas the many varieties of oak, so crucial for sustaining Himalayan agriculture, only merited a single bibliography. Actually, the broad-leaved forests of the upper Himalaya, condemned by the British as valueless, have in recent years attracted the attention of the commercial sector, after work at the

Forest Research Institute established their suitability for use as badminton rackets (Guha, Ramachandra 1989a). Again, the continuing raw-material crisis of the paper industry has encouraged research on agricultural residues as possible substitutes for scarce forest raw material. In both cases, scientists have unwittingly deprived local populations of a resource earlier earmarked exclusively for their use.

Fourth, in the system of forest management which has now prevailed for over a century, no social group has any stake in the long-term husbanding of forest resources. This is especially true of the three segments of our society most closely connected with forest utilization: the rural poor, forest-based industry, and the forest department itself. Unfortunately, the self-interest of none of these sectors is served by a proper maintenance of the tree cover. The rural poor can at best derive very low wages by working as forest labourers or by selling firewood, and since they do not share in the profits made from the forest produce these wages always remain low, at the subsistence level. While the rural poor gain little by protecting the tree crop, so far they have always succeeded in establishing their ownership over a patch of land by cutting down the trees and putting it to the plough. This is one reason why 2.5 million hectares of forest land has been lost to cultivation between 1951 and 1976 (see Table 6.3).

TABLE 6.3

THE LOSS OF FOREST AREA FOR VARIOUS PURPOSES BETWEEN 1951 AND 1976

| *Purpose* | *Area (thousand ha)* |
| --- | --- |
| River valley projects | 479.1 |
| Agricultural purposes | 2506.9 |
| Construction of roads | 57.1 |
| Establishment of industries | 127.2 |
| Miscellaneous purposes | 965.4 |

Forest-based industry and timber merchants both gain least in the short run by preserving forest stock. Instead, they can maximize immediate profit by concentrating on minimizing the present cost of resources. This is what they have done, in the process frequently resorting to violations of forestry regulations. Further, they have been given little incentive for investing in the preservation of resources. The very low prices, well below replacement cost, at which they had access to resources meant that they had no motivation to invest substantially in resource regeneration. Of course in recent years, when the forest department has failed them, industry has turned to helping private farmers grow wood, by providing them a remunerative price and loan facilities. But by and large the industrial sector has so far concentrated only on exploiting newer and newer species in more remote areas, as the currently used resources have been exhausted one by one (see below).

Meanwhile, the bureaucratic apparatus, with its diffusion of responsibility and lack of any accountability, provides no motivation to a good officer for the proper management of resources under his charge, nor disincentives for those who mismanage. Forest officers are frequently transferred, never spending more than three or four years in one assignment. As a consequence, there is little commitment within the forest bureaucracy towards good husbanding of the resources over which they are temporarily in charge.

### Sequential Exploitation: A Process Whereby a Whole Flock of Geese Laying Golden Eggs is Massacred One by One

During the British regime timber remained the main commodity in demand—teak, initially for the construction of ships and gun carriages, later for building-construction and furniture; deodar, pine, sal and some of the evergreen tree species for railway sleepers. There were of course other demands, such as wood for charcoal-making in the early days of the railways. But by and large the thrust of forest management was on the gradual conversion of India's forests into single species stands of teak, sal and conifers. The working plans ultimately aimed at bringing about this change, and if the natural mixed forests were overexploited, this was a matter of little concern. The pace of such overexploitation was especially great during the two world wars, and the forest working plans written immediately after World War II uniformly speak of the need to give the ravaged forests a long period of rest.

But no such rest was forthcoming, because India achieved independence soon after the end of World War II and launched on a course of industrial development. Forest-based industry in India had its hesitant beginnings with the use of bamboo for paper manufacture only in the 1920s. In 1924-5 just 5800 tonnes of bamboo were so used; by the beginning of the World War II this had risen to 58,000 tonnes. By 1987, however, paper and board production had risen to 2.66 million tonnes, correspond-ing to a raw-material consumption of over 5 million tonnes, a thousandfold increase over a period of sixty years. There has been an even more rapid expansion in the capacity of a number of other forest-based industries: plywood, polyfibre, rayon, and matchsticks to mention only the most important ones.

The stated policy in independent India was to enhance the forest stocks by 50 per cent, raising the overall area under forest from 22 per cent to 33.3 per cent. The existing forest stocks were then meant to be used sustainably, while additional areas were to be brought under tree cover. There was, however, no ap-parent attempt to match the growing capacity of forest-based industry with sustainable yields from existing forest stocks. Firstly, there was a wholly inadequate data base to assess sustainable yields from any forest area. Even for *Bambusa arun-dinacea*, which had been established as an excellent raw material for paper in 1912, the nature of the growth curve of an in-dividual bamboo and the effect on growth of extraction and grazing remained to be understood (Prasad and Gadgil 1981). And there was no understanding at all of the complex rain-forest ecosystems of the Western Ghats, the north-eastern hill states, and the Andaman and Nicobar islands, which provided raw material for the exploding plywood industry (FAO 1984; Saldanha 1989). At the same time, the growth in industrial capacity followed the economic rationale of a continuing increase, so long as investment was perceived to be profitable enough in the short run by the entrepreneur. Indeed such growth in industrial capacity parallels the logic of growth in the fishing fleet explored by Dasgupta (1982). Dasgupta shows that fishing boats will continue to be added so long as the net gain, i.e. the balance of gross profit minus the harvesting cost, remains above the threshold of the minimal acceptable level of profitability. This is likely to lead to overexploitation, if the market price of the produce is high relative to the harvesting and processing costs. Economists therefore recommend a tax to hike costs to a level that would avoid the excessive growth of harvesting effort. In the case of forest-based industry in independent India, exactly the opposite situation prevailed. The industries were subsidized so heavily, and could hike up the prices of their produce so freely in a seller's market, that their profitability has remained high, even as forest stocks have plummetted. Thus in the 1950s the paper industry was provided bamboo at the throwaway price of Rs 1 per tonne, when the prevailing market price was over Rs 2000 per tonne. Indeed, as early as 1860 Cleghorn records that local basket weavers were being charged at the rate of Rs 5 per tonne. Even in the 1980s, bamboo prices were raised only to Rs 200 to 500 per tonne, when market prices were well over Rs 5000 per tonne. The result of this state-subsidized profitability of forest-based industry has been an explosive growth in industrial capacity, and a non-sustainable use of forest stocks that may be summed up in the words of a forester speaking of the evergreen forests of Karnataka's Western Ghats: 'Most of the areas which were worked during the past have deteriorated considerably and it is doubtful if ever they would regenerate to their original structure' (Rai 1981).

It is worthwhile to follow the pattern of the over-use of

forest resources over the last thirty years. This pattern provides a chilling illustration of *sequential overexploitation,* i.e. the exhaustion of resources in a sequence of continuing decline in quality as well as in quantity. Such overexploitation can be discerned along six dimensions: (i) the mode of acquiring the resource, whether by gathering or production; (ii) control over the land from which the resource is acquired; (iii) the locality from which the resource is acquired, whether close to or far from where the industry is situated; (iv) the kind of terrain from which the resource is acquired, i.e. easy or difficult of access; (v) the plant species used, more or less suitable for the purpose; and (vi) the size of the tree used.

Thus the West Coast Paper Mill in northern Karnataka was set up in 1958 in the belief that the neighbouring state-owned reserved forest would supply *in perpetuity* all the requirements of the mill from the natural growth of bamboo, essentially at bare harvesting cost. By 1978 the mill had expanded to 200 tonnes per day, although less and less bamboo was available from the forests originally allotted to the mill. The mill gradually replaced bamboo with hardwood, also from the natural forests, and a large proportion of plantation-grown eucalyptus. Today only 10 per cent of the mill's raw material is bamboo, obtained at about Rs 650 per tonne (after the government has revised royalty rates) from nearby forests, while the rest is mainly eucalyptus (65 per cent)—a significant proportion from farm land—casuarina (13 per cent) and mixed hardwood (10 per cent) costing about Rs 1000 per tonne at mill site. There has thus been a continuous shift in raw material, from that merely extracted from natural forests at little cost, to that produced on plantations or fields with substantial inputs and at a far higher cost (Guha, S.R.D. 1989).

The ownership of land from which the resources are being supplied has also shifted in sequence. Initially all resources for a mill like Harihar Polyfibres in Karnataka came from state-controlled reserved forest land, which was used to supply raw materials at highly subsidized prices. As these supplies were no longer adequate, the mill shifted to purchasing eucalyptus grown on private farmlands. Here the element of state subsidy was restricted to the supply of free seedlings to farmers; thus the mill had to spend substantially larger amounts for the raw material. More recently, as we have seen, the mill has begun shifting the focus of resource production to erstwhile village common lands to be planted up by a joint-sector company, Karnataka Pulpwoods Limited.

Third, forest-based industries were started off with catchments close to the factory site from which the raw materials would come, supposedly in perpetuity. Thus the West Coast Paper Mill at Dandeli was expected to derive its resources from the districts of Uttara Kannada and Belgaum, while the Mysore Paper Mill at Bhadravathi was expected to do so from Shimoga district. As Figures 6.1 and 6.2 show, however, the resources of these areas have long been exhausted. While the joint sector Mysore Paper Mill then had to tap the entire state, the private sector West Coast Paper Mill is scrounging for bamboo and other woods from all over the country (Gadgil and Prasad 1978).

Fourth, the industrial demand has initially been met from the flat, more easily accessible, terrain. As the resources of such a terrain are exhausted through the process of selection felling, these areas are clear-felled and brought under the so-called conversion circle to develop teak or eucalyptus plantations. At the same time, the much steeper terrain earlier constituted as protection circle (from the point of view of watershed conservation) is brought first under selection and then conversion circle. Figure 6.3 schematically shows how this process has gone on in the Quilon forest division of Kerala (FAO 1984).

A fifth dimension of sequential overexploitation is the acceptance of less and less suitable species, as the more desirable ones are exhausted. As Table 6.4 shows, the Indian Plywood Manufacturing Company at Dandeli has increased the species utilized from 36 in 1946 to 61 in 1975, continually adding new localities for exploitation as well.

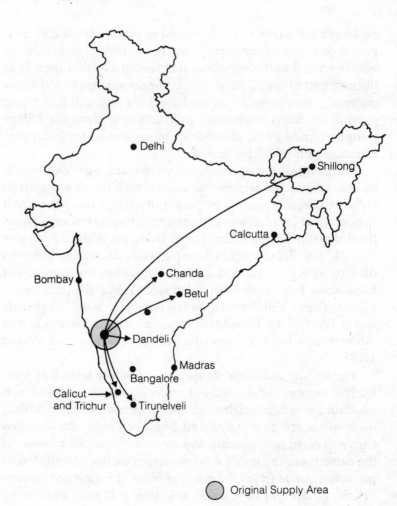

**Figure 6.1**
Areas from which the West Coast Paper Mills, Dandelli, began to draw
bamboo in the late 1970's (after Gadgil and Prasad, 1978).

**Figure 6.2**
Areas from which Mysore Paper Mills, Bhadravati, began to draw bamboo in the late 1970's (after Gadgil and Prasad, 1978).

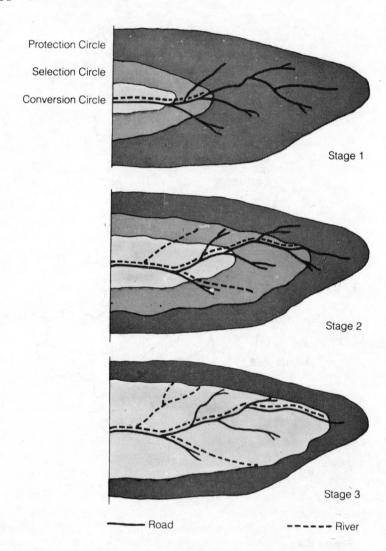

**Figure 6.3**

Successive changes in designation of areas under protection, selection and conversion circles in the Quilon Forest Division of Kerala over the period 1950–1980 (after FAO, 1984).

TABLE 6.4

SPECIES AND AREAS ALLOTTED TO THE INDIAN PLYWOOD
MANUFACTURING CO. LTD, DANDELI AND ITS ALLIES

| Year | Number of species allotted | Concession areas added |
|------|---------|------------------------|
| 1946 | 36 | Gund, Virnoli, Kulgi, Sambrani & Dandeli Ranges |
| 1948 | 45 | Nil |
| 1959 | 48 | Anshe - Ulve, Supa plateau series 2 & 3, Dandeli R. series 6 and 7. More areas of Yellapur, Sonda, unorganized forests of Mundgod, Nirsol slopes, Sonda high forests, Yekkambi |
| 1961 | 52 | Nil |
| 1974 | 53 | |
| 1975 | 61 | Nil |

SOURCE: Gadgil and Subash Chandran (1989).

Finally, there is a general though not overwhelming tendency to decrease the minimum girth for exploitation through selection felling. Table 6.5 shows the fluctuations in the exploitable girths under a succession of working plans for Ankola in coastal Karnataka. Rather than go to very low-girth sizes under selection felling, however, the tendency is to clear-fell forests being exhausted under selection felling regimes to raise plantations.

In the case of Indian forestry, one can thus discern a whole complex of processes leading to resource exhaustion. These processes vary greatly, depending on the part of the country and the industry concerned. However, four broad phases are discernible. In the first phase, lasting till the 1960s, selection fellings from natural forests were the mainstay of wood-based industry. When this was found inadequate, large-scale programmes of clear-felling and the raising of plantations of fast-growing species such as eucalyptus were launched with the

## TABLE 6.5
### EXPLOITABLE SIZE OF TIMBER UNDER WORKING PLANS OF ANKOLA HIGH FORESTS

| Name of the tree | Coppleston 1901 | Pearson 1918 | Miller 1918 | Millet 1927 | War felling 1940-45 | Mundkur 1955 | |
|---|---|---|---|---|---|---|---|
| | | | | | | WC I | WC II |
| Tectona grandis | 6½-7' | 6' | 6' | 6' | 4' | 6' | |
| Terminalia tomentosa | – | 7-6' | 6' | 5' | 5' | 3' | 6' |
| Terminalia paniculata | – | – | 7' | 7' | 5' | 5' | 6' |
| Lagerstroemia microcarpa | – | 7-6' | 7' | 7' | 5' | 5' | 6' |
| Dalbergia latifolia | – | 6' | 6' | 7' | 5' | 5' | 6' |
| Adina cordifolia | – | 8-7' | 7' | 7' | 5' | 5' | 6' |
| Other species | – | 6' | 7' | 7' | 5' | variable | variable |

WC I — Working Circle I — Plains Working Circle
WC I —Working Circle I I — Teak bearing areas

SOURCE: Gadgil and Subash Chandran (1989)

backing of special plan schemes by the central government. While clear-felling provided large amounts of wood, the productivity of plantations so raised was well below what was expected. Notably enough, this large-scale plantation effort had been launched without any careful trials of eucalyptus, or indeed any other species such as tropical pines. Consequently, by the mid 1970s came the next phase—of 'social' forestry, whose single remarkable success has been the raising of industrial softwoods on agricultural lands, and whose major failure relates to biomass for rural needs. The latest phase is a three-pronged attack, beginning in the 1980s, to fully open up the resources of untapped areas such as Arunachal Pradesh and the Andaman and Nicobar islands, liberalize the import of pulp and timber, and take over village common lands for industrial wood production.

## The Profligacy of Scientific Forestry

The strategies of resource use introduced by the British and continued by the government of independent India make a sharp contrast with earlier traditions of ecological prudence. As Parts I and II of this book document, the traditional patterns organized sustainable resource use by the implementation of a number of thumb rules such as the provision of refugia like sacred groves and sacred ponds, protection to keystone resources like *Ficus* trees, protection to specially vulnerable or critical life-history stages such as nesting birds, certain (probably conservative) quotas on how much fuelwood or leaf manure may be extracted from a village forest, and the diversification of the use of biological resources by endogamous groups. These thumb rules were bound to be arbitrary, and were implemented on the basis of magico-religious sanctions and social conventions.

A very different belief system was crafted on the European continent and brought to India by the British. The central tenet of this European belief system was the primacy of the objective of making maximum profit on the market. The currency in which this profit was measured was money; so that the whole diversity of resources that was earlier of significance to humans could now be transformed into money. This happened of course because there were now technologies available to change resources from one form into another, and into products that would fetch money; and new technologies were being continually invented to transform them in newer ways. All this meant that the goal of immediate profit from the most valuable resource of the moment was far more attractive than the goal of long-term conservation of a diversity of resources. The world dominance of European culture was a result of the European discovery of how to acquire mastery over nature, as well as how to translate this knowledge into technologies which manipulate nature. Along with the market, therefore, it was science and technology that became the most respected elements within the belief system of Europe. This science and technology had explicitly stated assumptions about the working of nature: it recognized only those natural forces whose working could be observed empirically. At the physical and chemical levels, where one is dealing with relatively simple phenomena, this methodology worked very well—Newton's laws of motion, the steam engine, and cannons that could be fired accurately from ships were among its many triumphs. The same methodology was then sought to be applied to more complex phenomena, including the working of ecosystems and human societies.

There were serious difficulties in applying the 'scientific' method at the level of these more complex phenomena. Consider, for instance, the functioning of a tropical humid forest, which is an interacting system of hundreds of plant species, along with thousands of species of micro-organisms as well as animals of varying sizes. To this day we do not even have a complete inventory of all the living organisms from a single hectare of any one tropical humid forest. We are also quite ignorant of how the populations of these thousands of species interact with each other. Only within the last few years have we realized, for instance, that insect predation on seeds or fungal infections on seedlings may profoundly influence these systems. Even if we were to leave aside ecosystem dynamics and concentrate on the population of a single species, we are able to say little about its population dynamics. For instance, we know very little about the 'regeneration niches' of most tropical-forest tree species. Forgetting even the population and focusing on individual plants, we know little about the growth patterns of tropical forest trees, or of the timing and periodicity of their seed production.

How can one then apply this supposedly 'scientific' method in decisions on how to manipulate a resource base such as a tropical humid forest? Obviously, all that is possible are a few crude prescriptions based on rules of thumb. If the object was the conservation of resources, these rules would differ scarcely at all from pre-scientific prescriptions of earlier times, precisely because our empirical knowledge of the workings of such a system has progressed only marginally beyond the empirical knowledge possessed by people who observed these systems, albeit informally, for generations. In our 'scientific' system sacred groves may be replaced by preservation plots, but the notion is the same, and the establishment of preservation plots has exactly the same arbitrariness.

However, if the object is not the conservation of a diverse resource base but the maximization of profits in the short run, rule-of-thumb prescriptions are obviously likely to be quite different. Their orientation, as we know too well, is quick harvests, and the replacement of a diversity of commercially less-valuable species by commercially valuable species—as perceived at that point in time. Add to this the fact that the maximization of profits has to accrue to a state apparatus and not to local people, and you get an additional rule of thumb, namely that as much land as possible should be brought under as complete a control as is possible by the apparatus of state. This is precisely what British forestry management attempted to do in India. While the British had radically different objectives from the traditional forestry management of village populations, they were hardly better informed about the working of forests as a result of 'scientific' investigations: such knowledge simply did not exist. It was therefore quite correct to term it 'forestry in the commercial interests of the state'; it was quite erroneous to call it 'scientific forestry in the interest of resource conservation'—as it was in fact incorrectly called.

Why was this misnomer applied? It was clearly a double misrepresentation of both the objectives and of the methodology employed to achieve those objectives. The objectives were represented as 'sustained yield' in the 'national interest', although in practice the yield was rarely sustained and the interests served were much narrower than society as a whole. The methodology too was misrepresented as 'scientific', an ideological device related to the high regard which science enjoyed in European society because of its success in dealing with simpler phenomena.

Let us take the objectives first. Did the consequences match the professed objective of sustained yield? Indeed not. The forest resource base was not used sustainably; rather, the pattern of its utilization is best described as successive overexploitation. For example, forests taken over by the state were classified into two categories: protected forests, taken over by the state but available to villagers for their requirements of gathered material; and reserved forests, fully at the disposal of the state and supposedly managed on a sustained-yield basis. (This classification itself acknowledges that there was no intention to manage the protected forests on a sustained-yield basis). Reserved forests were further categorized into protection forests and production forests. Protection forests were on sites such as steep hillsides or banks of streams, where an undisturbed vegetation cover was considered necessary for maintaining soil cover and the hydrological cycle. If resource conservation was the genuine objective of forest management, the delineation of such protection forests would have been independent of the commercial feasibility of their exploitation. In

fact, protection forests were originally declared in areas difficult of access; as these areas become accessible, they have been
reclassified as production forests. This, for instance, has been
the case with hill-slope forests near the Kudremukh iron ore
project in Karnataka. The terrain near the crestline of the
Western Ghats where this project is located was earlier without
any roads, and was declared 'protection forest' on the grounds
that it covered steep hill-slopes. It was reclassified as 'production forest' as soon as the project led to the construction of roads
into the area. Similarly, the high-level oak, fir and spruce forests
of the Himalaya have been converted into production forests
with the expansion of the road network into those areas.

The second issue relates to the methodology of scientific
forestry: can it be appropriately termed scientific? The basis of
modern science is its careful grounding in empirical fact. It
proposes a model of how nature works, based on systematic
observation; it then suggests that certain predictions follow if
this model is adequate; it verifies these predictions, and if
falsified, makes appropriate changes in its model. This model
suggests further predictions, which are then sought to be verified in an ongoing process. Now, the 'scientific' working of
forests ought to start with a model based on empirical observation, and make predictions about the yields being of such-and-
such quantitative level, and that they would change in such-
and-such fashion with time, given a certain system of
management. These predictions would then be verified, to
check if they have worked in practice; and if not, the model
would be altered accordingly. This cycle would then become an
ongoing one.

None of the elements of these processes occur in the exercise
of 'scientific' forestry. There is a wholly inadequate data base
which underpins the initial model of how the forest works;
there is no proper verification to see if the prescriptions have
worked, nor why they have or have not worked; and there is
no appropriate adjustment of the model in the light of any
attempted verification. Tropical forests are exceedingly diverse

species-rich systems, quite unlike the species-poor temperate forests. The nutrient cycle in tropical forests is quite distinctive, too, with the bulk of nutrients being held in the standing biomass, rather than in the soil. These facts imply that ecosystem and community-level interactions—for instance soil–plant relationships and inter-species interactions—are of far greater significance in tropical-forest dynamics. These are essentially ignored in the models of the working of temperate forests, such as single-species stands of pines that temperate-region foresters deal with; they have been equally ignored in the 'scientific' forestry introduced into India. Temperate foresters also ignore population dynamics, since they work with single-species stands maintained at a certain density, and Indian forestry thus also ignored population dynamics. Temperate foresters treated the dynamics of growth of an individual plant as almost the only necessary empirical data base, and this was also the approach adopted by 'scientific' forestry in India. The model then was that every tree species has a characteristic growth curve, such that the increment to biomass increases till the tree reaches a certain size and then begins to come down. The main prescription asked for was the right size for cutting the tree, so that some function of the timber yield is maximized. But the elementary data required for this, namely the growth pattern, is hardly known for any Indian species or for any exotic species in Indian conditions.

The bewildering variety of species also made the tropical forest very different from the temperate forests with which the German foresters were familiar. Dietrich Brandis drew attention to this signal difference when he called the low proportion of marketable species in any forest 'a peculiarity of forestry in India' (Brandis 1884, p. 12). But far from applying themselves to an understanding of the complexities of the forest ecosystem, colonial foresters were under considerable pressure to immediately change the species composition of the forest in favour of commercially valued trees. Yet increasing the proportion of commercially valued species was itself a 'most difficult problem . . . ' (Ribbentrop 1900, p. 163). It was therefore hardly surprising that existing attempts at inducing reproduction were often unsuccessful—and in many areas the department was 'living as yet to a certain extent on capital' (Ribbentrop 1900, p. 183). Quite apart from the ignorance of community-level interactions, the estimates of growing stock on which 'sustained yield' prescriptions were based were themselves flawed. But if Indian forest management deviated from the European model in the extreme paucity of data, the government did little to remedy the gap—indeed, foresters were told explicitly by the inspector-general of forests that constraints of time and money made it 'utterly impracticable and futile to try and elaborate working plans on advanced European models' (Smythies and Dansey 1887, p. 9). In other words, the proper enumeration of growing stock—an essential prerequisite for sustained-yield prescriptions—was out of the question. But even a century later, when in fact the scale of intervention in forest ecosystems has increased dramatically, there is the same attitude towards empirical data. Thus, while special experimental plots have been set up to collect information on species growth rates, the data on these have hardly ever been maintained or analysed properly.

This disregard for empirical data was revealed most strikingly by the Bastar pine plantation project. In the 1970s a project was proposed to grow the tropical pine *Pinus carribbeana* over some 40,000 hectares in the tribal district of Bastar in Madhya Pradesh. It was proposed that a 500 hectare plantation be grown initially for a five-year period to assess the performance of the species at this site—the decision on whether to pursue the project further was to be based on this experience. At the end of five years, when the larger project was being pushed, there was opposition to it from a number of quarters, including tribals, for whom the replacement of 40,000 hectares of natural forest (from which they gathered a number of resources) would have meant serious deprivation (see chapter 7). In view of this opposition, the central government appointed an expert group under the chairmanship of the inspector-general of forests to examine the project. One of the present authors served on that committee. When the committee requested data on the ex-perimental plantation, none materialized. In fact no data had been properly maintained at all. This attitude to empirical rigour explains the widespread failures of 'plantation' forestry, narrated earlier in this chapter, to realize more than a small percentage of expected yields.

The forestry practices introduced by the British and con-tinued thereafter are thus neither scientific nor conservation oriented. Their objectives have been threefold, namely (i) the demarcation and consolidation of forest land taken over by the state and alienated from access to the local people; (ii) the imposition of certain restrictions on the rate at which harvests are made from the forests. These restrictions do not ensure sustainable harvests; rather they have served to regulate the quantity of the material harvested to match commercial de-mands; (iii) the conversion of natural forests with a wide variety of resources valued by the local population into plantations of a relatively small number of species of maximal commercial value. While the environmental consequences of these objec-tives have been little short of disastrous, the unpopularity of 'scientific' forest management with the mass of the Indian population is a powerful indicator of its social inappropriate-ness. We have already dealt with the opposition to colonial forest management; the next chapter narrates the continuing resistance to state forest management in independent India.

CHAPTER SEVEN

# *Competing Claims on*
# *the Commons*

The continuing march of commercial forestry has greatly intensified conflicts between those whose interests it seems to serve
(chiefly the state and the commercial–industrial sector), and
those whose interests it seems to deny (chiefly the poorer
sections of the rural population). These conflicts apparently
came to a head with the debate around the 1980 Draft Forest
Act. The Act, which aimed at greatly strengthening the already
considerable powers enjoyed by the forest bureaucracy, was
opposed by grassroots organizations who argued that its punitive sanctions would be used against powerless groups such as
tribals and poor peasants. The controversy around the Act and
around the pine plantation in Bastar in Central India helped
fuel a larger debate on forest policy as a whole (Gadgil, Prasad
and Ali 1983; Guha, Ramachandra 1983).

This debate has brought in its wake a series of administrative changes which reflect the contending pulls on forest management. One set of changes works to strengthen the existing
system of state control, with exceptions made only for the
industrial sector. Thus, many states are entertaining proposals

to allot land to large industry as 'captive plantations'—though, significantly, the forest department is unwilling to hand over degraded reserved forests for the purpose, urging instead that industry be allotted other land owned by the state, including areas presently used as village commons. At the same time, another set of changes seems better disposed to ecological and social needs. These changes include the encouragement of voluntary organizations to take up afforestation in the countryside, and the ban on clear-felling imposed in several states. The government has also made important administrative changes. A new ministry of forests and environment was set up in 1985. In the same year, the National Wastelands Development Board was created as a nodal agency for afforestation. Severely hampered by the lack of access to land, in 1989 the Board was wound up and converted into a Technology Mission for Wastelands Development.

Only a decisive shift towards a more ecologically conscious and socially aware forest policy, however, is likely to diminish the severity of conflicts around the management and utilization of forest resources. This chapter (modelled on chapter 5, on the colonial period) provides a preliminary mapping of the conflicts over living resources in independent India. We distinguish between three generic types of forest-based conflict. To begin with, we follow the framework of chapter 5, successively analysing conflicts between the state on the one hand, and hunter-gatherers, shifting cultivators, settled cultivators, and artisans on the other (here we also deal with conflicts *within* the agrarian population). Well known from the colonial period, these conflicts all reflect a more basic clash between the commercial orientation of state forestry and the requirements of forest produce for subsistence on behalf of different sections of agrarian society. At the same time, while inheriting the framework of colonial administration and legislation, forest policy in independent India has been significantly influenced by the rapid expansion of the forest industries sector. This intensification of forest operations has brought in its wake new forms of conflict that were absent or at least not very visible in the earlier period. We then describe the characteristics of such conflicts *within* the forest, conflicts concerned with the distribution of gains from commercial forestry. Finally, the chapter examines a conflict which is the consequence not of intensive forest use but its obverse, i.e. the sharp curbs in certain areas on commercial *and* subsistence uses of the forest. We refer here to the massive network of wildlife sanctuaries, almost all of which have been established after 1947. Widely hailed as examples of successful conservation, these sanctuaries have often had a negative impact on the lives of the surrounding population. Our analysis of the conflicts around their management suggests that they rest, at least in part, on a narrow definition of biological conservation; enlarging this definition may perhaps be necessary to minimize future conflicts.

## Hunter-gatherers

Even during the colonial period, hunter-gatherers in the Indian subcontinent had begun to lose control over their means of subsistence. The takeover of large areas of forest by the state and the expansion of the agricultural frontier have continued to threaten both hunter-gatherers and their natural environment.

Medical advances have also taken their toll, for example the reclaiming of the Terai, the malarial tract at the foot of the Himalaya. Till 1947 inhabited only by the hunter-gatherer communities of Bhoksa and Tharu, the Terai is now the centre of a prosperous agrarian economy, created mainly by refugees from Pakistani Punjab.

Many hunting and gathering communities have had only one survival strategy open to them—to put their enormous knowledge of flora and fauna at the service of the new owners of their habitat, the forest department. Thus the Jenu Kurubas of Mysore, with a tradition of helping the state in the capture of elephants, now collect honey and other forest produce on behalf of the forest department and merchants. In one taluk, Heggada Devan Kote, the department in the nineteen seventies was earning a revenue of Rs 2.5 million per annum from minor forest produce, collected mostly by Jenu Kurubas working for a small daily wage. These tribals have very little freedom of choice with respect to work; living in the forest, they either work for the department or face the threat of eviction. This near total dependence led one anthropologist to characterize the forest department–Jenu Kuruba relationship as 'feudalistic' (Misra 1977). The Hill Pandaram in Kerala, who also depend heavily on the collection of minor forest produce, are equally fearful of the forest department. Residing in what are now 'reserved forests', they live in mortal fear of government officials who often extract favours from them (Morris 1982). Even where hunter-gatherers are comparatively autonomous of the forest department, ecological changes have forced new adaptive strategies, for example the sale of small animals, honey and plants to nearby markets (cf. Sinha 1972 on the Birhors).

## The Continuing 'Problem' of Shifting Cultivation

While hunter-gatherers have been powerless to effectively resist the forces of 'modernization', this has not always been the case with shifting cultivators who are, on the whole, a stronger and better organized group. However, in addition to economic and demographic pressures, shifting cultivators have had to contend with deep-seated official prejudice against this form of agriculture. Government reports have repeatedly affirmed the state's desire to do away *completely* with shifting cultivation. Such a policy rests on the unshakeable belief, expressed here by two economists, that shifting agriculture is invariably harmful for forest regeneration: '[Jhum] destroys the ecological balance and results in substantial soil erosion which subsequently leads to flooding of rivers [and] drying of hill springs' (Gupta and Sambrani 1978).

Recent anthropological work sharply contests this interpretation, providing abundant documentation that under conditions of stable population growth shifting agriculture is in fact a highly efficient and ecologically sustainable use of resources (cf. Geertz 1963; Conklin 1969). A major study of the Hill Marias of Abujmarh, one of the few surviving tribal communities in peninsular India who still depend exclusively on jhum, demonstrates that under the long fallow (twelve years) system practised by the Marias, both soil fertility and forest vegetation have sufficient time to recuperate (Savyasachi 1986). And as von Fürer-Haimendorf and his colleagues observe, 'some of the largest natural forests exist in areas inhabited by slash-and-burn cultivators for centuries, whereas intensive plough-cultivation has destroyed forests wherever it is practised' (Pingle, Rajareddy and von Fürer-Haimendorf 1982).

It is beyond our scope to adjudicate between these two conflicting interpretations. We would, however, like to highlight two points. First, one must remember that jhum is not merely an economic system with certain ecological effects; to its practitioners it is a way of life, the core of their material as well as mental culture. Secondly, many of the vocal opponents of jhum have not been at all critical of commercial forestry. In fact, most areas earlier used for shifting cultivation and taken over by the state have been subject to *greater*, not diminishing, forest exploitation, mostly for the market.

It is against this backdrop that we may view the continuing resistance of jhum cultivators to government forest policies. In Madhya Pradesh, jhum was banned in all areas covered by the Indian Forest Act, including large tracts earlier under princely states. Undeterred, tribals continued to follow the traditional rotations, although these areas had now been designated 'reserved' forests. When prosecutions and monetary fines failed to stop the cultivators, the forest department turned to the police, who made several arrests. The release of the arrested was then made conditional on a promise that they would take to plough cultivation. The transparent unwillingness to give up jhum was further illustrated by the Baigas, still confined to the *chak* in which Verrier Elwin had found them a quarter century earlier. While the forest authorities were 'hopeful of persuading [the] younger generation of Baigas to give up *bewar* altogether', the older generation was unyielding. They believed 'the Baigas were born to be kings of the jungle and the soil' and did not at all want to give up bewar, which provided 'the link with the past and with their ancestors' (NCAER, 1963, pp. 77-8).

Over many areas, the specific issue of jhum has been submerged in the larger question of tribal rights in the forest. Elwin and other anthropologists were hopeful that the new nation would redress the injustices the adivasis had suffered at the hands of the British. While their concerns found their way into important policy documents (including the Indian Constitution), there continued to be a wide gulf between policy and implementation. A special commission set up in 1960 to enquire into tribal problems was everywhere 'flooded with complaints from the tribals and their representatives against the forest administration'. Despite nearly a century of government control, the tribals were unshakeable in their belief that 'the forest belonged to them' (GOI 1961).

Viewing the forest department as a comparatively recent interloper, and undeterred by the provisions of the forest act, many tribal groups mounted a sustained challenge to the continuing denial of their rights. In 1957, a movement broke out among the Kharwar tribals of Madhya Pradesh, which called upon the people 'to stop payment of rent to revenue-collecting agents, utilise timber and forest produce without making any payment, defy magistrates and forest guards, and flout the forest laws which violated tribals' customary rights'. The movement slogan, *jangal zamin azad hai* (forests and land are free) succinctly expressed the opposition to external control and commercial use. For a time the movement brought forest operations to a standstill, dissipating only after the arrest of its leaders (Singh 1983).

More recently, attempts by the state to convert the mixed forests of Central India into monocultural plantations have met stiff resistance. The Bihar Forest Corporation's policy of replacing sal and mahua forests with teak has been sharply opposed by Ho, Munda and Santal tribals in the Chotanagpur area. In August 1979 the tribals, armed with bows and arrows, began cutting down the teak forests, asking simultaneously for their replacement with trees of species more useful to the local economy. The opposition to teak dovetailed with a wider movement of self-assertion which has demanded a separate tribal state of Jharkhand. A slogan of the movement, 'Sal means Jharkhand, Sagwan [teak] means Bihar', captures these links between the economic and ecological exploitation of the area. Outside Bihar, a World Bank aided project for raising Caribbean pine, as raw material for a new paper mill in the Bastar district of Madhya Pradesh, was stopped following opposition by tribal groups. And in the Midnapur district of Bengal, adivasis had opposed both the auction of forests and attempts to clear-fell sal in order to raise eucalyptus (PUDR 1982; Sengupta 1982; Anderson and Hubner 1988; Rana 1983).

This widespread and continuing resistance has, however, been no more than a holding operation. With the odds stacked against them, in most areas tribals have been forced to accept the new systems of forest working. Indeed the loss of their land has made thousands of tribals dependent on commercial forestry for a living. This integration of tribals into a capitalist system of wage labour has generated its own set of conflicts, dealt with later in the chapter.

## The Changing Ecology of Settled Agriculture

However unsuccessful, the resistance of jhumiyas to state forest control seems to have been continuous, spilling over into the post-colonial period without any perceptible break. The situation has been somewhat different with respect to caste society, settled in villages whose core activity is plough agriculture, with the allied (and in a sense dependent) occupations of animal husbandry, craft production and services. Of course, in the three decades immediately preceding independence the peasants, who were the backbone of the Congress, had availed of the nationalist movement to advance their claims in the forest. The available evidence suggests that in the decades *following* independence, however, access to forests ranked rather low on the list of peasant grievances—i.e. there appears to have been a diminution of forest-based conflict between the peasantry and the state.

How can we explain this attenuation of conflict? To begin with, peasants were hopeful that the victory of the Congress would inaugurate a new era. As peasants, ever since the advent of Gandhi had been the mainstay of the Congress Party, it was reasonable to hope that, with their government in power, their interests would be looked after. Second, peasants were to a large extent preoccupied with consolidating the gains that were readily forthcoming after 1947. Zamindars (large landlords) had been among the most loyal supporters of the Raj; the confiscation of land under their control became one of the most urgent items on the national agenda. Thus, access to *land* became an overriding concern, and in the coming years substantial tenants (mostly from the middle castes) benefited greatly from the abolition of landlordism, now having clear titles to land on which they had previously enjoyed only a rather tenuous right of usufruct. Lastly, with the early sixties, and with the onset of the so-called Green Revolution, farmers in many parts of India switched to a new mix of agricultural technologies which in fact *reduces* their dependence on forest resources. With the state providing water, electricity, fertilizers and machinery at highly subsidized rates, the country's landscape has been dotted with pockets of fossil fuel agriculture, and the production of food and other cash crops for the urban market. Ironically, while chemical agriculture has (for its practitioners) reduced dependence on living resources, it has at the same time provided a powerful impetus for the destruction of forests through the construction of large dams for irrigation and power generation.

Chemical agriculture is feasible only in certain parts of the country; in other areas, a healthy forest cover continues to be crucial to the practice of subsistence agriculture. One such region, the Central Himalaya, gave rise to perhaps the best known of all forest conflicts, the Chipko movement. In this hilly terrain, the possibilities for both intensive and extensive agriculture are strictly limited; at the same time, the Himalaya contain the best conifer forests in the country. Control over forest (and over its changing species composition, with peasants preferring broad-leaved species and the forest department conifers) thus became a paradigm case of the conflict between subsistence and profit-oriented uses of the forest. In the past, it had given rise to some of the most bitter and intense conflicts, chronicled in chapter 5.

With the improvements in communications after 1947, especially after the 1962 India–China war, commercial exploitation of the Himalayan conifers for both timber and resin has greatly accelerated. Simultaneously, Gandhian workers tried to organize forest labourers into co-operatives which then bid, with mixed success, for timber and resin contracts. These co-operatives also began small-scale processing of forest produce — for example, resin distillation and the manufacture of agricultural implements and bee boxes—hoping to generate local employment. Here the intention was not so much to oppose commercial forestry as to enable villagers to gain a substantial share from its operations. These activities enjoyed only a limited success, with the government preferring to give contracts to large merchants from outside the hills and greater subsidies to large-scale industry in the plains.

Chipko was sparked off by the government's decision to allot a plot of hornbeam forest in the Alakananda valley to Symonds, a sports-goods company from faraway Allahabad. A few months before this, the Gandhian organization in the forefront of the co-operative movement, the Dashauli Gram Swarajya Sangh (DGSS) had been refused permission by the forest department to fell trees from the very same forest (which they wanted for making agricultural implements). This transparent favouritism provoked the villagers, led by the DGSS, to threaten to hug the trees (Chipko means to hug in Hindi) and prevent them being felled by Symonds' agents.

This was in 1973. In the next decade, Chipko spread rapidly to other parts of the Uttar Pradesh Himalaya. In over a dozen separate incidents, villagers successfully stopped felling operations. The auction of forest coupés was also disrupted. The movement has received wide publicity, and its two main leaders, Chandi Prasad Bhatt and Sunderlal Bahuguna, have emerged as among the best-known environmentalists in India. Notwithstanding its public image as an 'environmental' movement, however, Chipko is best viewed as a *peasant* movement. It resembles movements in the colonial period which defended customary rights in the forest—so crucial for subsistence—from encroachment by the state and the commercial sector. Chipko participants are very conscious of this long history of protest. Their basic demands are captured in this pithy statement by a peasant of Badyargarh: 'We got only a little food from our fields; when we could not get wood to cook even this paltry amount, we had to resort to a movement' (Guha, Ramachandra 1989a).

A decade after it began, Chipko's echoes were picked up in northern Karnataka, where a similar conflict between villagers and the forest department gave rise to the Appiko movement (Appiko means 'to hug' in Kannada). In this district, Uttara Kannada, the forest department had for several decades been promoting teak plantations after clear-felling the existing mixed semi-evergreen forests. Inspired by Chipko, in August 1983 the villagers of Sirsi taluk requested the forest department not to go ahead with selection felling operations in the Bilegal forest of Hulekal range. When their requests were unheeded, villagers marched into the forest and physically prevented the felling from continuing. They also extracted an oath from the loggers (on the local forest deity) to the effect that they would not destroy trees in that forest (Mani 1984).

The Chipko and, indeed, Appiko cases are somewhat atypical. The peculiar ecological characteristics of the regions in which they arose explain the intensity of social conflict (and popular concern) over forest resources. Deforestation in the Himalaya, while undermining the basis of the agrarian economy, has unleashed a series of landslides and floods that have taken a heavy toll of human life. At the other end of the country, the spice-garden farmers who dominate the economy of Uttara Kannada are critically dependent on leaf manure from the forest. These farmers, mostly from the influential caste of Havik Brahmins and far richer than the average Himalayan peasant, have also taken the lead in initiating programmes of eco-restoration in collaboration with scientists (Gadgil *et al.* 1986).

Elsewhere, as we have suggested, the lack of concern among the traditional social base of agrarian movements—the middle and rich peasantry—has contributed significantly to the absence of forest-based *movements*. This is not to say that other sections of agrarian society are not seriously affected by biomass shortages, especially women and the rural poor (with women from poor families being doubly affected). The crisis in the availability of woodfuel, and the impact of such shortages on the health of women and children, has been richly documented. It is less well known that shortages of fodder are equally pervasive, and in some areas even more serious. According to one estimate, while the annual demand for fuelwood in Karnataka is 12.4 million tonnes, the annual production is 10.4 million tonnes, a shortfall of 16 per cent. In the case of fodder, however, the corresponding figures are 41.5 and 13.5 million tonnes respectively—a shortfall of nearly 70 per cent (Gadgil and Sinha 1985).

The earlier chapters have argued that state control over large areas of forest has significantly restricted, though not completely eliminated, one source of biomass for rural populations. Another source that has been seriously affected in recent decades consists of grazing land and forests held in common by villagers, land which does not come under the purview of reserved forests. Seventy-five years ago, one scholar noted that in parts of Bengal common pastureland was being wrested by the landlord from the villagers (Mukerjee 1916). However, such threats to common land have greatly intensified since independence. A major study of common property resources (CPRs) in seven states shows the rapid decline in their area and physical productivity. Population pressure and the privatization of CPRs (partly by rich farmers but also due to their distribution to landless families by the state) are chiefly responsible for this decline. As it happens, poor families are the hardest hit, for large landholders have access to both fuel and fodder (for example, from agricultural wastes). Attempting to quantify this dependence, the study estimates that the income generated for each households of the poor from CPRs ranges between Rs 530 and 830 per annum. These figures are substantially higher than the income generated by several anti-poverty programmes. The restoration of the status and productivity of CPRs may therefore be a most effective anti-poverty strategy (Jodha 1986).

A case study of Rajasthan suggests that official policies have considerably hastened this process of privatization and degradation. Thus the supersession of customary law by codified law—in other words, the shifting of the locus of decision-making away from actual users—has created great uncertainty concerning the management of common lands. As traditional institutions which earlier regulated the use of CPRs are not recognized in law, there is a powerful incentive for individual farmers to claim possession of parcels of land previously held in common, and get such encroachments 'regularized'. As richer farmers have more influence with both politicians and administrators, the apparently inexorable decline of CPRs is intimately connected with the declining fortunes of the rural poor (Brara 1987).

While the commons dilemma testifies to the growing fissures within agrarian society, as it happens these lands are under threat from outside forces as well. A particularly important case in Karnataka involves the transfer of land previously under the control of the revenue department to a joint-sector company, formed to grow industrial wood for the exclusive use of one factory, Harihar Polyfibres (owned by the powerful industrial house of the Birlas). Contending that these areas have been traditionally used by villagers for fuel and fodder collection, a voluntary group has challenged the formation of the new company in the highest court of the land. While the case is being heard in the supreme court, villagers have organized several 'pluck and plant' satyagrahas, symbolically uprooting eucalyptus (a species of very little use to the local economy) and planting saplings of more 'useful' local species instead.

At the time of writing, the final judgement on the KPL case is awaited from the supreme court. Whichever way it goes, it is likely to be a landmark judgement, determining the principles whereby private industry can or cannot be granted public land for exclusive use. However, peasant opposition to the takeover of common lands by KPL draws on long-standing traditions in defence of customary rights. For example, in the village of Baad in Dharwad district, uncultivated land belonging to the family of a former *jagirdar* (land grantee) was leased to KPL for growing eucalyptus. Peasants challenged this decision in court, arguing that the land leased to KPL had traditionally been used by them for grazing cattle. They contended that by enclosing the plot, harrowing the land in strips, and planting eucalyptus saplings, the operations of KPL would adversely affect the supply of grass for their cattle. The villagers of Baad invoked earlier court cases (dating back to 1943) wherein they had successfully defended their collective right to graze cattle on the contested area without paying any fee to the jagirdar. In those cases, judges had upheld the villagers' claim that this land was *mufat gayaran* (free pasture): consequently the jagirdars were barred from levying a fee. Citing these precedents, villagers asked that these lands be retained as pastures and that KPL operations be stopped (see plaintiff's appeal in suit no. 133 of 1988, court of the civil judge, Dharwad; suit no. 128 of 1943, court of the extra joint sub-judge, Dharwad).

Biomass shortages have also intensified conflicts between settled cultivators and pastoral nomads. Traditionally, a significant proportion of India's population (perhaps as much as 5 per cent) depended on nomadic grazing. With large tracts of land earlier left uncultivated, nomads had available abundant grazing for their flocks in the monsoon season. As natural grass growth declined after the rains, they turned to stubble left after the harvest, and to fallow fields. The farmers themselves typically kept only a few livestock, primarily for draught power. They thus enjoyed a mutualistic relationship with the graziers, exchanging stubble for manure and dairy products, meat and wool for grain. In recent decades this relationship has come under great stress. On the one hand, the acreage of non-cultivated lands open for grazing has shrunk dramatically, not least due to the state takeover of forests. On the other hand, a growing market for dairy products and meat has encouraged many cultivators to increase their own livestock holdings, thereby limiting access to crop residues for pastoralists. These pressures have forced many nomadic communities to shift from grazing to cultivation, mostly on marginal and unproductive lands. Elsewhere, growing competition for shrinking fodder resources has led to bitter conflicts between graziers and cultivators. These sometimes violent conflicts between groups that earlier enjoyed a co-operative relationship have been reported from the states of Madhya Pradesh, Uttar Pradesh, Rajasthan and Maharashtra (Gadgil and Malhotra 1982; CSE 1985; Wade 1988).

A brief mention, finally of one other set of castes deprived of biomass, namely artisans. Undeniably, the major reason for their decline has been competition from organized industry— i.e. from the mass production of utensils, containers, furniture, etc., which were earlier manufactured within the village. Simultaneously, shortages of raw material have also been a contributory factor. Basket weavers dependent on reed in Kerala and on bamboo in Karnataka, and hemp workers in Uttar Pradesh, are among many artisanal communities which have faced deprivation in recent years, as the resource traditionally used by them has either been depleted or diverted to industrial uses.

## Claiming a Share of the Profits

Changes in the proprietary status of the forests, as well as changes in its ecology, have clearly undermined the capacity of forest-dependent modes of subsistence—for example hunting, gathering, shifting cultivation, and some forms of plough agriculture and pastoralism—to reproduce themselves. As commercial forestry has continued its apparently unstoppable march, many forest-dependent communities have had to take recourse to alternative survival strategies. The most ready option, namely wage employment in commercial forest operations, ironically, concedes the inevitability of the new systems of forest working. One can see a classic process of proletarianization at work: divorced from the means of production (forest and land), forest dwellers are forced to accept a subordinate place in the new 'capitalist' system of production. This process has generated a set of conflicts characteristic of capitalism, in which labourers seek to improve their wages while capitalists seek to maintain high profit margins.

The basis of such conflict lies in the very mode of forest working in India, the so-called 'contractor' system. Although the state owns the forests, until very recently it has played little part in the actual extraction of timber and other forest produce. The procedure most widely followed is to mark, according to working-plan prescriptions, the trees to be felled in any given year. These trees are then collectively sold and auctioned to the highest bidder, who is now given a contract by the state. This 'contractor' is responsible for organizing labour, conducting felling operations, and transporting and supplying the converted logs directly to the actual users. A similar procedure is followed with respect to minor forest produce, wherein yearly contracts are offered to the highest bidder for the collection of any particular item—for example sal seeds—from a designated area of forest.

Corruption and waste are inherent in the contractor system. It is well known that at departmental auctions the contractor often bids a price far higher than the actual value of the marked trees, and then goes on to disregard silvicultural prescriptions by felling both unmarked and marked trees. The methods of extraction—for instance hurling logs down the hillside to a floatable stream—also cause great damage. By depressing collection wages, contractors realize staggering profits, often exceeding several hundred percent. Similar profit margins are realized by the processing industry, especially in the case of low bulk, high value items (such as perfumes) processed from minor forest produce (for illustrations, see Guha, Ramachandra 1983).

Recognizing these tendencies, policy documents have talked repeatedly of eliminating contractors, who, it is admitted, exploit both forests and labour. The need to replace contractors by forest labour co-operatives (FLCs) has been stressed in all the five year plan documents, in the 1952 forest policy, and by several commissions. A committee set up in 1967 to look into the tribal–contractor–forest nexus cynically observed that in this case they did not 'envisage the need for any changes in the wording of the existing [forest] policy' (GOI 1967, p. 13). The need to abolish intermediaries in forest working has been reiterated by the high-level committee for environmental protection, and in several meetings of the Central Board of Forestry (GOI 1980).

What has been the actual experience of the FLCs? Maharashtra is one state where the administration has taken some initiative in their formation. The districts of Thane and Dhulia in the state had been the centre of a major tribal upsurge in the 1940s, whose targets were the moneylender–forest contractor nexus (Parulekar 1976). Conceding the justice of these demands, the new Congress government tried to promote FLCs. However, their progress has been slow and halting.

Despite the presence of 444 FLCs in 1969-70 in the tribal parts of the state, co-operatives were in charge of only 40 per cent of the felling operations, with an equal share for contractors, and the remaining (20 per cent) in the hands of the forest department itself. Only one co-operative had taken the next step and started a sawmill to process timber. Lately, even these gains have been offset by the formation of the state's forest development corporation, whose brief is to augment the 'productivity' of forest lands by growing species such as teak. The new corporation has not recognized the earlier terms evolved between the government and the FLCs, and the number of forest blocks allotted to the latter has slowly dwindled (Muranjan 1974 and 1980).

In other states the experience has been even less encouraging. In Uttar Pradesh, the forest department was itself keen to encourage co-operatives, only to be told by the governor that since 'the forest department is a sort of a commercial department, it cannot be expected to extend concession in the transaction of its business, even to co-operative societies'. The government of Maharashtra (then Bombay) was actually invoked as an example not to be followed, for their schemes had 'cost the Bombay government quite dearly and considerable amounts were lost to the exchequer' (Guha, Ramachandra 1983). The favours shown to outside contractors and the processing industry were a major factor behind the rise of the Chipko movement. However, the most militant forest labour movements have been in tribal areas of Central India, concerned mainly with the terms of collection of minor forest produce. Protesting at their low wages, tribals have repeatedly struck work, refusing to deposit bundles at collection centres. In heavily forested districts, such as Srikakulam in Andhra Pradesh and Gadchiroli in Maharashtra, the affected tribals are being organized by Marxist revolutionaries, the so-called Naxalites (PUDR 1982; Calman 1985).

More recently, the controversy over the contractor system has resurfaced in Madhya Pradesh (MP). Responding to criticisms of exploitation of tribals by middlemen, in 1988 the MP government decided to bring the lucrative trade in tendu leaves (used for making *bidis*) under the co-operative fold. The collection rate for tendu leaves was fixed at Rs 15 per hundred bundles of a hundred leaves each, which compares favourably with the prevailing rates of Rs 12.90 in Maharashtra and Rs 9 in Orissa. At the same time, the MP government announced the formation of 2000 co-operative societies. It was claimed that these measures would benefit more than one million forest labourers, mostly tribals. The state takeover of the tendu leaf trade has been bitterly opposed by merchants and by politicians aligned to them. Scepticism has also been expressed about the feasibility of fostering co-operatives from above in the absence of local-level organizations (*The Economic Times*, 20 November 1988; *Economic and Political Weekly*, 18 March 1989). And with the change in government in MP in early 1990, moves have been afoot to abolish the co-operatives and give concessions to the merchants who earlier controlled the tendu leaf trade.

Despite successful struggles for wage increases, in most areas tribals continue to get only an infinitesimal share of the gains from commercial forestry. Nor have they benefited substantially from the abolition of the contractor system and the takeover of forest operations by newly-created government corporations. The state is not a model employer either, and since the contractor continues to dominate the trade in forest produce even where he does not anymore supervise extraction, resistance continues (Gupta *et al.* 1981).

## Wild Life Conservation: Animals Versus Humans?

Amid the continuing ecological decline of the Indian subcontinent, the massive network of parks and sanctuaries constructed after 1947 apparently stands out as a magnificent exception. As chapter 4 suggests, the British, as proponents of shikar on a large scale, had very little interest in wild-life conservation. The consequences of record-breaking shikar sprees and habitat destruction were apparent by the time India gained independence. The tiger population, estimated at 40,000 at the turn of the century, had slumped to 3000. The cheetah was extinct in 1952. Other large mammals, such as the elephant and rhino, had disappeared from areas in which they were formerly quite numerous, while the Asiatic lion survived only in the Gir forest.

The initiative for wild-life preservation came from the erstwhile princes, who had a rather better record than the British in maintaining their hunting preserves. The Indian Board for Wildlife was set up in 1952; since then, a steady stream of parks and sanctuaries has been constituted. A major conservation effort, Project Tiger, was launched with the help of international agencies in 1973, concerned exclusively with the protection and enhancement of tiger populations (more than fifteen sanctuaries covering 25,000 sq. km. come under this project).

Judged on their own terms, these programmes have been quite successful in stabilizing the population of some endangered species and in enhancing the population of others. Yet this success has not been without its costs. Immediately after independence, a well-known botanist had in fact warned the proponents of wild-life sanctuaries of possible adverse effects on the surrounding population, saying the interests of nature lovers and the interests of cultivators were not easy to harmonize (Randhawa 1949). His warnings were prescient, for villagers living on the periphery of sanctuaries face serious hazards against which they are insufficiently protected, especially crop damage and manslaughter. Destruction is particularly severe in areas adjoining forests inhabited by large animals such as the rhino and elephant. A study conducted in ten villages along the Karnataka–Tamil Nadu border estimates that the damage done to food crops by elephants was about Rs 1.5 lakhs per year. While on the average 10-15 per cent of the crop was destroyed, in some months fields were attacked by elephants almost every day. Over a period of two years, eighteen human kills were also reported (Sukumar 1989). One area in India where manslaughter by tigers has reached serious proportions is the Sunderbans delta. While the tiger population has increased from around 130 to 205 in the last decade, it is believed that Sunderban tigers have taken a toll of perhaps a thousand human lives in the past twenty years. National parks already cover 3 per cent of India's land surface, and there are proposals to double this area by the end of the century. In displacing villagers without proper rehabilitation, prohibiting traditional hunting and gathering, and exposing villages on the periphery to the threat of crop damage, cattle lifting and man-slaughter, the parks are, as they stand, inimical to the interests of the poorer sections of agrarian society. In some cases, villagers have responded by setting fire to large areas in the national parks. In the Kanha National Park of MP—a Project Tiger area also famous for swamp deer—the dry season of 1990 saw a wave of fires sweep through nearly half the park. Denied access to the park for their requirements of forest produce, the villagers, aided by political extremists, have in this manner taken 'revenge' on the forest department (see *Sunday Observer*, 13 May 1990).

These conflicts come out sharply in a recent status report on national parks and sanctuaries in India. The report cited frequent clashes between villagers and park authorities over access to natural resources. Between 1979 and 1984, fifty-one such clashes were reported as having occurred in national parks, and sixty-six clashes as having taken place in sanctuaries. Commenting on incidents of death or injury to humans on account of attacks by wild animals, the report observed that the states of Andhra Pradesh, Arunachal Pradesh, Himachal Pradesh, Manipur and Rajasthan do not pay any compensation in such cases. Among the states that did report payment, compensation for fatalities was quite meagre, varying from Rs 200 to Rs 10,000. The day-to-day management of the parks and sanctuaries is also biased against local users, there being sharp curbs on customary use. Indeed, a greater proportion of parks allows the manipulation of habitat by plantations than allows grazing by the cattle of surrounding villages (Kothari *et al.* 1989).

This is not to say that the preservation of biological diversity is not important, only that it should follow different principles. For in India, as in other parts of the Third World, national park management is heavily imprinted by the American experience. In particular, it has taken over two axioms of the Western wilderness movement: that wilderness areas should be as large as possible, and the belief that *all* human intervention is bad for the retention of diversity. These axioms have led to the constitution of massive sanctuaries, each covering thousands of square miles, and a total ban on human ingress in the 'core' areas of national parks (See Guha, Ramachandra 1989b).

These axioms of 'giganticism' and 'hands off nature', though cloaked in the jargon of science, are simply prejudices. When it is realized that the preservation of *plant* diversity is in many respects more important than the preservation of large mammals, a decentralized network of many small parks makes far more sense. The widespread network of sacred groves in India traditionally fulfilled precisely those functions. Yet modern wilderness lovers and managers are in general averse to reviving that system: apart from rationalist objections, they are in principle opposed to local control, preferring a centralized system of park management. The belief in a total ban on human intervention is equally misguided. Studies show that in fact the highest levels of biological diversity are found in areas with some (though not excessive) intervention. In opening up new niches to be occupied by insects, plants and birds, partially disturbed ecosystems often have a greater diversity than untouched areas. The dogma of total protection can have tragic consequences, as illustrated by the case of the famous bird sanctuary in Bharatpur. Here, villages were abruptly told that they must stop grazing cattle in the sanctuary, a right they had enjoyed for several decades. When they refused to agree, an altercation between them and officials culminated in firings and the death of several villagers. Ironically, scientific studies now suggest that grazing was not adversely affecting bird life in the sanctuary; on the contrary, it was essential for keeping down excessive growth of the *paspalum* grass, which otherwise choked the shallow marshes, rendering them unsuitable for the migratory species of waterfowl which are one of the sanctuary's main attractions (Vijayan 1987).

## The Changing Profile of Forest Conflicts

It would be appropriate to conclude this chapter with a brief comparison with the colonial period. While the framework of forest management introduced by the British is very much in place, the much higher rates of industrial growth have informed radical modifications in the systems of forest working. Simultaneously, the pressures of demographic expansion and ecological decline have forced many forest-dependent communities to look for alternative modes of subsistence. With respect both to the hardware and software of forest resource use, therefore, the post-colonial period is marked by both change and continuity. How is this reflected in the patterns of social conflict over living resources?

Table 7.1 summarizes what appear to be the salient features of this process. We can draw four major conclusions from the table.

First, there has been a diminution in the scale and spread of some conflicts, notably nos. 1, 2, and 3. However, this does not mean that these conflicts have been *resolved*. On the contrary, one party has been forced to accept defeat, and even in some cases (for example hunter-gatherers) virtual extinction.

Second, there has been the intensification of some conflicts (nos. 4 and 5) as well as the emergence of forms of conflict almost wholly absent in the colonial period (nos. 7 and 8). Conflict 7 is the consequence of the greater exploitation of forest resources to keep pace with a growing market demand. Conflict 8, on the other hand, is the result of schemes (however narrowly conceived) to *protect* certain habitats in their pristine state.

## TABLE 7.1

CHANGE IN INTENSITY OF DIFFERENT FORMS OF RESOURCE-
RELATED CONFLICT BEFORE AND AFTER INDEPENDENCE

| No. | Form of conflict | Comparative intensity (post-47 / pre-47) |
|---|---|---|
| 1. | Hunter-gatherers *vs* the state | Sharply reduced on the main-land; intensifying in areas like the Andamans |
| 2. | *Jhumiyas vs* the state | Reduced in peninsular India; intensifying in the north-east |
| 3. | Settled cultivators *vs* the State | Reduced in some areas (where forests are no longer important), constant in other areas (Himalaya, Western Ghats) |
| 4. | Conflicts within village society (decline of CPRs) | Sharply increased |
| 5. | Cultivators *vs* nomads | Sharply increased |
| 6. | Artisans *vs* the state | Diminished for some categories (charcoal makers), increased or constant for others (basket weavers) |
| 7. | Labourers *vs* contractors | Sharply increased |
| 8. | Wildlife *vs* villagers | Largely new conflict |

Third, the state, as one party to conflicts 1, 2, 3 and 6, is really acting as a proxy for the commercial–industrial sector. All these conflicts have the same underlying cause—the preference shown in forest management to commercial, profit-oriented uses at the expense of small-scale, subsistence-oriented uses.

Finally, conflicts 4 and 5 point to perhaps the most disturbing consequence of these processes—increasing conflicts between different sections of the agrarian population.

As in other former European colonies like Indonesia (see Peluso 1989), popular resistance to state forest management cuts across four axes: those of land control, labour control, species control, and ideological control. Thefts and conflicts with forest officials challenge the state's control over non-cultivated land; recent years have seen a growing militancy among forest labourers concerning their terms and conditions of work; efforts to manipulate the forest in favour of commercially valued species have been bitterly resisted by local populations, and the option of decentralized, community-based management has been put forward as an alternative to the ideology of state control. In 1913, the government of the Madras Presidency had bracketed the forest department with the salt department as the most unpopular among state agencies. Sixty years after Gandhi's Dandi march hastened the abolition of the state monopoly over salt, the forest department remains a largely unwelcome presence in the Indian countryside.

CHAPTER EIGHT

# *Cultures in Conflict*

This book has traced the broad contours of eco-cultural evolution in India, moving from the hunter-gatherer mode through the peasant mode and finally to the continuing attempt by the industrial mode to impose its definitive stamp on the ecological history of the subcontinent. The Indian experience is decisively marked by the colonial encounter; in this sense it differs from the paths traced by the two other Asian giants, Japan and China. Fortuitously escaping European colonialism, ecological change within Japan followed a more or less autonomous path, and it is as yet the only major Asian country to make a successful transition to industrial society. The Chinese experience is no less interesting: although the pace of industrialization has been noticeably slower, the dominantly socialist character of the state and existing property relations give its ecological history a distinctive twist. While Japan draws heavily on the natural resources of South East Asia and South America, China, like India, has to largely rely on its own resources in its industrialization strategy.

The ecological histories of Japan and China are finding their chroniclers (for sample writings in English, see Totman 1985 and 1989; Menezies, in press) and will doubtless display some interesting comparisons with the Indian case. We are more concerned here with the lessons that can be drawn from the two processes of eco-cultural change most extensively studied by historians—the European 'miracle' of successful industrialization and the imposition of Neo-Europes in the New World. While the environmental impacts of the neolithic revolution in Europe may have been relatively neglected by historians, many impressive volumes have been written on the socio-ecological dimensions of the clash between the peasant and industrial modes in that continent. To summarize a well-known story, it seems this conflict was resolved in two major ways:

1. By the discovery of new (chiefly inanimate) sources of energy, which when combined with the revolution in science and technology fuelled an unprecedented expansion in productive forces. Thus industrial expansion was able to absorb much of the surplus population from the countryside. The celebration of these European achievements in science and technology has, however, tended to obscure the critical role played in this transition by the substitution of plant by non-plant materials. The discovery of more efficient ways of using coal in the early eighteenth century led to coal quickly replacing wood in the critical industry of iron smelting. And as fodder became scarce for horses, the chief locomotory force till then, the steam engine was invented. Finally, iron replaced timber as building material as well as for textile machinery (i.e. in the leading industry). As a French observer put it in 1817, 'this continuous substitution of wood by iron is not at all the result of a craze or a passing fancy: it stems from a comparison of the low price for this metal with the dearness of timber which is excessive throughout Britain' (quoted in von Tunzelman 1981, p. 162). Having devastated its own forests and those of its Irish colony by 1700, England was saved by the timely discovery of non-wood substitutes.

2. By colonial expansion, involving (a) the extraction of raw materials, and (b) the settling of surplus population in

newly-conquered territories. Especially in its early stages, colonization was closely linked to the requirements of the domestic economy. As Karl Polanyi (1957, p. 179) observed many years ago, colonialism, following upon the commercialization of the soil and the increases in food production to meet the needs of a growing domestic population, was the third and final stage in the subordination of the resources of the planet to the imperatives of European industrialization. In time, however, the fabulous natural resources of these lands and the enterprise of the settlers worked synergistically to create an autonomous process of industrialization through much of the New World.

New World colonialism therefore provides an organic link between these two patterns of eco-cultural change. Fundamental to the making of the modern world, both processes have been strikingly successful when judged on their own terms: indeed, they are frequently held aloft as models for other cultures to emulate. This is not to say they were harmonious. New World colonization led to the extermination of indigenous cultures and populations, while the Industrial Revolution was not exactly a painless process either. Whatever these costs, however, both processes have ultimately resulted in the establishment of prosperous, relatively egalitarian and harmonious societies. Moreover, none of these societies are faced with the prospect of impending ecological collapse. Of course, outside their borders their behaviour is not always characterized by ecological restraint, as witness the large-scale logging of virgin forests by Japanese multinationals in Papua New Guinea and Kalimantan, and the equally notorious 'hamburger connection', wherein millions of acres of Amazonian rain forest are being destroyed to raise beef for the American market. Within their borders, however, many (though not all—as the case of acid rain in Europe testifies) industrial societies are doing a good job of protecting their forests. In some parts, for example north-eastern United States and Japan, vigorous tree growth has actually colonized land deforested for centuries.

The theme of this work bears a curious relation to the two

processes outlined above. For India is akin to Europe in having had for centuries a complex agrarian civilization, yet more like the New World in having been subjected to the ravages of European colonialism. Sidestepping for the most part the question of the extent to which the colonization of India actually enabled the European 'miracle', we have tried to document the other side of the coin, namely the impact it had *within* the subcontinent. British imperialism could not wipe out the population of India—ironically, it set in motion a process of demographic expansion—but it did certainly disrupt, perhaps irrevocably, the ecological and cultural fabric of its society. And after it formally left Indian shores, the tasks it had left unfinished were enthusiastically taken up by the incoming nationalist elites, whose unswerving commitment to a resource-intensive pattern of industrialization has only intensified the processes of ecological and social disturbance initiated by the British.

From an ecological perspective, the clash of pre-industrial and industrial cultures in India may be represented in terms of the closure and creation of niches. In India, as elsewhere, the British usurped the ecological niches occupied by the hunter-gatherers, many of whom also practised shifting cultivation, and diminished substantially the niche space occupied by food producers, by alienating them from access to non-cultivated lands. The resource processors and transporters of European civilization had by the nineteenth century a tremendously greater access to resources, largely because of their technological ability to tap additional sources of energy and materials. They out-competed and usurped the niche space of Indian handicraft workers and artisans, as well as of itinerant traders. This tremendous shrinkage of the niche space available to the Indian population was only marginally compensated by new niches which opened up to collaborators of the British, in the usurpation and transport of resources, as their clerks and trading partners.

The literate castes of pre-British India, involved in priesthood and administration, filled the clerical jobs, with merchants and shopkeepers in the role of trading partners. These groups prospered as time went by, and moved into the modern resource-processing industry. But the others—hunter-gatherers, peasants, artisans, and pastoral and non-pastoral nomads— had all to squeeze into the already diminishing niche space for food production. And they, we have seen, suffered great impoverishment.

While the British ruled India, they discouraged Indians from taking up resource-processing and transport on the basis of modern technology and with access to fossil fuel and other modern energy sources. With time, however, this resistance was broken down, and India began to industrialize. The emerging Indian capitalist class, in fact, provided financial support to the national movement, aware that in an independent country they would face less competition. Following independence, industrialists were able to steer the course of development on a path beneficial to them, namely as an all-out state-subsidized effort to intensify the use of resources such as land, water, vegetation, minerals, and energy.

Such an effort has to function within severe constraints, for unlike the Europeans who gained access to the resources of new lands in a comparable phase of industrial development, Indians are confined to a land already suffering from many kinds of resource depletion. Further, it is once again the Western world which has access to newer and newer resources as technological advances render useful various forms of energy and materials that were earlier of little value. India finds itself falling ever further behind in this race for technological substitution and resource creation. This disadvantage is reflected in the net outflow of resources from India to the West, be it fish, iron ore or scientifically trained manpower.

Within the country, meanwhile, efforts at the intensification of resource use have further stepped up the outflows of resources from cultivated as well as non-cultivated land to resource processors—industry, and the urban agglomerations built around it. Existing levels in the disruption of energy and material cycles, which ultimately must be closed, cannot be sustained indefinitely. They are in fact leading to a continuous depression of the productive potential of cultivated and non-cultivated land (cf. CSE 1985). The situation has been saved from serious disaster by an inflow of irrigation water, agro-chemicals and high-yielding varieties to the agricultural sector. This inflow is restricted to only 20 per cent of the land under cultivation. Nevertheless, it has succeeded in enhancing food production to a level adequate for subsistence by the entire population—though serious disparities in the availability of food for different social classes remain.

There has thus been a significant expansion in the niche space of food production in tracts of intensive agriculture. There has also been expansion of niche space for resource processing and transport, information processing and resource usurpation. However, these have been more than offset by the continuing contraction of niche space in tracts of subsistence agriculture, and for those dependent on foraging for resources—landless and small peasants, fishermen, and traditional resource processors (artisans and nomads). These difficulties are compounded by the overall growth in numbers of people. The consequence has been a scramble for resources and intense conflict, in the countryside and in the cities where people who have been driven out from elsewhere are flocking. While traditional relationships based on low levels of niche overlap between different endogamous groups have broken down, the barriers of endogamy persist. Endogamous caste groups therefore remain cultural entities, but have no common belief system to hold them together. No longer functional entities in the present scenario of shrinking niche space, castes and communities are set up against each other, with frighteningly high levels of communal and caste violence being the result.

In India the ongoing struggle between the peasant and industrial modes of resource use has come in two stages: colonial and post-colonial. It has left in its wake a fissured land, ecologically and socially fragmented beyond belief and, to some observers, beyond repair. Where do we go from here? There seems no realistic hope of emulating European or New World modes of industrial development. There is no longer a 'frontier' available with which to easily dispose of our population. Nor are there readily available substitutes for energy or construction material, enabling us to prevent our forest resources getting depleted. On both these counts the Western world has pre-empted the two-thirds of humanity which is lumped under the label 'Third World'. Through most of the Third World, the transition from the peasant to the industrial mode is very incomplete and, indeed, likely to remain incomplete for a very long time to come.

Not surprisingly, the Indian environmental debate has taken an altogether different track from its Western counterpart. Western environmentalists, contemplating the arrival of the 'post-industrial' economy and for the most part unaware of the damage its industrial economy is doing to other parts of the globe, are moving towards a 'post-materialist' perspective in which the forest is not central to economic production but rather to the enhancement of the 'quality of life'. In India, by contrast, the debate around the forest, and the environment debate more generally, is firmly rooted in questions of production and use. The issues in contention include the relative claims of the industrial and agrarian sector over natural resources (and within each, the claims of large *versus* small units), the uses of nature for subsistence or for profit, the respective proprietary claims of individuals, communities, and the state, and finally the role of natural resource management in an alternative development strategy.

But whether these debates will result in a new mode of resource use and a new belief system to hold our society together, it is too early to say.

# Bibliography

Aatre, T.N., 1915. *Gaon-Gada—Notes on Rural Sociology and Village Problems* (in Marathi). Mote Publishers, Bombay.

Adas, M., 1981. 'From avoidance to confrontation: Peasant protest in precolonial and colonial southeast Asia'. *Comp. Stud. Soc. Hist.*, 23.

Agarwal, A., 1986. 'Human-Nature interactions in a third world country'. *The Environmentalist*, 6.

Agarwal, D.P. and Sood, R.K., 1982. 'Ecological factors and the Harappan civilization'. G.L. Possehl (ed.), *Harappan Civilization*. Oxford and IBH, New Delhi.

Agulhon, M., 1982. *The Republic in the Village*. Cambridge University Press, Cambridge.

Aiyappan, A., 1948. *Report on the Socio-economic Conditions of the Aboriginal Tribes of the Province of Madras*. Madras Government Press, Madras.

Albion, R.G., 1926. *Forests and Sea Power*. Harvard University Press, Cambridge, Massachussets.

Allchin, B. and Allchin, F.R., 1968. *The Birth of Indian Civilization*. Penguin, Harmondsworth.

Allchin, F.R., 1963. *Neolithic Cattle Keepers of South India*. Cambridge University Press, Cambridge.

Ambedkar, B.R., 1948. *The Untouchables*. Amit Book Co., New Delhi.

Amery, C.F., 1875. 'On Forest Rights in India'. D. Brandis and A. Symthies (ed), *Report on the Proceedings of the Forest Conference held at Simla*, October. Government Press, Calcutta.

Anderson, R. and Hubner, W., 1988. *The Hour of the Fox*. Vistar Publications, New Delhi.

Anonymous, 1906. *A Manual of Forest Law Compiled for the Use of the Students at the Imperial Forest College, Dehra Dun*, Government Press, Calcutta.

Anonymous, 1913. *Report of the Forest Committee, Vol. I.* Government Press, Madras.

Anonymous, 1920. *Report of the Imperial Economic Committee : Tenth Report : Timber*. HMSO, London.

Anonymous, 1927. *Report of the Royal Commission of Agriculture.* HMSO, London.

Arnold, D., 1982. 'Rebellious Hillmen : The Gudem Rampa Rebellions (1829-1914)'. Ranajit Guha (ed.), *Subaltern Studies I.* Oxford University Press, Delhi.

Avchat, A., 1981. *Manase* (in Marathi). Granthali Publishers, Pune.

Ayers, R.V., 1978. *Resources, Environment and Economics. Applications of the Materials / Energy Balance Principle.* John Wiley and Sons, New York.

Badam, G.L., 1978. 'The quarternary fauna of Inamgaon'. V.N. Misra and P. Bellwood (ed.), *Proceedings of the International Symposium on Recent Advances in Indo-Pacific Prehistory*. Croom Helm, London.

Baden-Powell, B.H., 1874. *The Forest System of British Burma.* Government Press, Calcutta.

——, 1875. 'On the defects of the existing Forest Law (Act VII of 1865) and Proposals for a New Forest Act'. B.H. Baden-Powell and J.S. Gamble (ed.), *Report of the Proceedings of the Forest Conference, 1873-74.* Government Press, Calcutta.

——, 1882. *A Manual of Jurisprudence for Forest Officers.* Government Press, Calcutta.

——, 1892. *Memorandum on Forest Settlements in India.* Government Press, Calcutta.

Badgeley, C., Kelly, J., Pilbean, D. and Ward, S., 1984. 'The Palaeobiology of South Asian miocene hominoids'. J.R. Lukacs (ed.) *The People of South Asia*. Plenum Press, New York.

Bahro, R., 1984. *From Red to Green*. Verso Books, London.

Baker, C.J., 1984. *An Indian Rural Economy, 1880-1955*. Oxford University Press, Delhi.

Baker, D.E.U., 1984. 'A serious time : Forest satyagraha in Madhya Pradesh, 1930'. *Indian Econ. and Social Hist. Rev.*, 21.

Beckerman, W., 1972. 'Economists, scientists and environmental catastrophe'. *Oxford Economic Papers*, 24.

Beddington, J.R. and May, R.M., 1982. 'The harvesting of interacting species in a natural ecosystem'. *Scientific American*, 247.

Bennet, S., 1984. 'Shikar and the Raj'. *South Asia N.S.*, 7.

Berkes, F. (ed.), 1989. *Common Property Resources : Ecology and Community-Based Sustainable Development*. Belhaven Press, London.

Berkes, F. and Kence, A., 1987. 'Fisheries and the prisoners dilemma game : Conditions for the evolution of cooperation among users of common property resources'. *Metu Journal of Pure and Applied Sciences*, 20 (2).

Berlin, B., 1973. 'Folk systematics in relation to biological classification and nomenclature'. *Annual Review of Ecology and Systematics*, 4.

Best, J.W., 1935. *Forest Life in India*. John Murray, London.

Bhattacharya, S., 1972. 'Iron smelters and the indigenous iron and steel industry of India : From stagnation to atrophy'. S. Sinha (ed.), *Aspects of Indian Culture and Society*. Indian Anthropological Society, Calcutta.

Birrel, J., 1987. 'Common rights in the medieval forest'. *Past and Present*, no. 117.

Blanford, H.R., 1925. 'Regeneration with the assistance of Taungya in Burma'. *Indian Forest Records*, 11, part 3.

Blair, H., 1986. 'Social Forestry in India : Time to Modify Goals'. *Economic and Political Weekly*, 21 July.

Bloch, M., 1978. *French Rural History : An Essay on its Essential Characteristics* (1931 reprint). Routledge and Kegan Paul, London.

Borgerhoff Mulder, M., 1988. 'Behavioural Ecology in Traditional Societies'. *Trends in Ecology and Evolution*, 3 (10).

Brahme, S. and Upadhya, A., 1979. 'A Critical Analysis of the Social Formation of and Peasant Resistance in Maharashtra'. *Mimeo*, three volumes. Gokhale Institute of Politics and Economics, Pune.

Brandis, D., 1873. *The Distribution of Forests in India*. McFarlane and Erskine, Edinburgh.

——, 1875. *Memorandum on the Forest Legislation Proposed for British India (other than the Presidencies of Madras and Bombay)*. Government Press, Simla.

——, 1876. *Suggestions Regarding Forest Administration in the Central Provinces*. Government Press, Calcutta.

——, 1882. *Suggestions Regarding Forest Administration in the Northwestern Provinces and Oudh*. Government Press, Calcutta.

——, 1884. *Progress of Forestry in India*. McFarlane and Erskine, Edinburgh.

——, 1897. *Indian Forestry*. Oriental Institute, Woking.

Brara, R., 1987. 'Shifting Sands : A Study of Rights in Common Pastures'. *Mimeo*. Institute of Development Studies, Jaipur.

Brenner, R., 1976. 'Agrarian class structure and economic development in pre-industrial Europe'. *Past and Present*, 70.

——, 1978. 'Dobb on the transition from feudalism to capitalism'. *Cambridge Journal of Economics*, 2 (2).

Calman, L., 1985. *Protest in Democratic India*. Westview Press, Boulder.

Campbell, 1883. *Bombay Gazetteer : Kanara*. Government Central Press, Bombay.

Centre for Science and Environment, 1985. *The State of India's Environment 1984-85 : A Second Citizens' Report*. Centre for Science and Environment, New Delhi.

Champion, H.G. and Osmaston, F.C., (ed.), 1962. *E.P. Stebbing's The Forests of India*, Vol. IV. Oxford University Press, Oxford.

Champion, H.G. and Seth, S.K., 1968. *General Silviculture for India*. Manager of Publications, Delhi.

Chandra, G.S. and Srivastava, S.S., 1968. 'Forestry economirs in India, some areas of research'. *Arthavijnana*, 10.

Chandrasekhar, D.M., Krishnamurthy, B.V. and Ramaswamy, S.R., 1987. 'Social forestry in Karnataka : An impact analysis'. *Economic and Political Weekly*, 29 September.

Chatterjee, P., 1983. 'More on modes of power and the peasantry'. Ranajit Guha (ed.), *Subaltern Studies II*. Oxford University Press, Delhi.

———, 1984. 'Gandhi and the critique of civil society'. Ranajit Guha (ed.), *Subaltern Studies III*. Oxford University Press, Delhi.

Chattopadhyay, R., 1985. 'The Idea of Planning in India 1930-51'. Unpublished D. Phil thesis, Australian National University, Canberra.

Chaudhuri, K.A., 1977. *Ancient Agriculture and Forestry in Northern India*. Bombay.

Choudhury, L., Personal Communication.

Clark, C.W., 1985. *Bioeconomic Modelling and Fisheries Management*. John Wiley and Sons, New York.

Cleghorn, H., 1860. *Forests and Gardens of South India*. W.H. Allen, London.

Conklin, H., 1969. 'An ethno-ecological approach to shifting agriculture'. A.P. Vayda (ed.), *Environment and Cultural Behaviour*. Academic Press, New York.

Coward, M.P., Bulter, R.W.H., Chambers, A.F., Graham, R.H., Izatt, C.N., Khan, M.A., Knipe, R.J., Prior, D.J., Treloar, P.J. and Williams, M.P., 1988. 'Folding and imbrication of the Indian crust during Himalayan collision'. *Phil. Trans. Royal Society, London*, A 326.

Cronon, W., 1983. *Changes in the Land : Indians, Colonists and the Ecology of New England*. Hill and Wang, New York.

Crosby, A., 1986. *Ecological Imperialism : The Biological Expansion of Europe*. Cambridge University Press, Cambridge.

D'Arcy, W.E., 1910. *Preparation of Forest Working Plans in India*. Government Press, Calcutta.

Dasgupta, P., 1982. *The Control of Resources*. Oxford University Press, Delhi.

Dasgupta, S., 1980. 'Local Politics in Bengal : Midnapur District 1907-1923'. Unpublished thesis, School of Oriental and African Studies, University of London.

Dasmann, R., 1988. 'Towards a biosphere consciousness'. Worster, D., *The Ends of the Earth : Perspectives on Modern Environmental History*. Cambridge University Press, Cambridge.

Dharampal, 1986. 'Some aspects of earlier Indian society and polity and their relevance to the present'. *New Quest*, 57.

Dhareshwar, S.S., 1941. 'The denuded condition of the minor forest in Kanara coastal tract : Its history and a scheme for its regeneration'. *Indian Forester*, 67.

Dhavalikar, M.K., 1988. *The First Farmers of the Deccan*. Deccan College, Pune.

Digby, S., 1971. *War Horse and Elephant in the Delhi Sultanate*. Clarendon Press, Oxford.

Draz, O., 1985. 'The Hema system of range reserves in the Arabian peninsula : Its possibilities in range improvement and conservation projects in the near east'. J.A. McNeely and D. Pitt (ed.), *Culture and Conservation : The Human Dimension in Environmental Planning*. Croom Helm, Dublin.

Dumont, L., 1970. *Homo Hierarchicus : The Caste System and its Implications* (translated by M. Sainbury). Weidenfeld and Nicolson, London.

Eaton, P., 1985. 'Customary land tenure and conservation in Papua New Guinea'. J.A. NcNeely and D. Pitt (ed.), *Culture and Conservation : The Human Dimension in Environmental Planning*. Croom Helm, Dublin.

Edye, J., 1835. 'Description of the sea-ports of the coast of Malabar, of the facilities they afford for building vessels of different descriptions, and of the produce of the adjacent forests'. *Journal of the Royal Asiatic Society*, 11.

Ehrenfeld, U.R., 1952. *The Kadar of Cochin*. Madras University, Madras.

Elliott, J.G., 1973. *Field Sports in India, 1800-1947*. Gentry Books, London.

Elwin, V., 1939. *The Baiga*. John Murray, London.

——, 1942. *The Agaria*. Oxford University Press, Calcutta.

——, 1943. 'The Aboriginals'. *Oxford Pamphlets on Indian Affairs*, no. 14.

——, 1945. 'Saora Fituris'. *Man in India*, 25.

——, 1958. *Leaves from the Jungle* (1936 reprint). Oxford University Press, London.

——, 1960. *A Philosophy for NEFA*. Government Press, Shillong.

Engels, F., 1956. *The Peasant War in Germany*. Progress Publishers, Moscow.

Faith, R., 1984. 'The great rumour of 1377 and peasant ideology'.

R.H. Hilton and T.H. Ashton (eds.), *The English Rising of 1381.* Cambridge University Press, Cambridge.

Feldman, M.W. and Thomas, E.A.C., 1986. *Behaviour—dependent contexts for repeated plays of the prisoner's dilemma II : Dynamical aspects of the evolution of cooperation.* Working Paper Series : Paper no. 0002. Stanford University, U.S.A.

Fernandes, W. and Menon, G., 1987. *Tribal Women and Forest Economy : Deforestation, Exploitation and Status Change.* Indian Social Institute, New Delhi.

Fernandes, W., Menon, G. and Viegas, P., 1988. *Forests, Environment and Tribal Economy : Deforestation, Impoverishment and Marginalization in Orissa.* Indian Social Institute, New Delhi.

Fernow, B.E., 1907. *A History of Forestry.* Toronto University Press, Toronto.

Fisher, R.A., 1958. *Genetical Theory of Natural Selection.* Dover Publications, Inc., New York.

Food and Agricultural Organization, United Nations 1984 (FAO 1984), 'Intensive Multiple-Use Forest Management in Kerala'. *FAO Forestry Paper 53*, FAO, Rome.

Forde, C.D., 1963. *Habitat, Economy and Society : A Geographical Introduction to Ethology.* Dutton, New York.

Ford Robertson, F.C., 1936. *Our Forests.* Government Press, Allahabad.

Fortmann, L.P. And Fairfax, S.K., 1989. 'American forestry professionalism in the Third World—some preliminary observations'. *Economic and Political Weekly*, 12 August.

Freeman, M.M.R., 1989. 'Graphs and gaffs : A cautionary tale in the common property resources debate'. F. Berkes (ed.), *Common Property Resources : Ecology and Community-Based Sustainable Development.* Belhaven Press, London.

Gadgil, M., 1985a. 'Social Restraints on Resource Utilization : The Indian Experience'. J.A. McNeely and D. Pitt (ed.), *Culture and Conservation : The Human Dimension in Environmental Planning.* Croom Helm, Dublin.

——, 1985b. 'Cultural Evolution of Ecological Prudence'. *Landscape Planning*, 12.

——, 1987. 'Diversity : Cultural and Biological'. *Trends in Ecology and Evolution*, 2(2).

——, 1989. 'The Indian heritage of a conservation ethic'. B. Allchin, E.R. Allchin and B.K. Thapar (ed.), *Conservation of the Indian Heritage*. Cosmo Publications, New Delhi.

——, 1991. 'Conserving India's biodiversity : The societal context'. *Evolutionary Trends in Plants*, in press.

Gadgil, M. and Berkes, F., 1991. 'Traditional resource management systems'. *Resource Management and Optimization*. In press.

Gadgil, M., Hegde, K.M. and Bhoja Shetty, K.A., 1986. 'Uttara Kannada : A case study in hill area development'. C.J. Saldanha (ed.), *Karnataka State of Environment Report 1985-86*. Centre for Taxonomic Studies, Bangalore.

Gadgil, M. and Iyer, P., 1989. 'On the diversification of common property resource use by Indian society'. F. Berkes (ed.), *Common Property Resources : Ecology and Community-Based Sustainable Development*. Belhaven Press, London.

Gadgil, M. and Malhotra, K.C., 1982. 'Ecology of a pastoral caste : The Gavli Dhangar of Peninsular India'. *Human Ecology*, 10.

Gadgil, M. and Prasad, S.N., 1978. 'Vanishing Bamboo Stocks'. *Commerce*, 136 (3497), 17 June.

Gadgil, M., Prasad, S.N. and Ali, R., 1983. 'Forest management and forest policy in India : A critical review'. *Social Action*, 33.

Gadgil, M. and Sinha, M., 1985. 'The biomass budget of Karnataka'. C.J. Saldanha (ed.), *Karnataka State of Environment Report 1984-85*. Centre for Taxonomic Studies, Bangalore.

Gadgil, M. and Subash Chandran, 1989. 'Environmental impact of forest-based industries on the evergreen forests of Uttara Kannada district : A case study. Final report', submitted to Department of Ecology and Environment. Government of Karnataka.

Gadgil, M. and Vartak, V.D., 1976. 'Sacred Groves of Western Ghats of India'. *Economic Botany*, 30.

Gause, G.F., 1969. *The Struggle for Existence*. Hafner Publishing Company, New York.

Geertz, C., 1963. *Agricultural Involution*. University of California Press, Berkeley.

Goldschmidt, W., 1979. 'A general model for pastoral social systems'. *Pastoral Production and Society*. Cambridge University Press, Cambridge.

Government of India (GOI), 1929. *Report of the Forestry Committee.* Government Press, Simla.

——, 1948. *India's Forests and the War.* Ministry of Agriculture, New Delhi.

——, 1952. *The National Forest Policy.* Government Press, Delhi.

——, 1956. *The Second Five Year Plan.* Planning Commission, Delhi.

——, 1961. *Report of the Scheduled Areas and Scheduled Tribes Commission,* Volume I, 1960-61. Manager of Publications, Delhi.

——, 1964. *History of Indian Railways.* Government Press, Delhi.

——, 1967. *Report of the Committee on Tribal Economy in Forest Areas.* Department of Social Welfare, Delhi.

——, 1980. *Task Force Report on Taking Forestry to the People.* Ministry of Agriculture, Delhi.

Grigg, D.B., 1980. *The Agricultural Systems of the World : An Evolutionary Approach.* Cambridge University Press, Cambridge.

Grigson, W.V., 1938. *The Maria Gonds of Bastar.* Oxford University Press, London.

Grove, R., 1990. 'Colonial conservation, ecological hegemony and popular resistance : Towards a global synthesis'. J.M. Mackenzie (ed.), *Imperialism and the Natural World.* Manchester University Press, Manchester.

Guha, Ramachandra, 1983. 'Forestry in British and post-British India : A historical analysis'. *Economic and Political Weekly,* 29 October and 5-12 November.

——, 1989a. *The Unquiet Woods : Ecological Change and Peasant Resistance in the Himalaya.* Oxford University Press, Delhi and University of California Press, Berkeley.

——, 1989b. 'Radical American environmentalism and wilderness preservation : a Third World critique'. *Environmental Ethics,* Spring.

Guha, S.R.D., 1989. Personal Communication.

Gupta, R., Bannerji, P. and Guleria, A., 1981. *Tribal Unrest and Forestry Management in Bihar.* Indian Institute of Management, Ahmedabad.

Gupta, T. and Sambrani, S., 1978. 'Control of shifting cultivation : The need for an integrative approach and systematic appraisal'. *Indian Journal of Agricultural Economics,* 31.

Hahn, S., 1982. 'Hunting, fishing and foraging : Common rights and class relations in the post-bellum South'. *Radical History Review*, 26.

Halappa, G.S., 1964. *History of Freedom Movement in Karnataka*, vol. II. Government Press, Bangalore.

Hardin, G., 1968. 'The tragedy of the commons'. *Science*, 162, 1243-8.

Harris, M., 1980. *Culture, People and Nature : An Introduction to General Anthropology*. Harper and Row, New York.

Hawley, A.H., 1986. *Human Ecology : A Theoretical Essay*. Chicago University Press, Chicago.

Hay, D. *et al.*, 1975. *Albion's Fatal Tree*. Penguin, Harmondsworth.

Hays, S.P., 1958. *Conservation and the Gospel of Efficiency : The Progressive Conservation Movement 1880-1920*. Harvard University Press, Cambridge, Mass.

——, 1987. *Beauty, Health and Permanence : Environmental Politics in the United States, 1955-85*. Cambridge University Press, Cambridge.

Hearle, N., 1888. *Working Plan of the Tehri Garhwal Leased Forests. Jaunsar Forest Division*. Government Press, Allahabad.

——, 1889. *Working Plan of the Deoban Range, Jaunsar Forest Division, Northwestern Provinces*. Government Press, Allahabad.

Heske, F., 1937. *German Forestry*. Yale University Press, New Haven.

Hivale, S. and Elwin, V., 1935. *Songs of the Forest : The Folk Poetry of the Gonds*. Allen and Unwin, London.

Hutchinson, J., Clark, G., Jope, E.M. and Riley, R., 1977. *Early History of Agriculture*. Oxford University Press, Oxford.

International Institute for Environment and Development and World Resources Institute 1987 (IIED and WRI). *World Resources, 1987*. Basic Books, New York.

Jarrige, J.F. and Meadow, R.H., 1980. 'The Antecedents of Civilization in the Indus Valley'. *Scientific American*, 243.

Jodha, N.S., 1986. 'Common property resources and rural poor in dry regions of India'. *Economic and Political Weekly*, 5 July.

Johannes, R.E., 1978. 'Traditional marine conservation methods in Oceania and their demise'. *Annual Review of Ecology and Systematics*, 9.

Jones, E.L., 1979. 'The environment and the economy'. Peter Burke (ed.), *The New Cambridge Modern History*, vol. XIII. Cambridge University Press, Cambridge.

Joshi, N.V., 1987. 'Evolution of cooperation by reciprocation within structured demes'. *Journal of Genetics*, 66 (1).

Joshi, N.V. and Gadgil, M., 1991. 'On the role of refugia in promoting prudent use of biological resources'. *Theoretical Population Biology*. In press.

Kajale, M.D., 1988. 'Plant economy'. M.K. Dhavalikar, H.D. Sankhalia and Z.D. Ansari (ed.), *Excavations at Inamgaon*, vol. 1, part ii. Deccan College, Pune.

Kangle, R.P., 1969. *Arthasasthra*. University of Bombay, Bombay.

Karve, I., 1961. *Hindu Society : An Interpretation*. Deccan College, Poona.

——, 1974. *Yuganta. The End of an Epoch*. Sangam Books, Poona, and Orient Longman Ltd., New Delhi.

Karve, I., and Malhotra, K.C., 1968. 'A biological comparison of eight endogamous groups of the same rank'. *Current Anthropology*, 9.

Keer, D. and Malshe, S.G., 1969. 'Jotirau Phule, Shetkarya Asud : The Whipcord of the Farmer (1882-3)'. *The Collected Works of Mahatma Phule* (in Marathi). Maharashtra Sahitya and Sanskriti Mandal, Pune.

Khazanov, A.V., 1984. *Nomads and the Outside World*. Cambridge University Press, Cambridge.

Kjekshus, H., 1977. *Ecology Control and Economic Development in East African History*. University of California Press, Berkeley.

Kosambi, D.D., 1970. *The Culture and Civilization of Ancient India in Historical Outline*. Vikas Publishing House Pvt. Ltd., Delhi.

Kothari, A., Pande, P., Singh, S. and Variava, D., 1989. *Management of National Parks and Sanctuaries in India : A Status Report*. Indian Institute of Public Administration, New Delhi.

Krishnakutty, C.N. and Chundamannil, M., 1986. 'Are eucalyptus plantations fulfilling the goals in Kerala?'. *Eucalyptus in India : Past, Present and Future*. Kerala Forest Research Institute, Peechi.

Kumar, D. (ed.), 1983. *The Cambridge Economic History of India*, vol. 2. Cambridge University Press, Cambridge.

Ladurie, Le Roy, E., 1980. *Carnival in Romans*. George Brazillier, New York.

Lal, M., 1984. *Settlement History and the Rise of Civilization in the*

*Ganga-Yamuna Doab from 1500 B.C. – A.D. 300.* B.R. Publishing, Delhi.

Laptev, I., 1977. *The Planet of Reason.* Progress Publishers, Moscow.

Leakey, R.E., 1981. *The Making of Mankind.* Abacus, New York.

Leathart, S., 1982. 'Review of N.D.G. James, A History of English Forestry'. *Times Literary Supplement,* 8 January.

Leeds, A. and Vayda, A.P., 1965. *Man, Culture and Animals : The Role of Animals in Human Ecological Adjustments.* American Association for the Advancement of Science, Washington D.C.

Lefebvre, G., 1982. *The Great Fear* (1932 reprint). Princeton University Press, Princeton.

Lenski, G. and Lenski, J., 1978. *Human Societies : An Introduction to Macrosociology.* McGraw-Hill, New York.

Lewis, O., 1964. *Pedro Martinez.* Alfred Knopf, New York.

Linebaugh, P., 1976. 'Karl Marx, the theft of wood, and working class composition : A contribution to the current debate'. *Crime and Social Justice* n.s. no. 6.

Macleod, W.C., 1936. 'Conservation among primitive hunting peoples'. *The Scientific Monthly,* December.

Mahalanobis, P.C., 1963. *The Approach of Operational Research to Planning.* Indian Statistical Institute, Calcutta.

Malhotra, K.C. 1974. 'Socio-biological investigations among the Nandiwallas of Maharashtra'. *Bulletin of the Urgent Anthropological and Ethnological Sciences* (Austria), 16.

Malhotra, K.C., 1982. 'Nomads'. *The State of India's Environment, 1982-83 : A Citizens' Report.* Centre for Science and Environment, New Delhi.

Malhotra, K.C. and Gadgil, M., 1981. 'The ecological basis of the geographical distribution of the Dhangars : A pastoral caste-cluster of Maharashtra'. *South Asian Anthropologist,* 2.

Malhotra, K.C., Hulbe, S.K., Khomne, S.B. and Kolte, S.B., 1983. 'Economic organization of a nomadic community, the Nandiwallas'. P.K. Misra and K.C. Malhotra (ed.), *Nomads in India.* Anthropological Survey of India, Calcutta.

Malhotra, K.C. and Khomne, S.B., 1978. 'Social stratification and caste ranking among the Nandiwallas of Maharashtra'. *Proceedings of the Seminar on Nomads in India,* Mysore.

Malhotra, K.C., Khomne, S.B. and Gadgil, M., 1983. 'On the role of hunting in the nutrition and economy of certain nomadic populations of Maharashtra'. *Man in India*, 63.

Mani, A., 1984. 'Social Intervention in the Management of Forest Resources : A Report on the Appiko Movement'. *Mimeo*. Indian Institute of Management, Bangalore.

Martin, C., 1978. *Keepers of the Game : Indian-Animal Relationships and the Fur Trade*. University of California Press, Berkeley.

May, R.M. (ed.), 1984. *Exploitation of Marine Communities*. Life Sciences Research Report 32. Springer Verlag, Berlin.

McEvoy, A.E., 1988. 'Towards an interactive theory of nature and culture : Ecology, production and cognition in the California fishing industry'. Worster, D., *The Ends of the Earth : Perspectives on Modern Environmental History*. Cambridge University Press, Cambridge.

McNeely, J.A. and Pitt, D. (eds.), 1985. *Culture and Conservation : The Human Dimension in Environmental Planning*. Croom Helm, London.

Meher-Homji, V.M., 1989. 'History of Vegetation of Peninsular India'. *Man and Environment*, 13.

Mehra, K.L. and Arora, R.K., 1985. 'Some considerations on the domestication of plants in India'. V.N. Misra and P. Bellwood (ed.), *Recent Advances in Indo-Pacific Prehistory*. Oxford and IBH, New Delhi.

Menant, J.C., Barbault, R., Lavelle, P. and Lepage, M., 1985. 'African savannas : Biological systems of humification and mineralization'. J.C. Tothill and J.J. Mott (ed.), *Ecology and Management of the World's Savannas*. Australian Academy of Sciences, Canberra.

Menezies, N., (in press). 'A History of Forestry in China'. J. Needham (ed.), *Science and Civilization in China*. Cambridge University Press, Cambridge.

Merriman, J., 1975. 'The Demoiselles of the Ariege, 1829-31'. Merriman (ed.), *1830 in France*. New Viewpoints, New York.

Misra, P.K., 1977. 'The Jenu Kurubas'. Surajit Sinha and B.D. Sharma (ed.), *Primitive Tribes : The First Step*. Manager of Publications, New Delhi.

Misra, V.N., 1973. 'Ecological Adaptations during Stone-age in

Western and Central India'. K.R.A. Kennedy and G.L. Possehl (ed.), *Ecological Background of South Asian Prehistory*. South Asia Occasional Papers and Theses. Cornell University Press, Ithaca.

Mooser, J., 1986. 'Property and wood theft : Agrarian capitalism and social conflict in rural society, 1800-1850. A Westphalian case study'. R.G. Moeller, (ed.), *Peasants and Lords in Modern Germany*. Allen and Unwin, London.

Moosvi, S., 1987. *The Economy of the Mughal Empire c. 1595. A Statistical Study*. Oxford University Press, Delhi.

Morris, M., 1982. *Forest Traders : A Socio-Economic Study of the Hill Pandaram*. The Athlone Press, London.

Mukerjee, R.K., 1916. *The Foundations of Indian Economics*. Longman, Green and Co., London.

——, 1926. *Regional Sociology*. The Century Co., New York.

Muranjan, S.W., 1974. 'Exploitation of forests through forest labour cooperatives'. *Artha Vijnana*, 16(2).

——, 1980. 'Impact of some policies of the Forest Development Corporation on the working of the Forest Labourers Cooperatives'. *Artha Vijnana*, 22 (4).

Nair, K.K., 1967. 'Forestry Development in India'. *Indian Journal of Agricultural Economics*, 22.

Nandy, A., 1980. *At the Edge of Psychology and Other Essays*. Oxford University Press, Delhi.

Nash, R., 1982. *Wilderness and the American Mind*. Yale University Press, New Haven.

National Commission of Agriculture, 1976. *Report of the National Commission of Agriculture*. vol. IX. Ministry of Agriculture, Delhi.

National Council of Applied Economic Research, 1963. *Socio-Economic Survey of Primitive Tribes in Madhya Pradesh*. National Council of Applied Economic Research, New Delhi.

Nietschmann, B., 1985. 'Torres Strait islander sea resource management and sea rights'. K. Ruddle and R.E. Johannes (ed.), *The Traditional Knowledge and Management of Coastal Systems in Asia and the Pacific*. UNESCO, Indonesia.

Pandian, M.S.S., 1985. 'Political Economy of Agrarian Change in Nanchilnadu : The Late Nineteenth Century to 1939'. University of Madras, Unpublished Ph. D. thesis.

Pant, G.B. and Maliekel, J.A., 1987. 'Holocene climatic changes over north west India : An appraisal'. *Climatic Change*, 10.

Parulekar, G., 1976. *Adivasis Revolt*. National Book Agency, Calcutta.

Paul, G.P., 1871. *Felling Timber in the Himalaya*. Punjab Printing Co., Lahore.

Pearson, G.F., 1870. 'Deodar forests of Jaunsar Bawar'. *Selections from the Records of the Government of the Northwestern Provinces*, 2nd series, vol. II, Allahabad.

Peluso, N., 1989. 'Rich Forests, Poor People and Development'. Unpublished Ph. D. thesis, Department of Rural Sociology, Cornell University.

Peoples Union for Democratic Rights, 1982. *Undeclared Civil War : A Critique of the Forest Policy*. Peoples Union for Democratic Rights, Delhi.

Pimentel, D. and Pimentel, M., 1979. *Food, Energy and Society*. Arnold, London.

Pimm, S.L., 1986. 'Community stability and structure'. M.E. Soule (ed.), *Conservation Biology : The Science of Scarcity and Diversity*. Sinauer, Sunderland, Mass.

Pingle, U., Raja Reddy, N.V. and von Fürer Haimendorf, 1982. 'Should shifting cultivation be banned?'. *Science Today*, April.

Polanyi, K., 1957 (1944). *The Great Transformation*. Beacon Press, Boston.

Possehl, G.L., 1982. *Harappan Civilization : A Contemporary Perspective*. Oxford and IBH, New Delhi.

Prasad, S.N., Personal Communication.

Prasad, S.N. and Gadgil, M., 1981. 'Conservation of Bamboo Resources in Karnataka'. *Mimeo*. Karnataka State Council of Science and Technology, Bangalore.

Pressler, F.A., 1987. 'Panchayat Forests in Madras 1913-1952'. Unpublished Paper, Department of Political Science, Kalamazoo College.

Prucha, F.P., 1985. *The Indians in American Society : From the Revolutionary War to the Present*. University of California Press, Berkeley.

Rai, S.N., 1981. 'Protection forestry with a reference to ecological balance in the tropical forests of Western Ghats of Karnataka'. National Seminar on Forests and Environment, 2-3 Dec. 1981. Karnataka Forest Department.

Rajaguru, S.N., Badam, G.L. and Abhyankar, H., 1984. 'Late Pleistocene Climatic Change in India'. *Proc. Symp. Episodes.* Dept. Geol. M.S.University, Baroda.

Rana, S., 1983. 'Vanishing Forests'. *Frontier*, 16 (6).

Randhawa, M.S., 1949. 'Nature conservation, national parks and bio-aesthetic planning in India'. Presidential Address, Botany Section, 36th Indian Science Congress, Allahabad.

Rappaport, R.A., 1984. *Pigs for the Ancestors : Ritual in the Ecology of a New Guinea People.* Yale University Press, New Haven.

Raumolin, J. 1986. 'The Impact of Technological Change on Rural and Regional Forestry in Finland'. *Mimeo.* World Institute of Development Economics Research, Helsinki.

Rendell, H. and Dennell, R.W., 1985. 'Dated lower palaeolithic artefacts from northern Pakistan'. *Current Anthropology*, 26 (3).

Richards, J.F., 1986. 'World environmental history and economic development'. W.C. Clark and R.E. Munn (ed.), *Sustainable Development of the Biosphere.* Cambridge University Press, Cambridge.

Ribbentrop, B., 1900. *Forestry in British India.* Government Press, Calcutta.

Roy, S.L., 1925. *The Birhors : A Little Known Jungle Tribe of Chota Nagpur.* K.E.M. Mission Press, Ranchi.

Ruddle, K. and Johannes, R.E. (ed.), 1985. *The Traditional Knowledge and Management of Coastal Systems in Asia and the Pacific.* UNESCO, Indonesia.

Sagreya, K.P., 1979. *Forests and Forestry.* National Book Trust, Delhi.

Sahlins, M., 1971. *Stone Age Economics.* Aldine Publishers, Chicago.

Saldanha, C.J., 1989. *Andaman, Nicobar and Lakshadweep, An Environmental Impact Assessment.* Oxford and IBH Publishing Co. Pvt. Ltd., New Delhi.

Sankhala, K.S. and Jackson, P., 1985. 'People, trees and antelopes in the Indian desert'. J.A. McNeely and D. Pitt (ed.), *Culture and Conservation : The Human Dimensions in Environmental Planning.* Croom Helm, Dublin.

Sarkar, S., 1980. 'Primitive rebellion and modern nationalism : A note on forest satyagraha in the non-cooperation and civil disobedience movements'. K.N. Panikar (ed.), *National and Left Movements in India.* Vikas, New Delhi.

Sarkar, S., 1983. *Modern India, 1885-1947*. Macmillan, New Delhi.

———, 1984. 'The conditions and nature of Subaltern militancy : Bengal from Swadeshi to non-cooperation'. R. Guha (ed.), *Subaltern Studies III*. Oxford University Press, New Delhi.

Savyasachi, 1986. 'Agriculture and social structure : The hill Maria of Bastar'. *Mimeo*. World Institute of Development Economics Research, Helsinki, January.

Scott, J.C., 1976. *The Moral Economy of the Peasant*. Yale University Press, New Haven.

———, 1986. *Weapons of the Weak : Everyday Forms of Peasant Resistance*. Yale University Press, New Haven.

———, 1987. 'Resistance Without Protest and Without Organization : Peasant Opposition to the Islamic Zakat and the Christian Tithe'. *Comp. Studies in Soc. and Hist.*, 29.

Sengupta, N., 1980. 'The indigenous irrigation organization of South Bihar'. *Indian Econ. and Social Hist. Rev.*, 17.

———(ed.), 1982. *Jharkhand : Fourth World Dynamics*. Authors' Guild, Delhi.

Service, E.R., 1975. *Origins of the State and Civilization : The Process of Cultural Evolution*. W.W. Norton and Company, New York, London.

Shanin, T., 1986. *The Roots of Otherness : Russia's Turn of Century*, volumes I and II. Macmillan, London.

Sharma, G.R., 1980. *The Beginnings of Agriculture*. Allahabad.

Sharma, R.S., 1987. *Urban Decay in India*. New Delhi.

Shepard, P., 1982. *Nature and Madness*. Sierra Books, San Francisco.

Shiva, V., Sharatchandra, H.C. and Bandopadhyay, J., 1982. 'Social forestry : No solution within the market place'. *The Ecologist* 12.

Singh, K.S. (ed.), 1983. *Tribal Movements in India*, volume II. Manohar, New Delhi.

Sinha, A.C., 1986. 'Social Frame of Forest History'. *Social Science Probings* 3 (2).

Sinha, D.P., 1972. 'The "Birhors"'. M.G. Bicchieri (ed.), *Hunters and Gatherers Today*. Holt, Reinhart and Winston, New York.

Slobodkin, L.B., 1968. 'How to be a predator?'. *American Zoologist*, 8.

Smith, E.A., 1983. 'Anthropological applications of optimal foraging theory : A critical review'. *Current Anthropology*, 24.

Smythies, E.A., 1925. *India's Forest Wealth*. Humphrey Milford, London.

Smythies, A. and Dansey, E. (eds.), 1887. *A Report on the Proceedings of the Forest Conference held at Dehra Dun in October 1886*. Govt. Press, Simla.

Soule, M.E. (ed.), 1986. *Conservation Biology : The Science of Scarcity and Diversity*. Sinauer, Sunderland, Massachusetts.

Srinivas, M.N., 1962. *Caste in Modern India and Other Essays*. Asia Publishing House, Calcutta.

Stanhill, G. (ed.), 1984. *Energy and Agriculture*. Springer Verlag, Berlin.

Stebbing, E.P., 1922-27. *The Forests of India*, vols. I, II, III. John Lane, London.

Sukumar, R., 1989. *The Asian Elephant : Ecology and Management*. Cambridge University Press, Cambridge.

Terborgh, J., 1986. 'Keystone plant resources in the tropical forest'. M.E. Soule (ed.), *Conservation Biology—The Science of Scarcity and Diversity*. Sinauer Associates, Sunderland, Mass.

Tewari, K.M., 1981. 'Research Needs of Social Forestry'. *Indian Forester*, 107.

Thakurdas, P., *et al.*, 1944. *Memorandum Outlining a Plan of Economic Development for India*. Penguin, London.

Thapar, R., 1966. *A History of India—Vol. 1*. Penguin Books, Harmondsworth.

——, 1984. *From Lineage to State*. Oxford University Press, Bombay.

Thompson, E.P., 1970-1. 'The moral economy of the English crowd in the eighteenth century'. *Past and Present*, no. 50.

——, 1975. *Whigs and Hunters*. Penguin, Harmondsworth.

Totman, C., 1985. *The Origins of Modern Japan's Forests : The Case of Akita*. University of Hawaii Press, Honolulu.

——, 1989. *The Green Archipelago : Forestry in Preindustrial Japan*. University of California Press, Berkeley.

Trautmann, T.R., 1982. 'Elephants and the Mauryas'. S.N. Mukherjee (ed.), *India : History and Thought. Essays in Honour of A.L. Basham*. Subarnarekha, Calcutta.

Trevor, C.G. and Smythies, E.A., 1923. *Practical Forest Management*. Government Press, Allahabad.

Tucker, R.P., 1979. 'Forest management and imperial politics : Thana district, Bombay, 1823-1887'. *Indian Economic and Social History Review*, 16.

——, 1988. 'The depletion of India's forests under British imperialism'. D. Worster, (ed.), *The Ends of the Earth : Perspectives on Modern Environmental History*. Cambridge University Press, New York.

Tucker, R.P. and Richards, J.F. (eds.), 1983. *Deforestation and the Nineteenth Century World Economy*. Duke University Press, Durham.

USAID, 1970. *A Survey of India's Export Potential of Wood and Wood Products*, vol. I. United States Agency for International Development, Delhi.

Vail, D., 1987. 'Contract Logging, Clear Cuts and Chainsaws'. *Mimeo.* World Institute of Development Economic Research, Helsinki.

Vayda, A.P., 1974. 'Warfare in ecological perspective'. *Annual Review of Ecology and Systematics*, 5.

Vijayan, V.S., 1987. 'Keoladeo national park ecology study'. *Bombay Natural History Society, Annual Report 1987*.

Visveswaraya, M., 1920. *Reconstructing India*. P.S. King, London.

Voelcker, J.A., 1893. *Report on the Improvement of Indian Agriculture*. Government Press, Calcutta.

von Fürer Haimendorf, C., 1943a. *The Chenchus : Jungle Folk of the Deccan*. Macmillan, London.

——, 1943b. *The Reddis of the Bison Hills : A Study in Acculturation*. Macmillan, London.

——, 1945a. *Tribal Hyderabad : Four Reports*, Government Press, Hyderabad.

——, 1945b. 'Aboriginal rebellions in the Deccan'. *Man in India*, 25.

von Monroy, J.A., 1960. *Report to the Government of India on Integration of Forests and Forest Industries*. Food and Agricultural Organization, Rome.

von Tunzelman, G.N., 1981. 'Technical progress during the industrial revolution'. R. Floud and D. McCloskey (eds.), *The Economic History of Britain Since 1700*, vol. 1 : 1700-1860. Cambridge University Press, Cambridge.

Wade, R., 1988. *Village Republics : Economic Conditions for Collective Action in South India*. Cambridge University Press, Cambridge.

Wallerstein, I., 1974. *The Modern World System*. Academic Press, New York.

Walton, H.G., 1910. *British Garhwal : A Gazetteer*. Government Press, Allahabad.

Ward, H.C., 1870. *Report of the Land Revenue Settlement of the Mundlah District of the Central Provinces (1868-9)*. Government Press, Bombay.

Webb, W.P., 1964. *The Great Frontier*. University of Texas Press, Austin.

Webber, T.W., 1902. *The Forests of Upper India and Their Inhabitants*. Edward Arnold, London.

Whitcombe, E., 1971. *Agrarian Conditions in Northern India, Volume I : The United Provinces Under British Rule, 1860-1900*. University of California Press, Berkeley.

White, L., 1967. 'The historical roots of our ecological crisis'. *Science*, 155, 1203-5.

Wilkinson, R.G., 1988. 'The English industrial revolution'. Worster, D., *The Ends of the Earth : Perspectives on Modern Environmental History*. Cambridge University Press, Cambridge.

Womack, J., 1969. *Zapata and the Mexican Revolution*. Alfred Knopf, New York.

World Commission on Environment and Development, 1987 (WCED), *Our Common Future*. Oxford University Press, New Delhi.

World Resources Institute, 1987.

Worster, D. (ed.), 1988. *The Ends of the Earth : Perspectives on Modern Environmental History*. Cambridge University Press, Cambridge.

# Index

# Ecology and Equity

*To the memory of*

*Jotiba Phule*
*J.C. Kumarappa*
*Mira Behn*
*Salim Ali*
*Abdul Nazir Saab*

(The tragopan to the Bigshot)

'You say the town is short of water
Yet at the wedding of your daughter
The whole municipal supply
Was poured upon your lawns. Well, why?
And why is it that Minister's Hill
And Babu's Barrow drink their fill
Through every season, dry or wet,
When all the common people get
Is water on alternate days?
At least, that's what my data says,
And every figure has been checked.
So, Bigshot, wouldn't you expect
A radical redistribution
Would help provide a just solution?'

Vikram Seth,<br>
The Elephant and the Tragopan

*Right*
I will not stop cutting down trees
Though there is life in them
I will not stop plucking out leaves,
Though they will make nature beautiful
I will not stop hacking off branches,
Though they are the arms of a tree
Because –
I need a hut

Cherabandaraju<br>
(translated from Telugu by C.V. Subbarao)

# CONTENTS

# LIST OF ILLUSTRATIONS

# ACKNOWLEDGEMENTS

This book uses a variety of sources: aside from published materials, many of the illustrative examples of resource use and abuse come from twenty years of field experience on the part of one of the writers (Gadgil), working with ecosystem people, voluntary agencies, people's science movements and different levels of government. We have also benefited from conversations over the years with friends, too numerous to list individually, working in the spheres of academia, activism and administration.

Helpful comments on an earlier draft of the book came from Anil Gore, S. Bharath, Jayakumar Anagol, Keshav Desiraju, R. Sudarshan, Jessica Vivien, Michael Redclift, Jean Dreze, Dharam Ghai and Siddhartha Gadgil. Aside from typing drafts, Vijayageetha Gadagkar and Prema Iyer helped in many other ways in putting the book together; their assistance has been critical. The photographs were kindly supplied by the magazine *Frontline* (Plates 13 and 20) and Father Cecil Saldanha (all others). Finally, we would like to thank the Ministry of Environment and Forests, Government of India, for institutional support; and UNRISD and its Director, Dharam Ghai, for their patience and encouragement.

The United Nations Research Institute for Social Development (UNRISD) is an autonomous agency engaging in multi-disciplinary research on the social dimensions of contemporary problems affecting development. Its work is guided by the conviction that, for effective development policies to be formulated, an understanding of the social and political context is crucial. The Institute attempts to provide governments, development agencies, grassroots organizations and scholars with a better understanding of how development policies and processes of economic, social and environmental change affect different social groups. Working through an extensive network of national research centres, UNRISD aims to promote original research and strengthen research capacity in developing countries.

Current research programmes include: Business Responsibility for Sustainable Development; Emerging Mass Tourism in the South; Gender, Poverty and Well-Being; Globalization and Citizenship; Grassroots Initiatives and Knowledge Networks for Land Reform in Developing Countries; New Information and Communication Technologies; Public Sector Reform and Crisis-Ridden States; Technical Co-operation and Women's Lives; Integrating Gender into Development Policy; Volunteer Action and Local Democracy: A Partnership for a better Urban Future; and the War-torn Societies Project. Recent research programmes have included: Crisis, Adjustment and Social Change; Culture and Development; Environment, Sustainable Development and Social Change; Ethnic Conflict and Development; Participation and Changes in Property Relations in Communist and Post-Communist Societies; Political Violence and Social Movements; Social Policy, Institutional Reform and Globalization; and Socio-Economic and Political Consequences of the International Trade in Illicit Drugs. UNRISD research projects focused on the 1995 World Summit for Social Development included: Economic Restructuring and Social Policy; Ethnic Diversity and Public Policies; Rethinking Social Development and Social Policy; Social Diversity and Public policies; Rethinking Social Development in the 1990s; and Social Integration at the Grassroots: The Urban Dimension.

## UNRISD

A list of the Institute's free and priced publications can be obtained by contacting the Reference Centre at: the United Nations Research Institute for Social Development, Palais des Nations, 1211 Geneva 10, Switzerland; Tel (41 22) 798 84 00/798 58 50; Fax (41 22) 740 07 91; Telex 41.29.62 UNO CH; e-mail: info@unrisd.org; World Wide Web Site: http://www.unrisd.org

# INTRODUCTION

## I

Snapshots of our living earth relayed by a satellite above have undoubtedly been among the most dramatic images ever encountered by humans. This remarkable image has become commonplace only over the past decade. Let us for a moment imagine that such an image had been available for that momentous year 1947, when India attained freedom. If we were to contrast that fictional image with one captured for us by a satellite today, striking differences would be readily apparent. The earlier image would have much smaller areas of barren rocks and urban sprawl; fewer large bodies of water, but many smaller lakes freer of weeds at the peak of the monsoon; much more extensive tree cover, but a far lower extent of crop fields retaining their greenery even at the height of the summer.

Satellite pictures, real and imagined, help us understand the radical changes in the Indian landscape over the past four decades. It is a landscape in which the natural world is continually being replaced by a world of artefacts: where trees, shrubs and grasses are giving way to plantations and crop fields, roads and buildings; where rivers are being increasingly impounded with waters diverted through underground tunnels to turn giant turbines or merely being disciplined to flow along paths straight and narrow; where old wetlands are being drained and new ones created in the form of waterlogged fields.

For a fuller appreciation of the *human* consequences of this transformation, however, the modern bird's-eye view of the satellite might be supplemented by a more traditional worm's-eye view of life on the ground. A rapid walking tour through the cities and villages of India can help us understand how changes in the landscape reflect more subtle changes in the social fabric, and the ways in which India's intricate mosaic of human groups is coping with these changes.

A tourist walking through the Indian countryside (at rather more than a worm's pace) would be immediately struck by the chronic shortages of natural resources faced by every segment of Indian society. Fisherfolk are faced with the exhaustion of fish stock, shifting cultivators with the declining

1

availability of forest land. Mat weavers are running short of reeds and peasants are short of dung with which to manure their fields. Millions among the urban poor are shelterless and without adequate water supply. Irrigated farmlands are turning saline and whole coconut orchards are dying of disease. Paper mills are starved of their favourite raw material, bamboo, and textile mills are plagued by power cuts. City roads are clogged with traffic and city air is full of noxious fumes. The ever-growing numbers of Indians, their exploding appetite for consumption and their wasteful patterns of resource use have together conspired to ensure that all segments of society are in the midst of one resource crunch or another.

If the bird's-eye view revealed a picture of considerable ecological change, the worm's-eye view converts that image into one of a serious ecological crisis. This crisis is being translated into increasing social conflict, as different groups exercise competing claims on a dwindling resource base. India today is a veritable cauldron of social conflicts, many of which pertain directly to the control and use of natural resources. These conflicts are played out at different levels and with varying intensities. Within numerous scattered small villages, rich farmers and landless labourers fight for access to common grazing ground, while in city slums desperately poor households quarrel over the trickle of water that reaches them from a sole municipal tap. Such localized conflicts usually go unreported, and far better known are resource conflicts that involve large numbers of people and occur over extensive areas. These include, for example, the massive displacement of villages by a chain of dams being built on the Narmada river in central India, or the bitter fight between the politicians of the states of Karnataka and Tamil Nadu over the waters of the river Kaveri.

These conflicts, localized within villages or spread across large regions, provide the backdrop to the vibrant environmental movement in India. This is a movement that has grown rapidly in the past two decades. Indeed, it is the protests of environmentalists, rather than the concerns of the state or of the intelligentsia, that have generated a wide public awareness of the extent of environmental degradation in India, and (what is more important still) of its human consequences.

The numerous local groups comprising the environmental movement have been concerned, above all, with stopping economic activities that destroy the environment and impoverish local communities – be they large dams on the Narmada or magnesite mines in the inner Himalaya. By its very nature this has been a defensive movement, at times little more than a holding operation. In the circumstances it is hardly surprising that the environmental movement in India has not given sufficient thought to the larger processes that are contributing both to ecological deterioration and to social strife. Here some environmentalists have focused too narrowly on individual actors – for example, on forest managers in the case of the Chipko movement against commercial forestry, or on the World Bank as in the case of the Narmada agitation – while others have been content with identifying impersonal, abstract forces such as 'capitalism', 'materialism' or even 'modern Western patriarchal science' as being ultimately responsible for our present predicament. Universally lacking is a proper social-scientific analysis that might locate these individual actors in a wider context, or which would give flesh and bone to broad concepts such as 'capitalism' or 'science'.

## II

The first part of *Ecology and Equity* presents an original theoretical framework for making sense of what is, from an ecological point of view, undoubtedly the most complex society in the world. This is a society that contains within its ranks Stone-Age hunter-gatherers of the Andamans and white-collar *babus* of Delhi, nomadic shepherds of Himachal Pradesh and pavement dwellers of Calcutta, artisanal fisherfolk of Tamil Nadu and purse seine operators of Goa, shifting cultivators of Mizoram and sugar barons of Maharashtra, textile mill owners of Coimbatore and software exporters of Bangalore, fuelwood headloaders of Kumaon and engineers drilling the Bombay High for offshore oil. These varied constituents of Indian society differ greatly among themselves in their access to the resources of the earth. While bureaucrats in Delhi daily watch American soap operas on their Japanese television sets, the vast majority of women in villages of Bihar and Rajasthan can neither read, nor even listen to a cheap transistor radio. While citizens of Bombay have drinking water brought to their taps from rivers dammed tens of kilometres away, women in villages of Saurashtra must trudge long distances to bring home a pitcher of water. While obesity clinics sprout up in Madras and Bangalore, fully one-third of the Indian people cannot afford to buy enough food to keep their body and soul together.

The relentless transformation of the natural world into the world of artefacts, brought out so vividly for us by the satellite, has most asymmetric implications for these different constituents of Indian society. For the many who earn barely enough to fill their bellies, there is little left over to acquire the new goods on the market, be they soaps or blenders, mopeds or TV sets, apples flown in from the Himalaya or flats in high-rise buildings. The bulk of the poor, or even the not-so-affluent, must scratch the earth and hope for rains in order to grow their own food, must gather wood or dung to cook it, must build their own huts with bamboo or sticks of sorghum dabbed with mud and must try to keep out mosquitoes by engulfing them with smoke from the cooking hearth. Such people depend on the natural environments of their own locality to meet most of their material needs. Perhaps four-fifths of India's rural people, over half of the total population, belong to this category, which, following Raymond Dasmann (1988), we call *ecosystem people*.

As the natural world recedes, so shrink the capacities of local ecosystems

3

to support these people. Dams and mines, for instance, have physically displaced millions of peasants and tribals in independent India. Others have fled as forests and, with them, springs have vanished. These people constitute the *ecological refugees* who live on the margins of islands of prosperity, as sugarcane harvesters in western Maharashtra or farm labourers in Punjab, as hawkers and domestic servants of Patna or Hyderabad. As many as one-third of the Indian population probably live today such a life as displacees, with little that they can freely pick up from the natural world, but not much money to buy the commodities that the shops are brimming with either.

The remaining one-sixth of India's population are the real beneficiaries of economic development, which might be defined for the present purpose as the growth of the artificial at the cost of the natural. These beneficiaries are bigger landowners with access to irrigation, these are modern entrepreneurs in pockets of industrialization and the workers in the organized sector, these are the urban professionals – lawyers, doctors, investment bankers – rapidly gaining in wealth and prestige, and these are the ever-growing numbers of employees in government, semi-government and government-aided organizations. They have the purchasing power to buy cars and fly in airplanes, to dress in polyester clothes and feast on the fish, flesh and fruit brought to them from the four corners of the land. Not only do they have the money to pay for these commodities, but they have the clout to use the power of the state to ensure that the goodies come to them cheap, if not altogether free. As prosperous farmers they pay next to nothing for the electricity that runs their pump sets, as city dwellers they pay little for the water brought to them from hundreds of kilometres away. The news they read is printed on paper subsidized by the low rate at which bamboo is supplied to the mills, and the state builds and maintains at its own expense the highways on which ply the lorries that bring them all manners of commodities from great distances. Like their Western counterparts, whom Raymond Dasmann calls 'biosphere people', they enjoy the produce of the entire biosphere, in contrast to the ecosystem people, who have a very limited resource catchment. Devouring everything produced all over the earth, they might equally be termed *omnivores*.

*Omnivores, ecosystem people* and *ecological refugees*: three broad categories, to which we might assign all of India's huge population. These three classes might be distinguished by the size of their respective resource catchments, or by their relative ability to transform nature into artefact. They might be distinguished too by their widely varying powers to influence state policy, or by the degree of control they exercise over their own lives. These are thus categories at once ecological and sociological, and we use them in this book to provide a fresh interpretative analysis of the development experience in India, and of the social conflicts that have come in its wake.

Like all attempts to classify and interpret complex phenomena, this one too would run into the inevitable boundary problems. Is a village schoolteacher, with an assured salary from the state exchequer and perhaps some land, too, an ecosystem person, or is he a rural omnivore? The owner of a small garage in the city lacks the social power to qualify as an authentic omnivore, but can hardly be called an ecological refugee either. These examples could be multiplied; but while recognizing this difficulty, we are nevertheless convinced that a majority of the Indian population is covered by the three categories identified by us. More crucially, from a socio-ecological point of view we believe our categories to be a great improvement on the more conventional ones of class or interest group. This is not to say that the frameworks of class and interest group cannot be fruitfully applied in the study of other societies or other historical contexts. But as we hope to show in this book, our alternative, threefold classification provides a fuller and more convincing interpretation of political, economic and environmental change in contemporary India.

## III

In the second part of *Ecology and Equity* we turn from analysis to reconstruction, arguing for a new environment-friendly agenda for development that we believe to be in the interest of a vast majority of the Indian people. In their own, undoubtedly sincere, opposition to large projects, environmental groups have not thus far spelt out any concrete alternatives to present processes of destruction and deprivation. This might only be consistent with the defensive, almost siege-like position they find themselves in, but environmentalists have not always helped their cause by appearing to Just Say No to everything – be it eucalyptus, large dams or modern science. It has thus been easy for their opponents to dub them as anti-development, as backward-looking, retrograde rabble-rousers.

Despite its vitality and rapid rise to prominence, then, the environmental movement has been unable to contribute creatively to major debates on development policy in contemporary India. These debates have been conducted in the main between the proponents of economic liberalization, who favour a market economy and an outward-looking India fully integrated with the global marketplace; and those who wish to preserve the *status quo ante*, where the state has occupied the commanding heights of the economy and where a largely self-reliant India stands gloriously isolated, economically speaking, from the rest of the world. Yet this is a debate that might be greatly enriched, and given more meaning and relevance to the Indian context, by providing it with an ecological perspective. This does not mean simply balancing the sometimes conflicting objectives of economic growth and environmental protection, but rather, integrating into every step of policy formulation and execution the insights gleaned from an ecological interpretation of the Indian development experience. That, precisely, is what is attempted in the later chapters of this book.

5

We firmly believe that prudent, sustainable use of India's environmental resources is in the interests of a vast majority of India's population. It is therefore entirely possible to construct a development agenda that is at once in the interests of a majority of people and of the country's environment. Since most people appear to be primarily engaged in a pursuit of self-interest, such a programme ought to attract broad-based support. The snag, of course, is that the small minority of people who stand to lose from such a development process are the very people who are entrenched in power today. Therefore the agenda we advocate is bound to run into powerful opposition, at home as well as abroad. But we have faith that in the long run Indian society can move with vigour in the direction of the alternative future that we sketch in the second half of the book.

In this manner, *Ecology and Equity* hopes to provide the Indian environmental movement with a fuller analysis of the processes it has been fighting against; and, what is perhaps more important still, with a vision of what it should be fighting for. A truer picture, then, of the India that is, and a dream of the India that might be.

# Part I

# THE INDIA
# THAT IS

# 1

# CORNERING THE BENEFITS

## THE COLONIAL LEGACY

India is a natural geographical entity bounded by the great ranges of the Himalaya, the Thar desert and the Indian Ocean. Over the centuries, people have flowed in and out of it through passages to the northwest, the northeast and across the seas; but cultural exchanges have been especially intense within the boundaries of the subcontinent. While India has thus been a cultural entity for over two millennia, it was politically constituted as a state under the British colonial regime only about two hundred years ago.

The British who conquered and unified India were at that time the world's premier omnivores, drawing resources of the entire biosphere to their tiny island kingdom. The men presiding over the British Empire perched on chairs of Burma teak at tables of African mahogany, consuming Australian beef washed down with French and Italian wines. Their women were decked in Canadian furs and clothes of Egyptian cotton, dyed with Indian indigo, glittering with diamonds from South Africa and gold from Peru.

These levels of resource consumption among the British elite were attained by draining their many colonies, including of course India, of their natural resources. In order to accomplish this the pattern of land use within India had been so organized as to maximize the revenue it yielded to the British crown, and the commodities it could produce to feed the British economy. Since village communities could not conveniently be held responsible to pay taxes, land either became private property or was taken over by the crown. The privately held lands were primarily cultivated lands that were taxed heavily. In much of north and east India the ownership was handed over to feudal landlords, with the peasantry reduced to the status of much-exploited tenants and sharecroppers. In parts of south and west India cultivators were assigned lands, but, unable to pay the high taxes, quickly became chronically indebted, losing their lands to moneylenders. The peasants were forced to cultivate cash crops such as cotton, jute and indigo to feed the expanding British textile industry; but whatever the crops cultivated, they often could not make ends meet. While more and more land was brought under

9

cultivation, the productivity of agriculture remained utterly stagnant under the colonial regime. There were several disastrous famines under British rule with millions of deaths. With population beginning to grow after the First World War, per capita food production in India was at its lowest ever at the conclusion of the Second World War (see Guha 1992).

Peasants in India cultivate tropical soils, most of which are poor in nutrients. They compensate for this by maintaining a herd of cattle that convert the vegetation on the surrounding non-cultivated lands into manure for the crops. Peasant families must also cook their meals with fuelwood, and construct their huts and weave their baskets from the small timber and bamboo gathered from these lands. Such lands, as well as irrigation ponds, used to be managed for the most part by village communities. This involved restraints on overuse and contributions to maintenance, such as the periodic desilting of ponds by communal labour. However, the British had scant sympathy for community-based management systems. Where they contributed to higher levels of agricultural taxation as with irrigation ponds, they were permitted to continue. Otherwise, as with wood lots and grazing lands, the state took them over, declaring community control illegitimate. These lands were then either dedicated to producing timber on reserved forest lands or were constituted into open-access lands that suffered overuse and degradation. While in theory the reserved forest lands were supposed to be managed in a sustainable fashion, their takeover by the state merely amounted to confiscation, not conservation, as a dissenting colonial official eloquently put it in the 1870s. The tremendously diverse tree growth of India's forests was liquidated to build British ships and lay extensive railway lines. Mixed forests were replaced by single-species stands of a handful of commercially valued trees, such as teak, sal and deodar. This deprived the tribals and peasants of the forest produce they depended on, and even British experts pleaded that a certain relaxation of the wooden-mindedness of forest departments would greatly help the productivity of Indian agriculture and thereby permit higher levels of taxation (Voelcker 1893; Gadgil and Guha 1992).

The British wanted to retain India as a supplier of cheap raw materials and a market for higher-priced manufactured goods. Industry was therefore discouraged, as was scientific and technical education. A few industrial enclaves did develop regardless, such as the textile mills of Bombay and the Tata iron and steel mill in Jamshedpur. But these enclaves merely remained parasitic and did not serve as foci of modernization. The textile workers lived in miserable tenements in Bombay, leaving their families behind in the countryside, so that no broader, educated working class could emerge. The electric power that fed these mills came from hydroelectric projects built by summarily ejecting peasants, whose lands were submerged without due compensation. Little indigenous technological innovation accompanied this industrial development.

The fluxes of resources between India and Great Britain were highly asymmetrical under this regime: Out of India flowed large quantities of biological produce, rice and cotton, jute and indigo, tea and teak, as well as gold and precious stones. These commodities were produced cheaply; tea plantations in northeastern India, for instance, were set up by taking over tribal lands without any compensation; on them worked labour under conditions approximating slavery. There was little processing of these outgoing biological and mineral produce to add value to them. The return flows were of much smaller material quantities of value-added products of British manufacture.

As rulers, the British assiduously gathered and transmitted back home information on India's landmass, its plant wealth, its people and their customs. The topographic survey of India, the compilation of floras of British India, district gazetteers and ethnographic memoirs consumed a great deal of effort on the part of the colonial rulers (Viswanathan 1984; Desmond 1992). But very little technical information flowed from Britain to India. The Indian elite learnt the English language and read English literature. There was, however, a systematic attempt to keep scientific and technical information from reaching India. When J.N. Tata, one of the pioneers of Indian industry, proposed that an Indian Institute of Science be established to promote indigenous industry, he was consistently discouraged by the British administration. But the institute did come into being in 1909 and in its first decade actively helped establish indigenous technical enterprises. Its council then discouraged interaction between the institute and industry, forcing the faculty to concentrate on basic research. In the event, it accomplished hardly any innovative or original research either (Subbarayappa 1992).

The British omnivores were of course assisted in the task of mobilizing and draining the country's natural resources by representatives of the Indian omnivores of pre-colonial times. Indian society is an agglomeration of tens of thousands of endogamous communities, mostly Hindu caste groups, that observed a hereditary division of labour. In this traditional hierarchy the surplus, primarily of the agricultural produce of the countryside, was appropriated by the three upper strata of Indian society: the *Brahmans* (priests), *Kshatriyas* (warriors) and *Vaisyas* (traders). The two lower strata, *Sudras* (a category which included peasants, herders and fisherfolk, and the more skilled or prestigious service and artisanal groups) and erstwhile untouchables (i.e. the least skilled and prestigious service and artisanal groups), made up the bulk of the population and subsisted as ecosystem people. During colonial rule the upper strata helped serve the British objective of appropriating the surplus of the country's resource production, just as they had earlier collaborated with the many Indian chieftaincies and kingdoms in a similar endeavour, albeit at a less intense rate and on a more limited spatial scale. The task of collecting and transmitting agricultural land tax was assumed principally by the warrior and priestly castes, who became owners

11

of large tracts of land. The trader castes became partners of the British in the task of exporting natural resources and importing and distributing the goods of manufacture. The priestly castes came to man the lower echelons of the bureaucracy that ran the state apparatus.

At this time, the function of the state apparatus was relatively simple. It chiefly consisted in the collection of revenue and the maintenance of law and order. A small number of Indians were selectively educated to assist in these functions, but very few had access to science and technology. The Indian masses, whether peasants, fisherfolk, artisans, nomads or tribals, were left out, with very limited opportunities of education or of jobs in the bureaucracy, in the learned professions or in the slowly developing industrial sector. The tiny organized services and industries sector thus remained the monopoly of upper, literate and trader castes whose traditional divide from the primarily rural lower castes was further reinforced. Meanwhile, the colonial administration remained deliberately alienated from the people at large, with provisions such as the Official Secrets Act shielding it from any public scrutiny. This all-powerful state apparatus bore down heavily on the impoverished peasantry, with the once well-organized village-level systems of management and self-governance largely destroyed. Also gone was the traditional basis of subsistence of a large number of artisanal and service communities. Thus weavers were deliberately forced out of work to ensure a market for British textile mills, while river ferrymen went out of business with the building of bridges and nomadic traders with the expansion of roads and railways. A large populace therefore came to depend on cultivation of land and agricultural labour for subsistence, a populace that began to grow rapidly after the First World War, following the stagnation caused by serious famines and epidemics that characterized the period between 1860 and 1920. All this meant that the traditional network of mutual obligations characterizing Indian rural society was seriously disrupted. Of course, different communities in this network had always formed highly inequitable relationships with each other, but now the fabric itself was greatly strained.

One expects an imperial power to withdraw from a colony when the value of resources usurped is no longer attractive enough to offset the cost of such usurpation. When the British conquered India in the late eighteenth and early nineteenth centuries it had a substantial surplus of agricultural production. This surplus had largely disappeared by 1920. With the spurt in population growth that followed this period, a serious deficit developed. India's forest resources had also been seriously depleted by the Second World War (see Guha 1983), while its mineral resources were not very promising. At the same time the Indian elite had become increasingly conscious of the drain of the country's resources, and come to appreciate the possibilities of diverting these resource fluxes in their own interest. A section of the elite therefore took up the cause of Indian independence, a cause that came to be increasingly supported by the wider population as well (Sarkar 1983). For the British this meant that the costs of usurping the dwindling resource base of their colony were becoming excessively high.

## A NATION IN THE MAKING

It was at this low point that the British departed from India in August 1947. State power now passed into the hands of the landowning warrior and priestly castes of the countryside and the priestly and trader castes of the cities. This alliance was committed to the ideal of halting the drain of India's resources abroad. At the same time it was eager to refashion the pattern of resource use to serve its own interests. There were obvious limitations to what could be achieved in the framework of a low-input agrarian economy, one no longer capable of yielding much of a surplus. The solution obviously lay in industrialization; in tapping the energy of coal and petroleum, of hydroelectric power, in producing steel and cement and using the resources so generated to promote manufacture. The way forward also lay in the intensification of agriculture, by irrigating large tracts of land under river valley projects and supplying them with synthetic fertilizer and pesticides. Such a process of development could create a substantial new base of resources whose surplus could support the urban–industrial sector and the rural landowning elite.

This was the model of development India opted for, rejecting the alternative, once offered by Mahatma Gandhi and some of his followers, of crafting an agrarian society of village republics making low levels of demands on the resources of the earth by living close to subsistence (Kumarappa 1938, 1946). The latter path promised no surplus resources that could be channelled to the elite. On the contrary, it called on the central apparatus of the state to surrender its powers in favour of the masses in the countryside. It is possible that the rural masses would have opted in favour of the so-called Gandhian model. Under India's democratic system they had, in theory, an opportunity to do so. But people at the grassroots have so far ended up having very limited influence in India's parliamentary democracy. Under this system politicians are able to go back with impunity on most promises made at the time of election. Once elected they can vigorously pursue their self-interests without hindrance. When political power was quickly monopolized by the rural and urban elite after independence, this elite saw its interests as being far better served by the model of rapid industrial development.

'There is no free lunch in this world' is what the biologist Barry Commoner (1971) once termed the First Law of Ecology. Somebody had to pay for the wholesale intensification of resource use in independent India. In 1947, the state had substantial monetary resources, in part thanks to the contribution made by India in assisting the British in the Second World War. However, the nascent industrial and organized services sector had far fewer resources at its command. Thus the process of intensifying the indigenous use of country's natural resources would have taken off at a rather slow pace if it

13

had had to be paid for only by private enterprise. But there was already a well-established tradition of the British colonial state stepping in to subsidize private enterprise – albeit those units largely under British control. Then the state had readily assigned tribal land to tea plantations and conferred draconian powers over plantation labour to the British estate owners. The state had stepped in to acquire farmland cheaply to set up river valley projects and to quash peasant resistance to such projects. On independence the Indian state elaborated a new model of a mixed economy to take this process of state intervention much further.

This model was inspired in part by Soviet communism. Under Stalin the Soviet state had mopped up the surpluses of the countryside to build heavy industries. This was accompanied by tremendous suffering by the masses of people and degradation of the environment, but at that time the world was little aware of the magnitude of these problems (see Conquest 1968). The Indian political elite under Prime Minister Jawaharlal Nehru held the Soviet Union as the model to be followed, with state-run enterprises taking over the task of producing electricity and steel, fertilizers and radio broadcasts, and running trains and aeroplanes. But unlike in the Soviet Union, private enterprise was allowed to function under state regulation, with freedom to amass private property. Indeed, on the eve of independence leading Indian

*Plate 1* Ecological refugees in urban India are engaged in a desperate struggle for a few buckets of water

industrialists not merely supported but actively urged the acceptance of an interventionist role by the state apparatus (Thakurdas *et al.* 1944).

The process of the intensification of resource use in independent India thus became the charge of a bureaucratic apparatus inherited from the British. This was an apparatus fashioned primarily to better organize the drain of resources from the Indian countryside. The British were interested in acquiring these resources as cheaply as possible. They had no interest in the sustainable use of these resources, viewing with little concern destruction of extensive tracts of forests in the vicinity of railway lines or waterlogging of croplands in new irrigation projects (Howard 1940; Whitcombe 1971). This apparatus with its historical baggage was now put to the service of a new set of political masters.

In this framework, the process of development has come to be equated with the channelling of an ever more intense volume of resources, through the intervention of the state apparatus and at the cost of the state exchequer, to subserve the interests of the urban and rural elite. As a result, state subsidies have become a central element of the development process in independent India. These subsidies have served to lower the prices of many goods and services primarily for the more privileged segments of Indian society. They have also absorbed a large fraction of the costs of transporting these goods and services to the centres of concentration of the country's elite.

*Plate 2* India's elite now reach out to the remotest parts of the country, as on this island of the Andaman–Nicobar chain

# AN ECOSYSTEM PERSPECTIVE ON DEVELOPMENT

## The coast

The process of development in free India may be illustrated in terms of fluxes of material, energy and informational resources among the six major ecological regimes and urban centres of the country, as well as the outside world (see Table 1). Until independence the coastal waters of India were fished by artisanal fisherfolk employing rowing boats and sailing boats, with the fish consumed largely along the coastal towns and villages. Since fish was relatively inexpensive, part of the catch was employed as manure, notably in coconut orchards.

The fisheries development effort following independence supported mechanization, with the introduction of trawlers and purse seiners, the development of cold storage and canning facilities and the promotion of export of marine products. Other developments have included the establishment of a series of chemical industries on the coast, for instance a petrochemicals complex at Baroda and a rare-earth plant near Thiruvananthapuram, and the rapid growth of several coastal towns and cities.

All these developments have involved substantial investments of state funds. Thus purchases of trawlers have been heavily subsidized, and Indian Petrochemicals at Baroda is in the public sector. Calcutta has encroached on

*Plate 3* Mechanized fisheries and chilling and canning plants today support a major export industry of India

the salt marshes towards the coast with the costs of reclamation almost entirely borne by the government. The mechanization of the fisheries initially enhanced the fish catch by permitting fishing in waters further off the coast, and the development of cold storage and canning has greatly increased the movement of fish and shrimps away from the coast into inland towns as well as for export abroad. Indeed, the export of shrimps has become an important source of foreign exchange for the country. But mechanization, processing and transport also means a greatly increased dependence on petroleum and other external energy and material inputs. Furthermore, the mechanized craft have continued to fish close by the shore, although this zone is supposed to be reserved for countryboats. This has resulted in overfishing and a decline in total fish catch in several states. The extensive coastal pollution has further depressed fish catches, for instance around the chemical industries complex at Baroda. Unable to compete with mechanized craft, fisherfolk have been significantly impoverished. With processing and export the prices of fish have risen sharply so that the poorer segments of the coastal population are now denied access to this important source of protein; nor can fish manure any longer be applied to coconut trees. In summary, the state has invested in the transforming of fisheries so that they bring considerable profit to the traders who have newly entered the industry, through the enhanced supply of fish to urban centres in India and to countries such as Japan and the USA. These effects have been documented by Kurien (1978) and Kurien and Achari (1990).

## Inland waters

Fresh waters in India's streams and lakes have long been used for irrigation to enhance agricultural production, and the surplus available to the ruling classes. Evidence of the earliest irrigation works in India dates from the Mauryan Empire two thousand years ago. In south India chieftains promoted the construction of tens of thousands of village tanks in the medieval period (Ludden 1985). The Mughals built canals on the Yamuna. Most of these systems were maintained by local communities, arrangements which continued to some extent through the British period as well. It was under the British, however, that new technologies made possible construction of really large-scale irrigation works. Under the British too, water began to be used for hydroelectric power generation, a major step beyond the traditional water-mills to grind grain that were found in Himalayan villages. However, the colonial rulers had no interest in subsidizing Indian farmers, big or small. While the state stepped in to build dams, it collected irrigation levies that generated adequate levels of returns on the investment. The long-term health of farmlands was, however, ignored, and problems of waterlogging began to plague Punjab and Uttar Pradesh during the period of British rule (Whitcombe 1971; Agnihotri 1993). Power generation was in private hands and (as

Table 1 An ecosystem perspective on development: a summary

| To: | From: | | | | | | |
|---|---|---|---|---|---|---|---|
| | Sea | Inland waters | Forests | Grazing lands | Farmlands | Urban centres | Abroad |
| Sea | — | — | — | — | — | Sewage, industrial effluent, trawlers | Technology of mechanized fishing, fish processing |
| Inland waters | — | — | — | — | Pesticides, fertilizers | Sewage, industrial effluent, dams | — |
| Forests | — | — | — | — | — | Forest-based industry | — |
| Grazing lands | Chicken feed | — | — | — | — | Dairy industry | — |

| | | | | | | | |
|---|---|---|---|---|---|---|---|
| Farmlands | Fish manure no longer available | Irrigation, water, hydroelectric power | Manure, small timber, fuelwood | Dung as manure | — | Fertilizers, pesticides, high-yielding varieties, green revolution technology | Green revolution technology, diesel |
| Urban centres | Fish, shrimp | City water supply, hydroelectric power | Fuelwood, timber, paper, polyfibre, tea, spices, milk | Milk, meat | Grain, sugar, cotton, tobacco, bricks | — | Petroleum, electronic goods, technology |
| Abroad | Shrimp | — | Plywood, polyfibre, textiles, tea, spices | Leather goods | Cotton textiles, crop genetic resources | Trained manpower, software, pharmaceuticals | — |

*Note*: This table depicts the principal material, energy and informational flows within different ecological regions in India and the outside world described in Chapter 3) the British government did help the Parsi industrial house of the Tatas overcome peasant resistance and acquire land cheaply for its hydroelectric projects in the Maharashtra Western Ghats. But the power was sold by the Tatas to Bombay consumers at rates high enough to bring in a tidy profit. However, freshwater fishing was left largely alone under the British regime and problems of water pollution were modest given the comparatively low levels of urbanization and industrial development.

Following independence, the mobilization of water resources was seen as the key to stepping up agricultural productivity as well as the supply of electrical power. Water was also needed to service the rapidly growing urban centres. As urban centres grew and industries developed, water also had to serve to disperse the waste products. Even in the worst years of drought the Indian landmass receives on average 6,000 mm of rain; in years of good rain this climbs to 10,200 mm. Including the water received from the Himalayan watershed this comes to an average of 4,200 billion $m^3$ for the country as a whole. Given a population of 840 million people, this translates to nearly 5,000 $m^3$ per head. These are reasonably adequate levels; but in India this precipitation is largely confined to three to five months of the monsoon season. As a result serious water scarcities can and indeed do develop in the dry season even in a town like Cherapunjee in Meghalaya, which in an average year receives as much as 10,869 mm of rain (Centre for Science and Environment 1987).

Across India water must be impounded, preferably at a height, and transported as and when required to the point of use, whether for irrigation, industrial or domestic use, or power generation. In this old and densely settled country all such activity is bound to deprive some people of access to water or land, while improving resource access for others. For example, the westward diversion of the waters of the River Koyna from near its origin in the Western Ghats to a power station on the west coast has meant lower water supplies for villages lining the bank of the river on its eastward course towards the Deccan plateau. Other villages, mostly populated by small peasants and buffalo herders, were submerged under the dam, the inhabitants thus deprived of the only basis of subsistence they knew of. Subsequently, local people have also suffered from earthquakes triggered off by the impounding of this large body of water. Meanwhile, the power generated from the Koyna hydroelectric project has primarily fed industrial development in distant Bombay (Paranjpye 1981).

Other dams in the Maharashtra Western Ghats have taken water eastwards to irrigate lands (mostly under sugarcane), or to industry and cities. They have nurtured pockets of prosperity while at the same time depriving others of their livelihood. Such projects have also created access to hilly forested areas till then left untouched, thereby triggering deforestation. All these costs have been borne by small peasants, tribals, herders and rural artisans, who have been most inadequately compensated for the losses they have suffered

*Plate 4* River valley projects opening up access to remote areas have triggered deforestation in many parts of India

(Gadgil 1979; Sharma and Sharma 1981). Also ignored are the longer-term costs: for instance, the possibility of earthquakes triggered some decades later by the building up of strains in the geological substrate (Valdiya 1993). The benefits of large dams have gone primarily to the industrialists of Bombay, the sugar barons of western Maharashtra, and urban dwellers in general, who have received power and water at costs far below those incurred by the state exchequer (Singh 1994).

While promoting the construction of large dams, the state apparatus has overseen the collapse of traditional systems of smaller village tanks and distributaries. During the British regime these community-based management systems were sometimes permitted to continue, for they were efficient and made possible collection of land revenue at higher levels. In independent India, the state has no longer considered tax on agricultural land to be an important source of revenue. Instead, the state apparatus has wanted to enhance its control over the resource base. So all over India village tanks have been taken over by the Minor Irrigation Department, leading to a decay of local management systems and the rapid siltation of the tanks themselves (Somashekara Reddy 1988; Shankari 1991; Bandyopadhyay 1987).

The easiest solution to pollution is its dilution with the help of water and air, public goods in theory available to all. This is the solution adopted everywhere in India as growing industrial activity, exploding urban concentrations and the growing use of agrochemicals in intensive agriculture all

21

pump in more and more of polluting substances into India's environment (Alvares 1992). By and large little investment, private or public, has gone into treating industrial effluent or city sewage, and none at all into checking the discharge of agrochemicals. But this has not been without its attendant costs. These include the deprivation of drinking water to people and cattle dependent on natural watercourses, the loss of sources of irrigation water, a decline in fisheries, and, in some cases, even the poisoning of the underground aquifer. These costs have been widely dispersed, mostly borne by the ecosystem people, who must look after their own water needs. The benefits have accrued to those in the organized services–industries sector, the city dwellers, the better-off cultivators using fertilizers and pesticides. Here the pattern is very similar to that of the processes of exploitation of the resources of the sea. The state steps in to enhance resource availability to omnivores. If in the process the resource base of ecosystem people is attenuated or destroyed through pollution, the state barely intervenes. It certainly does not compel omnivores to bear these costs.

*Plate 5* Fly ash blankets the land in the vicinity of coal-fired power plants in many parts of the country

## Forests

The exploitation of India's forests provides a striking example of how state policies have favoured omnivores at the cost of ecosystem people, while at the same time promoting the exhaustive use of a renewable resource. The colonial state had already laid the foundation for this, by taking over land as reserved forests dedicated to the supply of cheap timber to build teak ships, to lay down railway lines, to put up British cantonments and provide for the two world wars. Apart from mills to saw timber for urban housing and furniture, little forest-based industry developed during British rule. These were, of course, only one side of the demands on forest produce, for the masses of India's ecosystem people needed large quantities of fuelwood, small timber and thatch. They built huts out of bamboo slats and stored grain in bamboo baskets. Their agricultural, fishing and hunting implements were all fabricated from plant material. Their livestock grazed extensively in the forest and in some parts tree fodder lopped by hand was important for their maintenance.

But the Indian Forest Department, founded during British rule, and in effect India's single largest landlord, viewed all these needs of the ecosystem people as a burden, as 'biotic' not 'anthropic' pressure, as if the people behind these demands were less than human. (Indeed, forest working plans classified 'man' as one of the 'enemies' of the forest.) Reluctantly some lands were set aside, as revenue 'wastelands', from which ecosystem people were expected to meet their substantial and vital biomass needs. However, ecosystem people had no longer any rights over these lands, only 'privileges', so that all these areas became no man's lands, overused without restraint by all and sundry (Gadgil and Guha 1992).

This framework has changed but little since independence, and such changes as have occurred have come about only in the past few years (see Chapter 7). At the same time the demands for forest resources have risen tremendously as the domestic consumption of paper, polyfibre yarn, plywood, etc. has soared, along with the continuing rural demand for fuelwood, fodder and small timber. Forest resources have been made available for industrial use at throwaway prices: bamboo at Rs 1.50 per tonne, when the prevailing prices were Rs 3,000 per tonne in 1960; at Rs 600 per tonne after much pressure, when the prevailing prices are Rs 12,000 per tonne. Indeed, it was the guaranteed supply of pine resin to a factory in Bareily at subsidized rates, while the locally set up cottage industry received erratic supplies at higher rates, that prompted the Sarvodaya workers of the Alakananda valley to question the working of the Forest Department in the late 1960s. Soon after this came the award of rights to fell trees, valued locally as a source of agricultural implements, to a sports goods factory in distant Allahabad. This precipitated the famous Chipko *andolan* in March 1973, with unlettered peasants threatening to hug trees to prevent them from being cut (Gadgil and Guha 1992; Gadgil 1989; Guha 1989a).

With the debate generated in the aftermath of Chipko, there has come about a shift in the way India's forest resources are being managed. This has included an official acknowledgement, in the National Forest Policy of 1988, that the biomass needs of ecosystem people must have primacy over the

23

commercial demands of omnivores. Some attempts have also been initiated to set up management systems involving local communities so that no man's lands could be brought under a more regulated regime of harvests (Agarwal and Narain 1990; Malhotra and Poffenberger 1989; and for a detailed assessment, see Chapter 7 of this book). But for the most part these policy initiatives have not translated into practice, with omnivores stonewalling all attempts at sharing power with ecosystem people. In the meantime the subsidized supply of forest raw material to industries goes on, while the large masses of ecosystem people continue to meet their biomass needs in a completely unregulated fashion from open-access lands. Thus the growing numbers of ecosystem people and the growing appetite of omnivores impose larger and larger demands on a dwindling resource base, contributing to widespread deforestation. This also means a serious impairment of the ecosystem services of the forests; in soil and water conservation, as a repository of biodiversity and as a means of sequestering carbon (Gadgil 1989).

### Grazing lands

Cattle and sheep, and to a lesser extent buffaloes, goats and pigs, have been major elements of India's rural economy. Grazing on the stubble and straw of crops and on the scrub and forests surrounding villages, they have been a very significant source of nutrients to replenish the poor soils that support much of Indian agriculture. Milk and milk products have been important components of Indian nutrition, and bullock power is critical to farm operations and transport. Traditionally, livestock was maintained by a majority of farmers as part of a mixed agriculture–animal husbandry system, as well as by specialist herders who moved with their cattle or sheep over large distances during the dry season. The farmers grazed their animals on village common lands subject to community control; the nomadic herders too had their own understanding as to the areas to be grazed by different groups and the relationships to be maintained with settled cultivators (Whyte 1968).

A prime focus of the British was on enhancing the revenue from cultivated lands; grazing lands were to them wastelands preferably to be brought under cultivation or converted into reserved forests. Irrigation was used as an important device to achieve the former objective – thus large tracts of grazing lands in Punjab were brought under the plough (Agnihotri 1993). Traditions of grazing cattle in the forests, as by nomadic Gujjars in the western Himalaya, were severely restricted by the constitution of reserved forests. Village grazing lands were treated as revenue wastelands converted into no man's lands subject to overuse.

This maltreatment of the grazing requirements of India's large livestock population has continued after independence. In particular, the roles of livestock in maintaining the fertility of farmlands and as a source of power for tilling and rural transport have been largely neglected. The emphasis has instead been on the use of synthetic fertilizers and the mechanization of agricultural operations and rural transport. The omnivores have, however, been interested in dairy products, in meat – especially of goat and sheep – and in leather goods. State interventions have therefore focused on the more efficient channelling of dairy products, meat and leather to the cities; and leather and leather products abroad. Substantial investments have gone into transporting, storing and processing milk, so that today the city of Bombay can draw its milk not just from much of its own state of Maharashtra but also from Karnataka, Gujarat and Madhya Pradesh as well. This of course implies much larger demands on petroleum and other energy sources. The goat population has increased at rates far exceeding those of other livestock as consumption of goat meat by the growing numbers of omnivores has been on the rise. But this has not been accompanied by attempts at enhancing the availability of fodder. Indeed, the efforts at enhancing agricultural production have invariably led to the introduction of high-yielding varieties with shorter stature and hence to a lower production of straw relative to grain. Nevertheless, the overall increases in productivity in tracts of high-input agriculture have been sufficient to offset this and to increase the total production of straw in these areas. The net result has been a greater dependence on crop residues for fodder, with the establishment of dairy industries in tracts of irrigated agriculture, while in other parts forests, scrublands and grazing lands are being increasingly overgrazed and degraded. As this process of degradation proceeds, the landless and poorer peasants shift to maintaining goats in preference to buffaloes or cattle, thus intensifying grazing pressure on the degrading vegetation (Gadgil, Pillai and Sinha 1989; Gadgil and Malhotra 1982).

With all this, the total livestock population of the country continues to go up, yielding increasing quantities of leather. These leather goods have become a very significant component of the country's export trade. Leather production is accompanied, however, by a serious problem of pollution from tanneries, for example in the cities of Madras and Kanpur (Alvares 1992). Little investment has gone into correcting this problem; and again the cost is borne by poorer people dependent on water and fish from streams and rivers. To sum up, then, state interventions to organize dairies, to provide cattle and goat purchase loans, or to promote leather exports have primarily helped to enhance the availability of dairy products and meat for city dwellers, to generate income for farmers with irrigated lands and to bring in foreign exchange, all at the cost of the degradation of India's forests, grasslands and water bodies.

### Farmlands

A simple approach characterized the British management of Indian agriculture. The British encouraged the bringing of increasing amounts of land under cultivation and under irrigation so as to extract as much revenue as possible from land cess. They also encouraged the production of crops like cotton and indigo to provide cheap raw material for British manufacture. In the process, the Indian peasant lost control over land to large, often absentee, landlords, and was perpetually in debt. Agricultural productivity remained largely stagnant and there were several disastrous famines. Small irrigation systems remained functional under community management and were supplemented by a few larger systems that enhanced agricultural productivity, for example in the dry tracts of Punjab. These large-scale dams were accompanied by problems of displacement of graziers, of waterlogging and salination of cultivated lands, and the outbreak of malaria (Whitcombe 1971, 1982, 1993).

State objectives in the farm sector shifted substantially after independence. A high priority was to enhance food production, along with the production of raw material for agro-based industries such as textiles, sugar and cigarettes. There was also a clamour from the peasantry for relief from debt and for land for the tiller. There were two sets of options open to the Indian state to meet these objectives. First, it could either spread inputs, enhancing agricultural productivity over wide tracts, or furnish concentrated inputs to restricted tracts. Second, it could either push through land reforms, or permit the continuation of large land holdings and share cropping.

In both cases the first of the two options would have favoured the large masses of ecosystem people, the second option the omnivores. But as it happens, by and large options favourable to omnivores have been followed over much of the country. Thus the strategy of the green revolution has been to pump in water, fertilizer, pesticides and high-yielding varieties to selected areas. It is this strategy that has successfully raised the productivity of Indian agriculture over the last quarter of a century. This success has taken the pressure off the need to enhance productivity on a broader base, a path that would have called for land reform. Since independence, the pressure for land reform has come both from Gandhians and from the political left of various hues ranging from Maoists reposing their faith in armed struggle, to the two mainline communist parties, the Communist Party of India (CPI) and Communist Party of India (Marxist) (CPM), which work within the framework of parliamentary democracy. Gandhian efforts have had little long-term impact, and land reform has gone furthest in West Bengal and Kerala, states with long years of communist rule. It has made least progress in the so-called BIMARU states: Bihar, Madhya Pradesh, Rajasthan and Uttar Pradesh (see Herring 1983; Joshi 1975; Jeffrey 1992).

The strategy of pumping in concentrated inputs in limited areas could never have been implemented if the farmers who benefited had to pay fully for inputs. Instead, the state has stepped in to massively subsidize water, power, fertilizer and pesticides. Ecologically speaking there are serious dangers in such an approach: dangers of waterlogging, of salination, of overuse and nitrate pollution of ground water, of the concentration of pesticides in the environment and the decimation of many non-target organisms, of a build-up of resistance in pest populations and thus of large-scale pest outbreaks in monoculture crops (Shiva *et al.* 1991a). Indeed, all these problems have materialized, along with problems of highly inefficient use of the subsidized resources. But in the process not only the chosen farmers, but those generating and directing the subsidized outputs – industry, bureaucrats and politicians – have done well. Moreover, concentrated inputs in restricted areas more easily generate surpluses that can be skimmed off to urban areas. All these subsidies permit the prices of agricultural produce to be kept at low levels. Further subsidies in food supply, primarily available in urban areas, have helped cope with the rapid growth in city populations.

The enhancement in the productivity of Indian agriculture is now shown to have been accompanied by a very low efficiency of resource use. As a result, while productivity per hectare has gone up substantially, productivity per unit of external energy input (for instance, in bringing water to the field, in manufacturing fertilizers, pesticides and so on) has sharply declined (Department of Science and Technology 1990). Especially since India has only a limited availability of petroleum, this has meant a greater dependence on imports. Foreign exchange must be earned to pay for such imports. Here textiles, along with tea, fish and leather, are critical to India's export earnings. The yarn for these textiles comes partly from polyfibre out of eucalyptus plantations that have replaced large tracts of India's natural forests, and partly from the cultivation of cotton. Indeed, there is an aggressive state-sponsored drive to enhance cotton production. But cotton, of all crops, demands very heavy doses of pesticides with all the attendant risks of environmental damage.

Farming in India has traditionally been a system of a rich mix of varieties of cereals and legumes. While wheat and millets and possibly rice were introduced to India from outside, a number of legumes, for example mung and pigeon pea, were domesticated there. Intercropping and rotation of cereals, nitrogen-fixing legumes and the extensive use of cattle dung as a manure were the key to the sustainability of traditional Indian agriculture. The cropping pattern also involved the use of thousands of locally adapted varieties. Modern intensification has destroyed this diversity – its emphasis is the creation, instead, of homogeneous stands of crop varieties that can perform well only if supplied with large amounts of water and heavy doses of nitrogen, potassium and phosphorus and protected by intensive application of weedkillers, insecticides and fungicides. Such systems have been perfected only in the case of a few cereal crops such as wheat, rice and sorghum. In particular, legumes, a most characteristic feature of Indian agriculture, are not a part of these systems of production characterized by high inputs and low diversity. Production of legumes in India has therefore remained more or less stagnant, even as cereal production has been growing along with human population (Bajaj 1982). The result is a great scarcity of pulses, a sharp rise in their prices and protein deprivation for the poor. A rise in the price of fish as this is diverted to the tables of omnivores in India and abroad of course further aggravates this problem.

While the rich diversity of India's crop varieties has been vanishing from the fields of the farmers, it has at least in part been preserved in agricultural research stations in India and abroad, and in seed banks owned by commercial companies. Much of this diversity is now under the control of foreign agencies that gained access to it entirely free of charge under a régime proclaiming it to be the common heritage of mankind. Indeed, India's crop genetic diversity has already contributed greatly to world-wide cereal production. Some years ago, for instance, an insect pest called the brown leafhopper inflicted billions of dollars' worth of damage on the rice crop of southeast Asia. A solution was finally found in the form of a gene conferring resistance to this pest in a rice variety from Pattambi in Kerala, located in the collection of the International Rice Research Institute in the Philippines. This Pattambi gene was obviously worth a great deal but there was no question of any payment under the common-heritage regime. This regime is now being scrapped under pressure from multinational seed companies based in the West, which see tremendous commercial potential in the exploitation of genetic diversity, especially with the advent of modern biotechnology. India has thus exported entirely free of charge a tremendous resource base built up through the efforts of millions of peasant households over thousands of years. It will now have to pay heavily for the new varieties incorporating these genes that the multinational seed companies are likely to produce in coming decades (Shiva *et al.* 1991a).

The developments in India's farm sector may then be summarized as a selective enhancement of agricultural productivity, which has augmented the agricultural surplus at the disposal of the omnivores while increasing the need for external inputs, especially those based on petroleum. These developments have left the bulk of India's peasantry, the dryland farmers and other weaker sections no better off than before as the prices of pulses, their principal source of protein, have skyrocketed. At the same time, the pumping in of large quantities of water and agrochemicals in selected areas has resulted in manifold environmental problems (Alvares 1992).

## Urban industrial enclaves

People obtain a subsistence in very many different ways. As hunter-gatherers, fisherfolk or miners they gather natural resources, partly for self-consumption, partly for exchange. As cultivators or herders they husband crops or livestock. As artisans or employees of an industrial corporation they process a variety of resources to add value to them. As shopkeepers or employees of a trading concern they arrange to distribute resources. As barbers, teachers, priests, bureaucrats, policemen, chieftains and politicians they provide a range of services. The variety of ways in which people thus subsist has multiplied and grown more complicated with technological advances. At the same time the terms of exchange have come more and more to favour activities adding value to resources in the organized industries and services sector in relation to those which simply gather or husband resources.

Britain gained control over India at a stage when the British were forging ahead in the development of technologies that could add value to resources. Their interest, therefore, lay in establishing a relationship with the colony where they could supply value-added commodities in exchange for biological and mineral resources gathered or husbanded within India. This in part required the active suppression of competition, for example of Indian shipbuilders or weavers, who had the competence to produce superior value-added products. In part, such a relationship was secured by ensuring that the new techniques percolated to India very selectively. Telegraph and railway lines were of course laid down in India to promote resource flows, but railway engines and telegraph machines continued to be manufactured abroad (Sangwan 1991). The introduction of these modern communication and transport devices meant that at a stroke a number of activities pursued by a substantial population of pre-colonial India became obsolete. These included the occupations of river ferrymen, itinerant traders and nomadic entertainers. Also suppressed in large measure were activities like shifting cultivation (Gadgil and Guha 1992). During British colonial rule therefore there was a shrinkage in the range of occupations that Indians could pursue, leading to a mounting pressure on cultivated lands. Since the productivity of agriculture grew only at a snail's pace, while population growth gathered pace after the First World War, there were large masses of underemployed people on the eve of independence.

It was in this setting that India began its own drive to industrialize, i.e. to add value to resources on its own, instead of merely exporting raw materials. In this setting labour was abundant, human-made capital, i.e. artefacts employed to add value to resources, scarce. With a large population and a base of natural resources that had been considerably depleted, the country was not especially endowed by nature either. It was clearly by no means easy for industrialization to take off. For it to gather pace industries needed to make profit, which could come from three routes: (a) by charging high prices, which would be possible if competition was suppressed and a sellers' market created; (b) by getting cheap access to power, water, land and raw materials, which would only be possible if the state stepped in to subsidize the cost of these resources which were by no means abundant, and indeed under considerable demand by a large population; (c) by taking advantage of cheap labour, since huge numbers were unemployed.

The soft options were evidently to get the state to create a sheltered market and permit the charging of higher prices than could stand international competition, as well as the passing on of a substantial fraction of the cost of the resources used to the state exchequer. The industrial sector vigorously canvassed for and successfully persuaded the state apparatus to implement such policies. Once such support was assured, it was easiest to go in for capital-rather than labour-intensive technologies, and to concentrate industries in places where the flow of subsidized resources had been well organized. So the possible option of labour-intensive industrial activities dispersed through the countryside, an option that would have been more favourable to the ecosystem people, was never pursued seriously. In deference to the heritage of Mahatma Gandhi, some support did go to village handicrafts, but that too was only in the form of government subsidies, and has slowly eroded over the years (Jain 1983).

The pattern of industrialization in India has therefore been one of a concentration of capital-intensive activities in relatively few areas. These centres, cities like Bombay, Baroda, Madras, Calcutta and Coimbatore, have in consequence been the loci of enormous government investments for

*Plate 6* Railway line overhanging the landslide it has caused. The extensive network of railways and roads in the country has often been a source of environmental damage

30

providing water, power, transport and communication facilities as well as subsidized food supplies. Indeed, in most cases the larger the city, the larger is the quantum of state subsidy flowing in (Maitra 1992). Naturally enough, the country's omnivores are concentrated in such centres. Even the rich landowners whose primary base is the countryside often have another base in the pampered urban areas.

In this scenario, it pays industry to manipulate and bribe politicians and bureaucrats, rather than to worry about technological innovation, efficient resource use or pollution control. So the soft options have been to import technologies and not worry at all about how efficient or environment-friendly these technologies are or were. The inevitable result is that (a) Indian industry is highly inefficient; (b) industry has no commitment to sustainable resource use or the maintenance of a clean environment; (c) employment in industry has grown very tardily. These are the hallmarks of India's high cost – low quality economy.

*Plate 7* Untreated sewage from the city of Bangalore poses a serious health hazard for villagers living downstream

This complex of processes favouring a narrow elite of omnivores at the cost of masses of ecosystem people has created, too, large numbers of ecological refugees in the hinterlands. These have flocked to the cities, to which the state has channelled resources, from all over the country. Of course, ecological refugees cannot become full partners in sharing this largesse. As Eduardo Galeano (1989) has written of their counterparts in Latin America, the ecological refugees in Indian cities 'sell newspapers they cannot read, sew clothes they cannot wear, polish cars they will never own and construct buildings where they will never live'. But they may still be better off in city slums than in declining villages, with marginal access to water and to opportunities of employment, at least in the unorganized sector as domestic servants, hawkers and recyclers of garbage. Or, as in the outskirts of Calcutta, they may learn to cultivate vegetables very efficiently using solid organic wastes and city sewage. In consequence city populations have grown rapidly, not really through growth of economically productive activities, but as parasites subsidized at the cost of the hinterlands.

*Plate 8* Shantytowns of ecological refugees are an inescapable feature of urban India

This, then, is the current social and environmental scenario in India. But such a parasitic development has its natural limits. Over the years the state apparatus has grown to levels at which it no longer has the capacity to subsidize the omnivores outside its fold; most of its resources are now being swallowed by the salaries and perks of government employees and their political bosses. But with no resources left over to patronize other omnivores, or to pacify ecosystem people and ecological refugees, the state apparatus is in trouble. The only recourse left to it is to beg for resources from abroad. But the global omnivores – represented by the World Bank, the International Monetary Fund and transnational corporations – would not make such resources available for nothing. Their interest is to ensure that Indian markets are opened up to their products. Under pressure the Indian state has little option but to concede these demands. Thus the last few years have exposed India's high-cost, low-quality economy to traumatic competition from abroad. As a result of international competition Indians are being alerted for the first time to the massively inefficient process of resource use that has characterized the country's development process. At the same time the protests of ecosystem people and ecological refugees, who have suffered most over the years, are becoming increasingly evident. To the liberal economist's clamour for *efficiency* these protests have counterposed the equally compelling slogans of *ecology* and *equity*.

It is, then, a propitious occasion for us to examine where Indian society and environment is, how it got there, and where it may proceed. It is our purpose to explore these processes further in the succeeding chapters of this book.

# 2

# PASSING ON THE COSTS

## ISLANDS OF PROSPERITY, OCEANS OF POVERTY

In nature, material and energy tend to flow down the gradient. When an iron rod is heated at one end and cooled at the other, heat tends to diffuse from the hotter end. When some coloured water is poured into a bucket full of plain water, the colour gradually spreads throughout the water body. These processes tend to reduce differences, to favour homogeneity. The processes of material and energy fluxes outlined in Chapter 1 do just the opposite. Once a pocket of irrigated cultivation is created, into it flow subsidized fertilizers, pesticides and electric power. Once a pocket of industrialization takes root, it likewise attracts subsidized water, power, raw materials and communication facilities. Significantly, these fluxes tend to be at the cost of the hinterlands. Thus people in the catchment of a dam are legally prevented from using the reservoir water for irrigating their own lands. Refugees from the Koyna dam, which supplies power to the cities of western Maharashtra, cannot themselves afford even kerosene to light their huts at night.

Forty years of planned development has created an India in which islands of prosperity peep out of a sea of poverty. The omnivores inhabiting these islands are securely on firm ground. The bulk of India's ecosystem people are submerged in the sea of poverty. The ecological refugees are hangers on at the edges of the islands of prosperity, somewhat like mud-skipper fishes hopping around on the muddy beaches fringing mangrove islands. From time to time the tide swallows them; they manage to clamber back on to the mud, but can never make it to dry land.

In a democratic society a small minority ought not in theory be able to thrive at the cost of a much larger majority. It does so in India, as in many other countries of the world, by perverting the spirit and subverting the workings of democracy, by managing to concentrate a great deal of power in its own hands. To accomplish this the omnivores have built up an alliance akin to an iron triangle – an alliance of those favoured by the state (industry, rich farmers and city dwellers); those who decide on the size and scale of these favours (politicians); and those who implement their delivery (bureaucrats and technocrats). To illustrate the working of this alliance, take the creation of Salt Lake City, the fashionable new extension of Calcutta. This was once a salt marsh, the northward stretch of the mangrove swamps of the Sundarbans. Enormous investments were needed to drain it, to create dry habitable land and to bring in sweet water. The state made all these investments and created prime real estate abutting the metropolis of Calcutta. The land was then sold to the politically influential and to well-paid bureaucrats at less than 10 per cent of the market price – never mind that the market price itself did not reflect the prior state investments in its reclamation and development. Part of what was so acquired was used by the beneficiaries personally; the rest was sold at huge profits. The land awardees undoubtedly kicked back part of the profits to the politicians and bureaucrats making the awards.

Analogous processes operate all the time: when alignments of new roads and irrigation canals are decided on; when forest raw material is assigned to the paper or plywood industry; when licences to import television sets or to manufacture cars are awarded.

Figure 1 illustrates the operation of this alliance of omnivores, the iron triangle. The prime beneficiaries of this system of state-sponsored resource capture are those in the organized industry and services sectors and the larger landowners in areas of intensive agriculture and horticulture. The state absorbs a large fraction of the costs of water, power, raw material, fertilizers, etc. provided to industry, services and intensive agriculture; it also organizes provision of these resources to the homes of omnivores, again with large subsidies from the state exchequer. This resource capture by omnivores is at the cost of the other five-sixths of the population: the landless labourers, small peasants, rural artisans, herders, countryboat fisherfolk, nomads and tribals; and such of these as have ended up in urban shantytowns.

The masses are persuaded to accept this system through three kinds of devices: by permitting a trickle of handouts to reach them; by ensuring that they remain assetless and uneducated; and, finally, by more active coercion. There is a whole range of rural development programmes that aim to take state-mediated benefits to the ecosystem people; so many indeed that they come to be known only by their acronyms or abbreviations – RLEGP, TRYSEM, IRDP, DPAP and what have you. Some offer straightforward patronage such as large-scale distribution of *dhotis* and saris to the poor. Others provide subsidies for digging a well or purchasing a cow or a goat. Still others generate employment in constructing roads or digging trenches around forest plantations. A portion of the handout does reach the intended beneficiaries, but a substantial fraction is misappropriated by the political-bureaucratic machinery. As a result the programmes often end up unproductively; the farmer may pocket merely half the subsidy due for purchasing a milch cow, the veterinarian the other half for certifying the existence of the mythical cow. There may be false muster rolls of payment for carrying out soil conservation

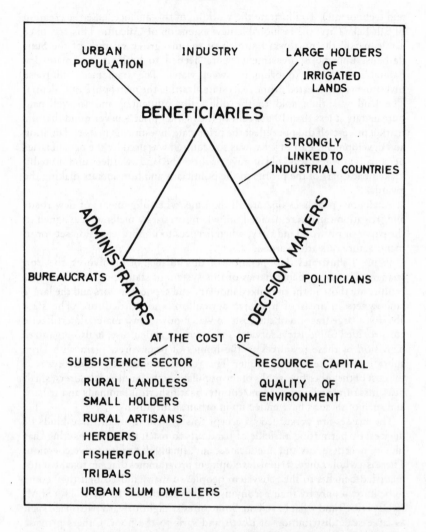

*Figure 1* The iron triangle governing resource use patterns in India. Large state-sponsored subsidies have created an iron triangle of components of Indian society benefiting from, administering and deciding upon state patronage. Constituents of this iron triangle are forcing the country into a pattern of exhaustive resource use at the expense of the environment and a majority of the people

work, itself never adequately implemented in the field. Nevertheless, all citizens of India, poor and rich, have come to look upon the state as a source of free handouts. All are involved in a scramble for a share – minuscule, moderate or large – of these handouts, with few questioning the pernicious system itself.

Through forty-five years of independence the masses of people have largely continued to remain assetless. The most significant asset in a predominantly agricultural country is land. Land reform, assigning ownership of the land to the tiller, has made hardly any progress over large parts of India. This has ensured that the poor cannot effectively make themselves felt in the political process over much of the country. It has also meant that efforts at enhancement of agricultural production have concentrated on very limited tracts of land, with a wide range of undesirable environmental and social consequences.

After forty-five years of independence too a majority of Indian citizens remain illiterate. Little money proportionately has been made available for primary education, although quite substantial amounts are invested on training engineers and chemists who permanently settle abroad. Again, the money actually made available is ill spent, since very many teachers do not attend to teaching in more remote villages, but simply draw their salaries. The villagers have no control over teachers, who report only to some education officer at the district headquarters.

However, there is every reason to believe that a very substantial proportion of India's rural people are motivated to absorb education if an opportunity presents itself. Such an opportunity arose in the form of the Literacy Mission, a programme to achieve minimal levels of literacy for all Indians that was initiated by the Government of India in 1988. For the first time this mission seriously attempted to involve the voluntary sector in a major national effort. It was rewarded by an excellent response, especially in a number of south Indian districts. The voluntary sector became organized and began to play an active role in bringing in more accountability to the development effort. Likewise, in the state of Karnataka, teacher attendance in village schools dramatically improved when local-level political institutions, known as *mandal panchayats* and *zilla parishats*, were given a measure of real autonomy between 1986 and 1990. However, these developments are evidently not to the liking of omnivores. Thus the *panchayats* and *parishats* were abolished when the Congress Party returned to power in Karnataka in 1989, even as the involvement of the voluntary sector in the National Literacy Mission has been watered down.

While the rural masses have been cheated of access to education on a broad scale, there have been attempts to buy their acquiescence by reserving a fraction of seats in higher education to the scheduled castes, scheduled tribes and some of the other backward castes. In the absence of serious land reform the assetless have no chance of acquiring good-sized parcels of irrigated land, but through such reservation they do have a slender chance of acquiring an education and a job in the organized industry–services sector. Higher education is the one avenue available for outsiders to join the ranks of omnivores. There has therefore been considerable political pressure to extend reservations for backward communities not only in institutions of higher education, but in the form of job quotas in the state (or omnivore) sector as well. Such reservation opportunities mean that a small number of ecosystem people and ecological refugees can make the transition into omnivory, but without in any way affecting the overall distribution of power and resources between the three classes.

## DEMOCRACY AND POWER

Indians can justifiably be proud of being citizens of the world's largest democracy. Except for a brief interlude between 1975 and 1977, there have been reasonably free and fair elections to the state legislatures and national parliament over a forty-five-year period. But this does not mean that the interests of the masses have really prevailed. Since these interests are far more congruent with those of a healthy environment, their subversion by omnivores has often resulted in environmental degradation.

Omnivores have subverted democratic processes on a variety of scales and in a variety of ways. The British initiated and the Indian state continues the assumption of draconian power in the acquisition of land for ostensibly public purposes. These powers, embodied in the Land Acquisition Act of 1894, have been misused to drive peasants off their lands without adequate notice and compensation in order to facilitate a variety of development projects. As early as 1921 the protest against the Mulshi hydroelectric project in western India was precipitated when a British engineer organized the digging of trenches in private fields even before any notice of intention to acquire land had been issued (see Chapter 3 for details). To this day farmers whose lands are being submerged by dams on the River Narmada are protesting that they have not been served due notice, nor have provisions been made for proper rehabilitation following their eviction.

Indian society has had rich traditions of decentralized governance, not least in the sphere of natural resource management. Down the centuries, while rulers came and went, taxing and plundering, India's villages managed their own lands, waters and forests. Recent historical research suggests that pre-colonial community-based management systems had elements of equitable sharing and democratic decision making, and on the whole functioned quite effectively (Guha 1989a; Sengupta 1985). The British dismantled the community management of farmland (where it existed) to facilitate taxation, and took over large areas of community forests as state property. However, they largely left intact the community management of small-scale irrigation works, which facilitated the collection of land tax at higher rates. The Gandhian ideal of reviving India as a land of village republics called for the continuance, and indeed revival, of such community-based management practices (see Kumarappa 1938). Instead, after independence the heavy hand of bureaucracy came down on all such remaining systems, proceeding to dismantle them.

The peninsular Indian Archaean crystalline shield areas of Andhra Pradesh, Karnataka and Tamil Nadu have an intricate network of tens of thousands of tanks constructed over hundreds of years. Over the centuries village communities have maintained these systems through regular desilting and the repair of canals through community labour, the observance of a variety of regulations relating to the maintenance of the catchment areas and the sharing of water which might include assignment of plots for cultivation depending on the availability of water. Use of water from such tanks was regulated by a village functionary authorized and paid for by the community. The systems continued through the period of British rule, although village communities themselves did drastically change through concentration of land in the hands of a small number of people.

On independence it would have been in the democratic spirit to have continued these decentralized management systems, perhaps correcting for excessive domination by large landholders. Instead, the machinery of the Minor Irrigation Department took over the management of virtually all these tanks. This is consistent with the overall drive of the bureaucracy in independent India to take on more and more responsibilities and to enlarge bureaucrats' power and job opportunities, regardless of whether the assigned responsibility can indeed be discharged effectively by them. The experience of the past four decades has been that tens of thousands of tanks so taken over by the state apparatus have almost all fallen into disrepair. The village communities have stopped contributing voluntary labour, while the state is quite incapable of paying for and efficiently executing the functions earlier performed voluntarily. Further, the subsidized supply of synthetic fertilizers, irrigation pumpsets and electrical power has reduced the interest of better-off agriculturists in the silt and water from the tank. So the irrigation authorities are now proposing simply to write off this huge and ancient network of tanks, while urging that the state instead go on constructing newer, bigger irrigation projects (Shankari 1991; Sengupta 1985; Singh 1994; Von Oppen and Subba Rao 1980).

Not all community-based tank management has yet been abandoned and in a number of villages these systems continue, albeit without the co-operation of the official machinery. A proportion of these community-based tanks have also fallen into disrepair, but a significant number of them continue to be managed well. Unfortunately this invites not encouragement but, rather, efforts by the official machinery to disrupt their functioning. Thus it is reported that the *phad* irrigation system of northwestern Maharashtra was deliberately destroyed by the authorities to ensure that peasants could not function in an autonomous, independent manner in this democratic nation.

Community-managed forest lands, which once existed over large parts of India, were steadily dismantled during the nineteenth century. The process of state usurpation was consolidated in the Indian Forest Act of 1878, under which the colonial government took over massive areas of forest. Over the years the extent of forest appropriated by the state steadily increased until it came to exceed one-fifth of India's total land area (Gadgil and Guha 1992).

As they have been destroyed for over a century, there is little concrete information on pre-colonial community-based forest management systems. Glimpses of how such systems functioned may, however, be obtained in the writings of colonial officials. An officer in the Garhwal Himalaya wrote in the 1920s of how, despite official apathy, customary restrictions on the overuse of forests operated over large areas, with village grazing grounds well maintained and fuel and fodder reserves carefully walled in (see Guha 1989a: 31). A more detailed account can be found in the report by G.F.S. Collins, a British revenue official assigned to review lands available for community use in the Uttara Kannada district of Karnataka. Between 1860 and 1890, these common lands had mostly been taken over as reserved forests, only very limited areas remaining available for community use. But here too, the communities were stripped of all rights of regulating the use of resources on lands retained for their use. These were treated as open-access lands, and anybody could come and harvest produce without local communities being

*Plate 9* A fuelwood depot in Karnataka. Exhaustion of open-access forests in the neighbourhood of villages is forcing the government to arrange for supplies of fuel from more distant forest areas

in a position to control or monitor use. This resulted in the overuse of these tracts of land, which was used in turn to justify the progressive conversion of community lands into reserved forests. Thus the extent of such lands was reduced from 7,185.9 km² to 353.3 km² between 1890 and 1920.

The resulting protests led to a reassessment by Collins in 1920 (Gadgil and Iyer 1989; Gadgil and Subash Chandran 1988; Collins 1921). Collins reported that he found three villages in the coastal Kumta district which had on their own initiative continued systems of community-based management, although such systems had no official backing. The inhabitants of Chitragi, Kallabbe and Halakar had been strictly regulating harvests from forest lands available for community use adjoining their villages, paying for a watchman themselves. As a result these forests had an excellent standing biomass, apparently being used on a sustainable basis. Collins commended these systems and suggested that they be formally recognized through promulgation of a vil.age forest act, so that the system could be implemented elsewhere as well. Such an act was passed in 1926, and the village forest councils given formal authority.

The management of these three village forest councils of Chitragi, Kallabbe and Halakar continued in excellent shape until the 1960s. At that time the provincial boundaries were redrawn and Uttara Kannada was included in the newly formed state of Karnataka. Promptly the Karnataka Forest Department served notices disbanding these village forest councils. By that time the town of Kumta had grown right up to the village of Chitragi, whose inhabitants accepted the dissolution of the forest council. This was followed by a quick spurt of felling and total destruction of that forest tract over a period of three months, the state authorities clearly being in no position to arrest this destruction. The people of Kallabbe and Halakar, on the other hand, went to court, challenging the dissolution of village forest councils, and continued to regulate harvests. In the mid-1970s, while the court case was still pending, a timber contractor for a plywood mill began to extract wood from the Kallabbe forest. When the secretary of the village forest council asked the help of the Forest Department in preventing these commercial harvests, the authorities instead backed the timber contractor. Thus have arms of the iron triangle, in this case the Forest Department, the plywood mill and the timber contractors, continued to work with each other, and against the interests of ecosystem people, sustainable resource use and the spirit of democracy (Gadgil and Iyer 1989; Gadgil et al. 1990).

Following independence, there have been sporadic attempts to decentralize political authority, to create democratic institutions at the level of *mandals* (= village clusters), *taluks* (= counties) and *zillas* (= districts). Traditionally a council of five or more village elders, the *panchayat*, used to manage community affairs. In consequence, village-level elected committees are called *mandal panchayats*, and the district-level committees *zilla parishats* (*parishat* = assembly, council). All over the country such institutions have been set up but almost always they have been undermined by an alliance of state and national-level politicians and government officials. This is because the politicians operating at higher levels can afford to ignore the concerns of poorer people at the grassroots, instead working in league with the bureaucracy to pursue the interests of the omnivores. But politicians at the *mandal* and *zilla* level have perforce to take account more directly of the interests of ecosystem people, which are often in conflict with those of omnivores. Hence decentralized political institutions have been systematically sabotaged by the omnivores. Even where they have been set up, they are subject to arbitrary dissolution and indefinite postponement of elections.

There are, however, forces, especially outside the dominant Congress Party, that have from time to time tried to take advantage of the potential support from the masses of people and have pushed for such decentralized institutions. Thus the Janata Dal government that came to power in the state of Karnataka in 1983 made a serious attempt to empower and establish *mandal panchayats* and *zilla parishats* in the state. On returning to power in 1989, Congress moved quickly to suspend the workings of the *panchayats*. However, in the period during which these institutions were functioning, local environmental issues had indeed become much more visible in the political process. Thus the Uttara Kannada *zilla parishat* passed a unanimous resolution opposing the setting up of an atomic power plant in its district, and the Tumkur *zilla parishat* supported the resolution by one of its constituent *mandal panchayats* that it did not want a cement factory to be built in its locality. In both cases the state and central authorities overruled the wishes of the local people.

The only state in which decentralized political institutions have been nurtured over a sufficiently long period is the state of West Bengal. It also happens to be the state with the longest period of stable rule by any coalition of political parties, having been governed at the time of writing for over seventeen years by an alliance of left parties. The leftists who initially came to power with the support of industrial labour and the urban middle class found their hold over this constituency of the omnivores to be rather tenuous. They then turned to building up a rural support base by pushing through agrarian reform (most notably, the recording and enforcing of the rights of sharecroppers) and creating genuinely powerful political institutions at the village and district levels. With the support generated by rural reform the left could capture control of the decentralized political institutions through their village-level political cadre. Once this system has been entrenched the ruling left parties have a strong interest in its maintenance (Kohli 1987).

This system is of course by no means free of corruption. But it is a system in which the interests of the ecosystem people have a significant measure of influence. Only in the state of West Bengal must officials respect and listen to members of the *mandal panchayat*. It is therefore not a coincidence that the involvement of people in management of forest resources has made greater progress in West Bengal than anywhere else in the country (Malhotra and Poffenberger 1989).

India is a federation of states with considerable powers devolving to the state governments. Over the past forty years there has, however, been a growing tendency towards centralization of power, progressively weakening state-level political institutions. There is a widespread belief that this trend has favoured the cause of environment and that state governments are far less likely to promote environmental degradation. But there is little solid evidence to support this view. It is true that in 1982–3, when Mrs Indira Gandhi was Prime Minister, she took the side of environmentalists and against the wishes of the Kerala state legislature halted the construction of the Silent Valley hydroelectric project, which would have destroyed some of the finest rain forest of the Western Ghats. But on a different occasion Mrs Gandhi also bowed to political pressure to permit encroachment over large tracts of rain forest by influential plantation owners in the same state of Kerala. Again, while the Uttar Pradesh Chief Minister, Mr Kalyan Singh, called for a suspension of the controversial Tehri Dam in the Garhwal Himalaya after a massive earthquake in October 1991, the central government refused to accept this suggestion. On the whole, it would appear that both state and central governments tend to be dominated by the same omnivore interests, and little will be gained or lost for the cause of environment if the balance shifts one way or another. On the other hand (as we shall demonstrate in Part II of this book), a great deal is likely to be gained if political power is actually transferred to the masses of ecosystem people through a system of decentralized political institutions fully operative at the village and district levels.

Those who maintain that a strong central government is in congruence with the interests of a healthy environment might also have to reckon with the experience of the Indian Emergency of 1975–7, when democratic processes were in abeyance. It was during this period that the Kudremukh Iron Ore project was initiated in Karnataka as a public-sector undertaking without the state government even having been consulted. This project uses ore poorer in iron content than many others available in Karnataka. The ore is then enriched by grinding it to the consistency of talcum powder, increasing the iron content and converting it into pellets that are marketed. This is an expensive process which has ensured that the company ran losses for more than a decade while other private mining concerns were all making handsome profits. The mines are located in heavy-rainfall areas, much of them covered by a pristine rain forest, from where originates the Bhadra, an important river of peninsular India. With huge amounts of waste of the consistency of talcum powder, the project has caused serious siltation of the Bhadra river and the reservoirs, and pollution of the sea near the site of the pelletization plant on the west coast. The mines have also opened up access to a rich rain forest tract and rendered possible its destruction. It is perhaps not an exaggeration to say

*Plate 10* Prior to initiation of extraction of the ore, natural grassy banks and evergreen forest graced the site of the Kudremukh mines in Karnataka's Western Ghats

*Plate 11* The Kudremukh iron ore mines have laid to waste one of the most picturesque parts of the hills of the Western Ghats

that the project is at once an economic and an environmental blunder of a major magnitude, a blunder perpetrated during the Emergency when public debate of such issues was almost totally suppressed.

The Emergency period also witnessed a fierce coercive drive for population control. This program of forced sterilization caused great resentment and was indeed a major factor in the defeat of the Congress Party when elections were finally held in March 1977. This coercive drive was clearly a major disservice to the goal of halting the growth of India's population, since no politician is willing any longer vigorously to pursue a policy of population control, even by democratic means. Thus there is little to suggest that the suspension of democratic processes during the Emergency made any positive contribution to the cause of the Indian environment. Neither is there evidence from other countries, be it Myanmar or the erstwhile Soviet Union, that dictatorships are anything but abusive of environmental interests.

## THE ECOSYSTEM AND ITS PEOPLE PAY

India has thus become effectively organized as a democracy of the omnivores, for the omnivores, by the omnivores. This is a system in which the interests of the huge numbers of ecosystem people and ecological refugees can be largely ignored. The omnivores can capture resources by using the state apparatus, while passing the costs of resource capture on to the rest of the population. This permits the system to tolerate massive environmental degradation and to use resources in an exceedingly inefficient manner.

The trends in the utilization of bamboo in India provide a graphic illustration of this process. Among the fastest-growing plants in the world, bamboos, with their long fibres and ease of working, find an enormous number of applications throughout the tropical world. Bamboo slats go to make walls, doors, windows of huts and large pillars and beams. Grains are stored in bamboo baskets and bamboo culms are fashioned to form cylindrical vessels and spoons. Bamboos are joined end to end to form irrigation pipes and used in seed drills. Bamboo shoots are a nutritious food and bamboo seeds as good as rice.

But for the early British forest managers bamboo was a pernicious weed. No matter that millions of India's ecosystem people extensively used it, no matter that hundreds of thousands of families of basket weavers depended on bamboos to earn a living. For the British wanted tropical forest lands to produce teak and little else. When forests were cleared to raise teak plantations, bamboo thrived, overtopping teak and reducing its growth. So for many years forest managers prescribed the eradication of bamboo from the forests of India. This policy began to change only when research in the 1920s showed bamboo, with its long fibres and soft wood, to be a very desirable raw material for paper making. But for over half a century after bamboo became a commodity of industrial use, it was still viewed as an inexhaustible resource.

45

It was made available to paper mills at a throwaway price, as low as Rs. 1.50 per tonne to the West Coast Paper Mill in the Uttara Kannada district of Karnataka in 1958. At that time, the market price of bamboo in the cities of Karnataka was of the order of Rs 3,000 per tonne. While poor basket weavers paid a hefty price for this vital raw material, industry was given virtually free access to stocks on reserved forest lands (Gadgil and Prasad 1978).

Economic theory tells us that entrepreneurs will always be interested in maximizing the returns on their investments. At any time there are many alternative avenues by which to accomplish this, and technology is opening up new avenues all the time. Modern enterprises are therefore not necessarily interested in long-term sustainability of a renewable resource, be it bamboo or fish stocks. Indeed, it is rational for them to use up the resource at a rapid rate and to switch to a substitute at a later date if the current profit margin is high enough. Economists therefore suggest that the state should step in to levy a tax and to bring down the profit margins of the use of renewable resources in danger of overuse, so that the users are motivated to use a renewable resource at a rate moderate enough to ensure sustainability (Dasgupta 1982).

But sustainable use of bamboo or fish stocks or ground water is truly important only for the ecosystem people who have no other options – options of moving elsewhere or substituting for a particular resource. Sustainability is of little interest to omnivores who might have plenty of such options. A state acting in the interest of ecosystem people may then be expected to price renewable resources so as to moderate the pace of their commercial exploitation. On the other hand, a state acting in league with commercial interests would subsidize renewable-resource supplies to omnivores, thereby increasing the profit margins and levels of availability to the privileged, while at the same time promoting overexploitation and exhaustion at the cost of the underprivileged.

Indeed, the Indian state apparatus has consistently underpriced renewable resources, or in other ways enhanced the profits realizable from their exploitation. To revert to the example of paper mills and bamboo, the West Coast Paper Mill (WCPM) was not only awarded bamboo at nominal prices, but also granted highly subsidized access to land and water. In theory the WCPM was expected to meet its bamboo resource needs in perpetuity from the district of Uttara Kannada. But this expectation was based on faulty, probably deliberately inflated, figures of resource stocks and a very incomplete understanding of bamboo ecology.

Furthermore, the paper mill contractors had little interest in carefully harvesting bamboo, a few culms at a time from each clump, so as to keep the clumps growing. Rather, they were interested in minimizing harvest costs, and tended to clear-cut the most accessible clumps. The result was that the WCPM very soon exhausted the bamboo resources of Uttara Kannada and then turned to the neighbouring state of Andhra Pradesh. As the bamboo resources of that state became depleted, the WCPM moved further afield to the states of Orissa, Assam and Arunachal Pradesh. As bamboo diminished all over, the mill began to use other softer woods such as *Kydia calycina* and eucalyptus. With a captive market ensured through strict restriction of imports the paper industry could hike up prices and maintain large profits despite the substitution of bamboo by more expensive raw material. While the paper industry has been doing very well, and undoubtedly kicking back part of its profit to politicians and bureaucrats who have favoured it, bamboo exhaustion has hit the rural people very badly. For them the quality of housing has significantly declined, and a valued food resource has all but vanished. More critically, for the basket weavers incomes have drastically diminished, turning many into ecological refugees (Gadgil and Prasad 1978).

This ability to pass on the costs of profligate resource use to the masses also means that there is little concern with obtaining adequate returns on the resources deployed. That is why there is scant concern within the iron triangle with dams that have silted up rapidly and with irrigated lands that have become waterlogged, with high levels of transmission losses in the supply of electricity, low efficiencies of boilers in factories and massive thefts of coal transported on railway wagons. For the users of water, power, coal and transport are not paying the full cost of services, and do not care if the real costs are unduly large. As for those in charge of delivering these services, the greater the tolerance of inefficient resource use, the greater the possibilities of their misappropriating public funds.

## DECOUPLING RESOURCES SPENT FROM SERVICES DELIVERED

So for the state apparatus, which had until very recently taken on the task of delivering everything from water, coal and electricity to classical music and technological innovations, what becomes important is not what the people get, but what the government spends. The state also has had the powers to regulate everything from cutting of trees and building of houses to marketing of liquor and manufacture of soap. Again, given that no influential segment of society really cares for what happens to the trees or who drinks the hooch, the apparatus has developed a vested interest in accumulating regulatory powers to harass and extort, rather than genuinely protect trees or maintain the quality of soap. As a distinguished scholar and civil servant has written, the implementation of development programmes in India is in the hands of 'an unholy alliance of the elite, the educated and the bureaucrat, incapable of eschewing coercion, corruption and fraud' (Mitra 1979). This alliance has then concentrated on accumulating regulatory powers, and spending a great deal of money while delivering very little. Indeed, the whole functioning of the government in independent India seems to have evolved so as to decouple resources expended from services delivered!

The Republic of India has been set up as a confederation of states; instead, the central and state governments behave as confederations of numerous ministries, departments and public-sector corporations. The efficiency of the overall development process obviously depends on a proper co-ordination of the activities of the different agencies. Thus the provision of irrigation must be related to soil conditions, and attempts at forest regeneration to the provision of fodder to livestock. Soil and water conservation activities in a watershed must treat land controlled by revenue, irrigation, forest and public works departments as well as by private holders in a co-ordinated fashion. But despite the existence of bodies like the district development councils, not a shred of such co-ordination exists. Instead, each department has developed a culture of a well-knit, highly organized group pursuing its own vested interests in an independent fashion. Of course, each department does interact with others to carve out the total share of the pie, but to no other useful purpose. So district development councils or state or central-level planning commissions are reduced to the function of accounting for the allocation of funds among different government agencies, and play little role in genuinely integrating the process of development.

There are innumerable examples in our experience of how the lack of such co-ordination leads to considerable resource wastage. All over India, surface irrigation has been developed without reference to soil conditions. So water flows along thousands of kilometres of unlined canals even where the soils are heavy clays such as the black cotton soils. These soils have poor drainage and become waterlogged very readily. With time, salts diffuse to the surface of waterlogged soils leading to salination (N.J. Singh 1992). Since none of this has been taken into account in the design of irrigation systems, despite the existence of government departments like the Soil Survey and Land Use Bureau, huge tracts of lands on either side of irrigation canals have become saline, waterlogged wastelands in many parts of India.

Mahatma Gandhi once called the goat a poor man's cow. Goats are hardy animals capable of grazing on all manners of plants that sheep or cattle or buffaloes will not put a tongue to. Goats can earn supplementary income for the otherwise assetless rural poor, and their numbers have been rapidly increasing with the exploding demand for goat meat from omnivores. The provision of subsidies for the purchase of goats is therefore something that can be readily justified as a part of rural development programmes, especially to help members of tribes and lower castes. Since government agencies love disbursing subsidies, goat subsidies are a popular component of programmes of the Animal Husbandry Departments, including those operating on the Western Ghats hill chain. These have been a regular component of the Western Ghats Development Programme, a special programme of assistance from the central government. But much of the grazing on the Western Ghats is on forest land, and this increased goat population is apt to increase the grazing pressure on forests. There has been no systematic investigation of the effects of grazing on forest regeneration or growth, as is indeed the case for most issues pertaining to forest ecology. However, just as government agencies love disbursing subsidies, they love powers to regulate people's activities. Thus the Forest Departments have been urging a ban on the maintenance of goats in areas with large forests, including all *taluks* (counties) falling under the Western Ghats Development Programme. In this manner, while one wing of the government has been disbursing subsidies to encourage the purchase of goats, another wing has been demanding that goats be banned.

Over the years there has been an increasing realization of the resource wastages that result from this lack of co-ordination between sectors. Here watersheds are evidently an appropriate unit for integrating land-based development activities such as agriculture, animal husbandry, fisheries and forestry. As a consequence, integrated watershed development programmes are coming into vogue in many parts of the country, the state of Karnataka having taken a lead in this matter. But the entrenched government departments are totally unwilling to work with each other in a co-ordinated fashion for this purpose. So the Karnataka government has created yet one more agency, the Watershed Development Board, to work in an integrated fashion in a few specially demarcated localities. There can be no more eloquent testimony to the complete unwillingness of the government machinery to work in a co-ordinated fashion, and to its interest in the proliferation of state agencies and departments (State Watershed Development Cell 1989).

The states of pre-colonial India, as well as the state in British India, invested little except in maintenance of law and order and the collection of taxes. There were, of course, some investments in roads, communications and irrigation, but the management of natural resources depended greatly on private and community effort. (The Indian farmer adapted many new crops brought in from outside by Europeans prior to conquest, such as chilli, potato, tomato, guava and cashew. The spread of chillies was by far the most remarkable; chilli had become a common ingredient of Indian cooking within half a century of its introduction in the early 1500s (Achaya 1993).) Village communities put up and maintained small-scale irrigation works with their own effort, and took care of health services, including vaccination against smallpox. After independence all such activities became a function of the state apparatus, calling for an investment of public funds. New crops and crop varieties are promoted by the state, and since the state cannot desilt irrigation tanks they have fallen into disuse. All development programmes have thus become utterly dependent on the availability of state funds.

Since there is little internal pressure to use this money effectively the state machinery has relentlessly pushed up the costs of delivering all services. A foremost component of this is civil construction: that is, the building of houses and offices, roads, bridges and dams. In charge of such activities is the PWD – the Public Works Department, also informally known as Public Waste

Department or Plunder Without Danger. Costs for construction supervised by the PWD are substantially higher than what private effort would require. Anecdotal evidence suggests that this is because around 30 per cent of the costs are misappropriated by the state machinery, shared among the concerned politicians and bureaucrats. In such a system the PWD has strong vested interests in inflating costs, and the departmental norms are indeed designed more to push up costs than to enhance quality.

As a consequence, the PWD consistently opposes all attempts at innovation that would cut down costs. One such noteworthy attempt was by Laurie Baker, an architect from Kerala. Baker has designed a number of elegant private and public buildings at costs well below PWD norms. In the early 1970s, he built the campus of the Centre for Development Studies in Thiruvananthapuram at costs well below those initially estimated. With the money thus saved, the centre was able to build one of the finest social science research libraries in Asia. At that time, Mr Achutha Menon, a progressive politician, was the Chief Minister of Kerala and thought it would be an excellent idea to use Baker's designs and techniques to economically construct a large number of primary school buildings. He called together a group of educationists and engineers along with Mr Baker to discuss this scheme. The story goes that Baker was virulently attacked by the engineers, while the others were convinced that his ideas were worthy of implementation. In the end the Chief Minister gave up the idea of using Baker's designs, apparently afraid that the government machinery would somehow sabotage the construction, which might even end up in the collapse of some school buildings.

Not that structures constructed at high costs following PWD norms and built under its supervision do not collapse. Several have done so, including the big bridge over the river Mandovi in Goa. Goa was under Portuguese control until 1961 and its citizens tell the story of how an engineer working in the Goa government was at that time sent to jail for several years when a building constructed under his supervision collapsed, leading to the death of four people. But there is no such accountability in independent India, and the collapse of the Mandovi Bridge with nobody held responsible is an unfortunate symbol of the nature of India's development efforts, which have erected a high-cost, low-quality economy plagued by environmental degradation and social inequity.

Given the extreme levels of inefficiency with which development projects are executed, it is not easy to rationalize their being taken up. To get over this problem, project planners tend to greatly overestimate likely benefits. Thus in the late 1950s foresters advocated giving up selection fellings in stands of natural forest, replacing them instead by aggressive plantation forestry, which called for large-scale clear-felling of natural forests to create plantations of fast-growing exotics such as eucalyptus and tropical pine. This approach was adopted in the absence of any careful trials of how successful such human-made plantations were likely to be. Without proper evidence it was estimated that the eucalyptus plantation yields would be between 14 and 28 t/ha/year. On this basis, large tracts of prime rain forest, even primeval sacred groves, of the Western Ghats were cut down, only for it to be discovered that in these high-rainfall areas eucalyptus falls prey to a fungal disease agent and may either die off or grow very slowly. The realized productivities of the eucalyptus plantations so raised were only between 1.5 and 3 t/ha – barely 10 per cent of the projected ones, and well below the productivities of the natural forest so replaced (Prasad 1984).

Many irrigation projects are similarly justified on the basis of greatly exaggerated estimates of the likely enhancement of agricultural productivity. It tends to be assumed, for example, that the farmers will receive reliable water supply, and will follow a recommended pattern of cropping involving use of moderate amounts of water over extensive areas. Neither of these assumptions ever holds in practice. Farmers in the upper reaches end up cultivating water-intensive crops such as sugarcane and rice and using too much water, often leading in the long run to problems of waterlogging and salination, while farmers in the lower reaches receive very inadequate supplies. The net result is a far lower enhancement of agricultural productivity, well below what has been projected. Nevertheless, despite the experience that the kind of cropping pattern prescribed can never be implemented, estimated benefits continue to be inflated, being computed on an imaginary basis.

Benefits of irrigation projects are also overestimated by underestimating rates of siltation and the life of the reservoirs. There has been very little careful research on the rates of erosion and siltation of streams under different patterns of land use in India, and the rates used at the project formulation stage are invariably arbitrarily assumed minimal rates. Several years ago one study by India's Planning Commission revealed that the realized rates were on average 2.17 times greater than the rates assumed at the time of project formulation. Among the worst cases of such an underestimation is the Tungabhadra project in Karnataka. The estimated life of this project, completed in 1953, was 100 years, that is, its dead storage capacity was expected to be fully silted up over such a period. However, by 1983 it was estimated that its live storage had already been reduced to 90 per cent after full exhaustion of its dead storage capacity through siltation.

Several features of the functioning of the state apparatus are consistent with this absence of any concern over resource wastage; indeed, they convey the distinct impression that resource wastage is welcome. These include the extremely irregular releases of funds, the chasing of paper targets and the frequent transfer of government officials. The government financial year runs from 1 April to 31 March, and typically, large amounts of funds are suddenly released towards the end of the financial year, even as late as 29 March, with the expectation that the money will be effectively utilized by 31 March. This is quite clearly impossible so that huge amounts of money are regularly wasted and misappropriated. Indeed, in official folklore the month of March is known as *Loot-mar-ka-mahina* – the month of plunder and robbery.

The gross inefficiency of this system of the sporadic use of funds is brought out by a study of the construction of fuel-efficient wood stoves in the state of Karnataka. The design being propagated through this programme in the period 1984–7 was known as Astraole, and was developed by engineers at the Indian Institute of Science in Bangalore. The fuel efficiency of the village wood stoves in Karnataka is between 12 and 17 per cent; under laboratory conditions the Astraole can generate efficiencies of the order of 42 per cent. In the field the efficiencies are lower, of the order of 32 per cent, since a variety of dishes are cooked and they cannot be so adjusted as to fully use up all the heat generated. The doubling of fuel use efficiency is still a significant enough gain. But its realization depends, while the stove is being constructed, on carefully maintaining the sizes and relative dispositions of the different parts. Careless construction can disrupt the proper flow of fresh and hot air and drastically bring down efficiency gains. There may even be situations where the new stove is worse than old ones if improper construction leads to reverse airflows from under the pots towards the fuel box.

Recent years have witnessed an attempt to propagate non-conventional energy sources such as solar energy and devices enhancing efficiency of energy use such as the Astraole. These drives have taken on the standard form of stimulation through free or subsidized supplies. Once subsidies come in, the government machinery develops a vested interest in pushing through the programme while misappropriating a part of the subsidy. The drive is also fuelled by the setting of targets to be fulfilled within some time limit – such as 10,000 Astraoles to be constructed before 31 March. All of this conspires to promote shoddy execution, which may in fact lead not to better, but worse, performance.

The most serious lacuna in the system is of course the near-total absence of any evaluation of the performance, especially by those who are supposed to benefit from the services generated. Thus villagers who end up using Astraoles in their huts have no authority to evaluate how useful their introduction has been. By and large nobody else evaluates the performance either. However, in the case of the Astraole programme in Karnataka the State Council for Science and Technology did conduct a post-programme evaluation, including interviews with users. Figure 2 sums up the results of these interviews, revealing that in a substantial number of cases the stoves turned in very poor performances. Figure 3 brings out one of the factors behind this, namely the hurried construction of a large number of stoves towards the end of the financial year (Ravindranath *et al.* 1989).

Another serious victim of *targetbaji* – the chase of paper targets at the cost of performance – is India's family planning programme. Individuals are given financial incentives to undergo vasectomy or tubectomy operations, and

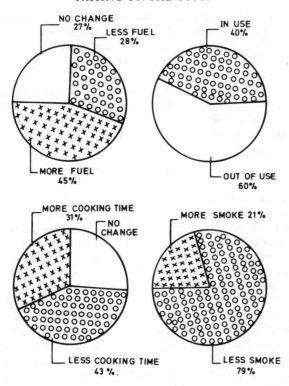

*Figure 2* Results of a study for evaluating the performance of a new fuel-efficient stove design, Astraole, based on a survey of 280 stoves in fourteen districts of Karnataka state in 1988 (after Ravindranath *et al.* 1989)

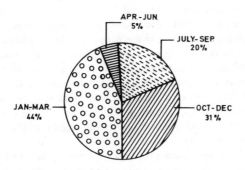

*Figure 3* Time of year of construction of a total of about 153,000 Astraole stoves over the three-year period 1984–7 in the state of Karnataka (after Ravindranath *et al.* 1989)

health centres are assigned targets of so many thousands of operations to be performed per year. A substantial proportion of India's masses are not really motivated to restrict the number of their offspring, who are seen as useful devices to add to the very low family incomes. Meeting targets thus necessarily ends in involving a measure of coercion. On top of this, the lack of performance evaluation and compulsions to meet targets promote careless surgical operations. Shoddy construction of bridges or wood stoves is bad enough but shoddy operations can be a hazard to health and even cause deaths, which can readily turn people against these programmes.

## GETTING AWAY SCOT-FREE

All over India people are of course quite aware of the thoroughly unsatisfactory performance of the state apparatus and the high levels of misappropriation by the iron triangle of politicians, bureaucrats and contractors working in league with each other. The failure to bring the situation under check relates to three factors: lack of evaluation of performance; the difficulty of assigning responsibility for poor performance; and the lack of appropriate machinery to reward good or punish poor performance. We will come back later to the vital issues of lack of performance evaluation and suitable incentives, positive and negative, to promote good performance. Equally significant in encouraging wasteful resource use is the difficulty of assigning responsibility. A main reason for this is the system of frequent and arbitrary transfers. The British introduced this system because they did not want officials to strike roots and develop sympathies with the subject population. The state in colonial times also performed the relatively simple task of maintenance of law and order and collection of revenue – tasks that did not require very much specialized locality- or society-specific knowledge. But the state machinery is now involved in a variety of interventions in complex local ecosystems and social systems, whether it be introducing silkworms, goats, groundnut or coconut, providing drinking water, eradicating guineaworms, eliminating sharecropping or freeing bonded labourers. Effective action therefore calls for a deep understanding of and empathy for local people and their environment. But frequent transfers make this understanding impossible to acquire. Indeed, by the time an official begins to appreciate the local situation he or she is invariably packed off. Thus government programmes become a shifting kaleidoscope reflecting the whims and fancies of a multitude of poorly informed even if sometimes well-intentioned administrators.

Unfortunately, a significant proportion of the government machinery is merely self-serving. The administrators are not responsible to the people they deal with, but only to their political bosses. The political masters have perfected the art of manipulating elections with the help of catchy slogans, communal politics, money power and hooliganism. They are themselves hardly responsible to anybody. Politics in India has thereby become the pursuit of public affairs for private profit. Yet it is notable that no state or national politician has ever been convicted and sent to prison for corruption, although a large number (perhaps even a majority) are certainly guilty on this count. Meanwhile, administrators are not rewarded for being honest, capable, or for doing the job they are assigned to do. Instead they are rewarded for conniving with politicians in extending patronage to people in favour and in misappropriating public resources. In this system transfers are used by politicians to post pliable administrators to where they want them to be, and to remove administrators who will not do their bidding.

Apart from organizing specific services such as delivering water or vocational training, the bureaucracy is also responsible for regulating resource use and abuse, be it the cutting of trees or the pollution of streams. As with delivering services, the self-serving political-bureaucratic apparatus can convert such regulation into an opportunity to further private ends, rather than to render public service. In popular parlance this is known as earning *aankhbandi* allowance – an allowance for looking the other way when a law or regulation is being violated. Thus, just south of the centre of the Western Ghats range lie the Nilgiris or Blue Mountains, a massif rising to an altitude of over 2,400 m. In recent times a Private Trees Protection Act (PTPA) has been promulgated, applying to the Nilgiris district in the state of Tamil Nadu. This act prohibits the cutting of any trees without government permission, even on private lands.

It happens that the Nilgiris are full of tea estates, estates with tens of thousands of silver oak, *Erythrina* and other shade trees. The landowners regularly harvest part of the crop, replacing it with new saplings. With the PTPA they have to run from pillar to post and grease palms to get the required permission. At least, so allege all landowners – from tribals to big estate owners. They say that despite the difficulties they face, the act has in no way reduced the rate of tree cutting. It is just that individual landowners are now incapable of handling the transactions involved. They have to go to timber merchants who arrange to get the necessary permits and of course swallow a good portion of the price, part of which is undoubtedly shared with the authorities. In this manner regulation results not in gains for the environment, but for the politician–bureaucrat–contractor iron triangle.

Pollution control too has apparently degenerated into a similar system of ill-gotten gains for the regulators. Harihar Polyfibres, whose factory discharges effluents into the River Tungabhadra, was embroiled in a long dispute with local people and voluntary agencies who strongly contended that the State Pollution Control Board behaved as if it was an arm of the industry (Hiremath 1987).

## THE MONOPOLY OF PRODUCTION, DISTRIBUTION, INFORMATION

Such an inefficient system can function only if it does not have to compete with other enterprises, Indian or foreign, delivering the same goods and services. Nor can it survive if there are alternative systems for accomplishing the same regulatory objectives. It is therefore only natural that the state apparatus had until very recently assumed monopolistic control over its various functions. Thus government electricity boards and power corporations have had monopolies over the production and distribution of electrical power over most of the country. These systems are notoriously inefficient, with constant breakdowns of power stations, huge transmission losses and rampant thefts of power. Moreover, until very recently they have focused on large-scale, centralized power generation and its distribution over countrywide grids. Such a system cannot effectively deliver power to small, remote hamlets, especially in regions like the Himalaya. Small-scale power generation for local use using hill streams can therefore be a superior alternative in such areas. Indeed, dwellers in such Himalayan villages have traditionally employed water power to grind wheat.

There have been some attempts to use these Himalayan hill streams to generate power on a small-scale, local basis. Not far from the town of Gopeshwar in the Alakananada valley, where many of the activists of the Chipko *andolan* are located, is the village of Tangsa. In Tangsa the villagers have set up a small-scale sawmill using water power to saw timber for local construction purposes. They have been interested in employing the same technology to generate electric power for lighting village streets and houses. But despite many attempts by the Chipko activists, the Uttar Pradesh electricity board stands in the way, preventing any violation of its monopoly over the generation and distribution of power.

The state similarly claims monopolistic rights over all streams, rivers, estuaries and sea waters. Only state agencies such as irrigation departments have therefore the right to dam and divert watercourses, or to auction sand from stream beds, and so on. There are a series of problems with such a management regime. First, decisions such as damming, or auctioning the sand tend to be taken without any reference to the aspirations or preferences of the people involved. Thus those whose lands are to be submerged or acquired for a canal are not advised of this fact even long after the decision has been taken, and may even be forcibly evicted at the eleventh hour without any proper arrangements for rehabilitation or compensation. Nor are the recipients of irrigation water consulted as to the course of distributaries, the amount of water they desire, or the cropping pattern they plan to adopt. The beneficiaries of irrigation by no means pay for the state investment in bringing the water to their fields, investment that may be as high as Rs 52,000 per ha. But by and large they welcome the irrigation water and try to grab as much

*Plate 12* Unregulated removal of sand from river beds has led to serious disruption
of water regimes in many parts of the country

of it as possible. Those in the upper reaches of command areas go in for
cultivating water-hungry crops such as paddy and sugarcane; those in lower
reaches often do not receive the water they were expected to get. The
resultant chaotic pattern leads to very inefficient resource use, which
worsens water scarcity. At the same time, since the beneficiaries do not pay
for the water that reaches them, nobody questions the heavy costs of
delivering this water. The project implementers take advantage of the
situation to hike up the project costs to exorbitant levels.

It is only now that this process of water misuse is beginning to be
questioned. Among the pioneers of such attempts are the farmers of
Tandulwadi in the Sangli distict of Maharashtra. In the 1980s these peasants
ran into serious problems of water shortages. Coming into contact with a
voluntary organization called Mukti Sangharsh, they began to search for the
causes, only to discover that the rainfall records did not indicate any trend of
decline. They then realized that scarcities were primarily due to the new
pattern of cropping involving heavy water demand.

At the same time the farmers realized that the flow of streams through their
village was being adversely affected by the indiscriminate removal of sand by
private contractors. The farmers decided to take matters into their own hands
– to build a dam on their own, using the money raised by selling sand removed
to promote dam construction; to share the irrigation water from the dam on
an equitable basis, even allotting a share to the landless families; and finally

to use the water effectively through a cropping pattern that kept out crops with a heavy demand for water. For the builders of this dam, called Baliraja by the peasants, this involved challenging the state monopoly on more than one count, for in theory they have no rights over the sand or water, except through the mediation of the state machinery. But they have succeeded in challenging this monopoly and the Baliraja dam stands today bringing water to parched fields. Its achievement brought forth this disgruntled, and most· revealing, remark by a minister in the Maharashtra government: 'What will the government have left to do if peasants go on building dams themselves?' (Singh 1994: 221).

Perhaps the most pernicious of all state monopolies is that over information. In the traditional caste society of India, based on a hereditary division of labour, information was compartmentalized in the extreme. The upper castes, especially *Brahmans*, had monopoly over formal knowledge, be it of astronomy, the Ayurveda system of medicine, or religious texts, chants and rituals. The many separate artisanal castes – potters, goldsmiths, weavers – each had exclusive knowledge of materials and techniques pertaining to their specialities. Peasants had knowledge of farming, herders had knowledge of sheep or camel rearing, specialized nomadic castes had knowledge of herbal medicine, and so on. Information was passed on from generation to generation and almost exclusively within kin groups, though occasionally through specialized teaching institutions such as *pathashalas* for training priests in religious texts and rituals. The monopoly was maintained by this tradition of the transmission of knowledge being limited to members of kin groups. But there was coercion as well to ensure monopolistic control. Thus Manusmriti, the Hindu text of the second century of the Christian era, which prescribes codes of behaviour, specifies that no lower caste or *Sudra* should even hear the sacred Vedas of the *Brahmans*. If a *Sudra* commits such an offence he could be punished by having molten lead poured into his ear. There is also the famous story in the Mahabharatha epic of Ekalavya, of a tribal youth who had acquired the knowledge of archery supposed to be limited to high-caste warriors or *Kshatriyas*, and who, once discovered, was forced to lose his right thumb, so that the monopoly would not be broken.

This fragmentation of knowledge, so that the *Sudras* – artisans, peasants and herders – with knowledge of the relevant techniques in their limited domains were prevented from access to the knowledge and technique of other caste groups, as well as from access to the formalized body of knowledge with the priests, was undoubtedly a significant reason why Indian society failed to take a lead in the development of science and technology, despite considerable advances on many fronts. When the British conquered India, they were helped in good measure by their mastery over the newly emerging scientific knowledge and its accompanying techniques. They arrogated to themselves the status of a superior isolated group, like the *Brahmans* of yore, and attempted to maintain a monopoly over science and technology. The British, however, needed to interact with the Indian population in order to maintain law and order, collect taxes and organize a flow of raw materials to Britain and of manufactured articles to India. For this purpose they required allies, who came almost exclusively from the upper castes; that is, from among the omnivores of pre-colonial times. These were provided with an education appropriate to serve their alloted functions, which had little scientific or technical content.

The function of the colonial establishment and its Indian allies was predominantly one of subjugation. It did not want to be questioned by the subjects, and therefore shielded itself behind an Official Secrets Act. The government did of course acquire such information about the subject population as it deemed desirable, as witness the proliferation of district gazetteers, land settlement reports and other official records. But the subjects had absolutely no access to information about what the state was up to. The government communicated with the public only through occasional gazette notifications and government orders.

After independence the state apparatus radically changed its function from that of subjugation to that of all-round development. But this all-round development was only in theory; in practice it became a mix of subjugation on behalf of the Indian omnivores, coupled with the disbursement of patronage trickling down to the ecosystem people and the increasing numbers of ecological refugees. This system can also perform best only by being kept out of the public gaze. So, despite the ushering in of a democratic government, and elected legislatures, the state apparatus continues the way of the colonial state in absolutely refusing to share any information with the public. The Official Secrets Act remains intact, all government papers are marked 'Official Use Only', and members of the public have no access to information of vital interest to them – pertaining to submersion areas of a proposed dam, or plans to clear-fell a particular piece of forest, or of who is eligible to receive subsidies for digging a well, or what were the findings of a government committee that looked into an incident of river pollution resulting in massive fish kills. This monopoly of information comes in very handy in favouring the allies of those in power, as well as in misappropriating public resources.

Over almost five decades now, the omnivores have thus been merrily pursuing the capture of India's resource base. Only fragmentary estimates are available of state resources devoted to enhancing resource supplies through subsidies – most of which ultimately go to omnivores – but the magnitude is staggering, perhaps over 15 per cent of the gross national product. This share has been steadily increasing; over the 1980s it grew at 18 per cent per year, well over the inflation rate (Mundle and Rao 1991). At the same time, the size of the state apparatus in terms of numbers employed in government, joint sector and government-aided institutions has been rapidly increasing, so that salaries and perks of employees are swallowing an ever greater proportion of state funds. All of this is financed to a significant extent by deficit financing creating inflationary pressures. Such inflation primarily hurts the masses outside the charmed circle of omnivores. In part the state operations are financed by external loans, so that the debt burden has also been continually going up. There is increasing pressure to exploit the natural resources of the country, be it iron ore, fish, leather goods or textiles, to service this debt and pay for the exorbitant levels of import of oil. The impact of this export of natural resources is again felt primarily by the ecosystem people, whether through overuse, through diversion of resources traditionally used by them or through pollution-related damage (Kurien 1993). It is little wonder, then, that India has become a cauldron of conflicts directly or indirectly triggered by the abuse of natural resources to benefit the narrow elite of omnivores. It is to a consideration of these conflicts that we now turn.

# 3

# A CAULDRON OF
# CONFLICTS

As the centre of power and patronage, the Indian city of New Delhi is the venue of year-round demonstrations by organizations representing different classes, castes and ethnic groups. Farmers demanding the provision of subsidized power and fertilizer, industrial workers campaigning for higher pay, upper-caste students fighting against job quotas for backward communities, and ethnic minorities fighting for a separate state all recognize the symbolic significance of a show of strength in the national capital. Assured of widespread coverage by the print media, these demonstrations are often held at the Boat Club lawns, a stone's throw both from the Houses of Parliament and from the government secretariat.

The month of May 1990 saw a phenomenon unprecedented even for New Delhi: a demonstration followed, within a week, by a counter-demonstration. First, villagers to be displaced by the massive Sardar Sarovar dam, being built on the Narmada river in central India, assembled in a peaceful yet joyous *dharna* (sit-down strike) on Gol Methi Chowk – in the heart of New Delhi, and very close to the residence of the then Prime Minister, Mr V.P. Singh. Mostly poor peasants and tribals, these ecosystem people of the Narmada valley sat there for several days, singing, dancing and listening to exhortative speeches by their leaders. Most of the demonstrators had come from Madhya Pradesh, the state containing a majority of the villages to be submerged by the dam. They dispersed only after Mr V.P. Singh met a delegation of the protesters, and assured them that the Sardar Sarovar project would be reviewed. Immediately, politicians in Gujarat, the state that stands to benefit most from the project, set about organizing a counter-demonstration on behalf of the omnivores. After a public meeting at the boat club, the Gujarat protesters also went to meet Mr V.P. Singh. The Prime Minister granted them an immediate audience – he had kept the Madhya Pradesh peasants waiting for days – and told them what they wanted most to hear, namely that he and his government were fully committed to the implementation of the Sardar Sarovar project.

A few months later, the two contending groups were involved in a face-to-face encounter hundreds of miles from New Delhi, on the Madhya Pradesh–

61

Gujarat border. On 25 December 1990, the Narmada Bachao Andolan (Save Narmada Movement), an organization working in the interests of the potential oustees of the dam, began a 250-km march from Rajghat in Madhya Pradesh to Kevadia colony, the site in Gujarat of the Sardar Sarovar dam. The marchers, several thousand in all, were stopped by the Gujarat police at the border village of Ferkuva, and prevented from entering the state. On the Gujarat side, a large group, including students and plain-clothes policemen, had assembled to heckle the marchers. A stalemate lasting several days ensued, with the pro-dam agitators – who were addressed by the then Chief Minister of Gujarat, Chimanbhai Patel, on 29 December – raising slogans in favour of the dam and against the Narmada Bachao Andolan and one of its leaders, the respected septuagenarian social worker Baba Amte. For their part, the protesters insisted on their right to march peacefully to the dam site at Kevadia.

On the second day of the new year, a group of twenty-five protesters, with their hands tied to emphasize the non-violent nature of their struggle, entered Gujarat, to be stopped by the police 150 m inside the state. Two more groups, again with their hands tied, joined them the next day. On 5 January 1991, Baba Amte and another group of twenty-five also entered Gujarat. After being allowed to cross the border but not proceed further, they began an indefinite *dharna* (strike) on the Gordah River bridge, a bare 30 m inside Gujarat. The next day, other anti-dam activists, including Medha Patkar, indisputably the movement's most important leader, went on hunger strike

*Plate 13* Medha Patkar leading a *dharna* by Narmada Bachao Andolan supporters and tribal people

on the Madhya Pradesh side of the border. With the Gujarat government unrelenting, the stalemate continued for several weeks. Finally on 28 January Patkar and her associates, with their own lives in danger, were persuaded by other social workers and voluntary organizations to give up their fast (Anon. 1991).

The Narmada controversy is a particularly charged example of a wide spectrum of social conflicts over natural resources in contemporary India. The past two decades have witnessed a rapid sharpening of these conflicts, although of course they have a long history stretching well into the country's past. Notably, these conflicts were muted in the first two decades of independent India, when the state was widely perceived as the authentic legatee of an all-class and genuinely mass-based national upsurge. On independence the state had also changed its character from being an instrument of subjugation and extortion of surplus on behalf of a colonial power, to one extending its patronage to the entire population. With time, however, the Indian state has lost much of its legitimacy, being increasingly seen as an instrument of a narrow elite of omnivores. At the same time, the democratic system has conferred on the growing numbers of ecosystem people and ecological refugees a modicum of political clout. These twin and somewhat contradictory processes, namely, resource capture by the elite and the creation of a space for social protest, have generated numerous conflicts between and among the three broad ecological classes of modern India: the omnivores, the ecosystem people and the ecological refugees.

The conflicts occur on a variety of scales, from encroachment on a plot of grazing land by families of local landless labourers, to widespread protests by thousands of peasants and tribals against displacement by a mega-project like the Narmada dam. They relate to many different natural resources: land for cultivation or grazing livestock; water for domestic, agricultural or industrial use; mineral deposits; fish stocks; woody biomass; or wildlife. Most dramatically they pit haves against have nots: trawler owners against artisanal fisherfolk, tribals against paper mills, peasants against irrigation authorities. But they also pit poor against poor, as village common lands are ruined in a scramble for fuelwood or pots broken in a fight at a public well. They pit rich against rich as in 1991, when Bangarappa and Jayalalitha, the Chief Ministers of Karnataka and Tamil Nadu respectively, engaged in a slanging match and precipitated violence in a dispute over the use of Kaveri waters for the rich farmers and city dwellers of the two states.

## LAND TO THE TILLER

In this chapter we propose to describe a range of such conflicts pertaining to access to and control over natural resources. In an agrarian country such as India the most basic resource is cultivable land, a resource that has become all the more critical as village-based crafts have been collapsing one by one (a

63

process that began with the deliberate sabotage of the Indian weaving industry under British rule, in order to open up markets for the mills of Manchester). This process has continued after independence, with government policies relating to synthetic yarn production and wool weaving further undercutting the viability of a cottage industry that once employed a very significant proportion of India's population (Jain 1983).

A fundamental failure over large parts of India has been that of land reforms. While all land in pre-British India belonged in theory to the crown, peasant communities controlled it very effectively. With the land : person ratio still quite favourable, peasants had the option of moving somewhere else and starting cultivation if a local ruler became particularly oppressive and demanded too large a fraction of surplus agricultural production. Agricultural land was not a commodity commonly bought and sold, so that a peasant could not easily be alienated from his land. Local chieftains collected the surplus of agricultural production, usually in kind, and passed on a fraction to the king. A good bit of the land was also assigned to temples whose controlling priests appropriated the surplus. In this system too an omnivore class of warriors, priests and bureaucrats lived on production appropriated from ecosystem people. But the latter still enjoyed a substantial degree of autonomy and control over their local resource base.

The British land settlements helped tie peasants down to particular pieces of land, demanded high levels of taxes in cash, converted agricultural land into a commodity that could be bought and sold, and created a large class of tenant farmers cultivating on rent land that was owned by a handful of landowning families. This was done more directly in the eastern and northern territories, which were conquered first, with the so-called *zamindari* settlements assigning all ownership rights over large tracts of land to landlords. It was achieved indirectly in the western and southern territories conquered later, when peasants unable to pay the heavy land revenue assessments in cash got into debt and lost their lands to moneylenders, who were chiefly from trading and priestly communities. The colonial landholding systems created a large body of impoverished peasantry, perpetually in debt, in the Indian countryside. Its numbers swelled as large numbers of rural artisans such as weavers, nomadic traders and river ferrymen lost their base of subsistence owing to the import of British goods and improvements in transport and communication. Their numbers grew further after the 1920s as the population, which had declined with a series of disastrous famines and epidemics between 1860 and 1920, assumed an upward trend.

Throughout British rule this mass of ecosystem people remained dependent on a largely stagnant system of agricultural production, many tilling land as sharecroppers in perpetual debt. They backed the national struggle for independence expecting fully that the land would come to the tiller after the British left. Indeed, the Indian National Congress promised such a reform; but the promise has been left substantially unfulfilled, especially in the *zamindari* territories (with the notable exception of the communist-ruled state of West Bengal), where the British had installed landlords on a large scale. Land reform has gone much further in territories where land was initially assigned to the peasants, though many of them later lost it to moneylenders. In these *ryotwari* states of Maharashtra and Karnataka, for instance, land reform has progressed substantially.

The political response to this failure has been twofold: the Gandhian, as represented by Vinoba Bhave and Jai Prakash Narayan, and the socialist, as represented by a whole spectrum of groups from social democrats to revolutionary communists. The Gandhian approach has been largely ineffective. While large amounts of land were voluntarily donated, even whole villages in the early days of the Bhoodan and Gramdan movements, the Gandhians never got down to organize rural society on a new, egalitarian basis. The leftists on the other hand have proved far more effective, albeit on a still limited scale. It is the communist governments of Kerala and West Bengal that have pushed through relatively complete land reforms. In other states like Karnataka and Maharashtra too it has been the pressure from the left that has been instrumental in persuading Congress governments to implement land reform.

The failure to implement land reform over much of the country has fuelled violent leftist movements, beginning with the communist-led peasant revolts of Telangana in the 1940s. More recently, groups working for armed struggle have come to be known as Naxalites, after the uprising in 1967 at Naxalbari in West Bengal. This extreme-left movement has had considerable influence in the tribal areas of West Bengal, Bihar, Madhya Pradesh, Orissa, Andhra Pradesh, Maharashtra and parts of Tamil Nadu and Karnataka. The Naxalites have been instrumental in tribal and poor landless farmers staking claim to agricultural land and forests for cultivation (Naidu 1972; Calman 1985). Although they do control some areas, and even extort taxes (for instance from forest contractors in Bastar), Naxalites have had little direct success in promoting land reforms or local control over forest resources. But their influence has undoubtedly been of significance in the recent moves towards greater involvement of local people in forest management (the history of agrarian movements in twentieth-century India is usefully surveyed in the volumes edited by the sociologist A.R. Desai (1979, 1986)).

This failure to implement land reforms, especially in the fertile agricultural tracts of the Gangetic plains, has meant that enhancement of agricultural production cannot be a broad-based effort. Instead, it has been pursued as an attempt to inject large quantities of outside inputs: water, fertilizers, pesticides, high-yielding varieties over limited areas to produce large surpluses of agricultural production that can be siphoned off to support a growing urban population and agro-based industries such as sugar and cigarettes. This is the 'green revolution', which has primarily occurred in the northern states of Punjab and Haryana. These are areas of relatively low rainfall where pastoralism had been the mainstay of the traditional economy, supplemented to a limited extent by agriculture. Since pastoralism, especially when it involves nomadic movements, is difficult to tax, no major landlords emerged in this region, which initially yielded little revenue for the British government. But the soils of these states are fertile alluvium and there was considerable scope for irrigated agriculture if the tributaries of the Indus and the Ganga could be dammed. A major thrust of British policy therefore was to bring these lands under irrigated agriculture. This was continued more vigorously after independence, converting Haryana and Punjab into the agriculturally most productive areas of modern India, along with parts of the Godavari, Krishna and Kaveri deltas.

But this strategy of boosting agricultural production in limited areas while permitting agriculture to stagnate on fertile lands owned by conservative landlords elsewhere has meant that the masses of ecosystem people of the Indian countryside remain impoverished, while a new class of omnivores has been created from among the large and medium-sized landholders of the areas of 'Green Revolution' agriculture. This is how the landless of Bihar have become migrant agricultural labourers in Punjab, where they have been massacred time and again by terrorists in yet another manifestation of social conflict flowing from inequitable development. This lopsided agricultural development is also accompanied by a series of undesirable environmental consequences, including waterlogging of overirrigated lands, depletion of soil fertility, pesticide pollution, nitrate pollution of the underground aquifer and so on.

The enhancement of agricultural productivity over the past three decades has thus left much of India's rural population out of its ambit. Nor has there been a significant enough increase in employment in the industrial sector. Together this has spawned enormous numbers of people in the Indian countryside with a hunger for ownership of land for cultivation. In the absence of land reform, these people have no recourse but to encroach on common lands, be they village grazing lands controlled by revenue authorities or reserved forest lands. Such pressures have led to a variety of endemic conflicts, each on a small scale, but brewing across the length and breadth of the country. Thus landowners wish to maintain the grazing grounds intact, while landless agricultural labourers, mostly of the lower castes, attempt to occupy these for cultivation. Perched atop the Nilgiri hills at an altitude of 2,400 m is a village called Nanjanad. Most of the village land here is owned by a community of commercially oriented farmers called *Badagas* who have been cultivating potato and tea. Just outside the village is a hamlet of scheduled caste labourers who were working on *Badaga* fields. In 1990 several among them occupied the village grazing lands, in part encouraged because the then District Collector was actively assigning land to lower-caste families. The *Badagas* rose as one and ensured that they were evicted, and the district collector transferred. Similar skirmishes have been reported from many other parts of the country as well.

With enormous inputs concentrated in urban areas, the value of urban land is far higher than that of agricultural land, even if the latter is irrigated. Buildings on urban land require bricks, which can be formed from the soil of agricultural fields, baked with wood from village trees. This makes a quick profit, although it drastically reduces agricultural yields. So on the fringes of all urban centres in India, old banyan and peepul trees, once revered as sacred and therefore never to be cut, or mango trees previously maintained for their fruit, are firing kilns with bricks made of topsoils from the field (Gadgil 1989). There are sporadic protests against these trends, as near the town of Honnavar in coastal Karnataka, where a voluntary agency called Snehakunj is trying hard to stop brick and tile factories from buying the topsoil of paddy fields.

With the price of urban land skyrocketing, there is a scramble to acquire pieces of it. New packages of urban land are continually created by the various urban development authorities, with the land being made available not only below the value commensurate with state inputs into its development, but even below the prevalent market prices. This creates a scramble for such land and an opportunity for politicians and bureaucrats to extend patronage while making a killing. There are scandals around land development in most major cities with accusations of corruption against politicians of virtually all parties. One such celebrated case pertains to housing development proposed in land surrounding the Thippegondanahalli reservoir, which provides 30 per cent of the total supply of drinking water to the city of Bangalore. Land around this reservoir was purchased by forty-two ex-army officers in 1972, when it did not come under the purview of the Land Reforms Act. A special commission granted conversion of land for non-agricultural purposes in 1979 after permission was obtained from the Pollution Control Board and the town planning authority to construct 771 houses. In 1982, however, the Bangalore Water Supply and Sewerage Board appealed against the project on the grounds of the threat of pollution to the reservoir. In response, a firm of Delhi-based promoters put up a proposal for construction of 270 country villas in 414 acres (168 ha) of land around the reservoir, arguing that villas, instead of group housing, would protect the green belt. A single judge quashed the order permitting the formation of the layout in 1987 and this judgment was upheld in 1991 by a division bench of the high court. During Mr Bangarappa's chief ministership of the state of Karnataka a government order was passed approving the formation of a township around the reservoir in contravention of these court judgments. Opposing this order, A.R. Lakshmisagar, a former Law Minister of the state, and several others filed a public-interest litigation. Questioning the validity of the government order, they submitted that this constituted contempt of court and argued that with the formation of the proposed township not only would the quantity of water flowing into the reservoir deplete, but the water would be contaminated as construction workers would flock into the area, forming slums around the reservoir. They also pointed out that the order had been passed in great haste, it having taken only ten days for the files to move from one department to the other. Mr B. Basavalingappa, Minister for Environment and Ecology in Mr Bangarappa's government, also strongly condemned the government order. He accused the promoters of processing their application through the Secretary of Housing and Urban Development, who had nothing to do with the clearance of the project. The high court now has set aside the government order approving the formation of the township, branding the state's action as 'arbitrary and high-handed'. The division bench of the high court remarked that by overruling the previous high court judgment, the government had not only thrown the laws to the wind, but also violated the writs issued by the court. In addition, the high court pointed out that the government order had also contravened the Land Revenue Act.

In all this, ecological refugees flocking to the cities have little chance of proper housing given the enormous prices of urban land. So they create shantytowns on land peripheral to the city without any organized transport, water or power supply. As the numbers of people in such slums mount – and they have come to constitute as much as half the population of the metropolis of Bombay – they form significant vote banks. Political bosses, often doubling as slum lords, then help organize some water, power, transport facilities to such settlements. But as cities expand these once peripheral lands become highly attractive for developing higher-income housing or office buildings and acute conflicts may develop over clearance of slums, so that omnivores can gain fuller control over such land. A major violent incident during the Emergency years of 1975–7 thus involved demolition of slums in Delhi, and real estate developers are suspected to be involved in the Bombay communal riots of 1992–3 which led to large-scale destruction of slum areas.

## DAMS AND THE DAMMED

Along with land, water is the resource in widest demand. Some of the most virulent conflicts have arisen when omnivores attempt to capture water resources by denying ecosystem people access to cultivated land. A major context in which ecosystem people are thrown off their lands is the construction of major dams – the most ambitious of which are the chain of dams being built on the River Narmada. The movements of dam-displaced people have gathered force in the past two decades. We shall come to these contemporary protests presently, but we must first note one important, though as yet little known, precursor. Known as the Mulshi *satyagraha*, this was the opposition to a dam being built on the Western Ghats south of Bombay by the flourishing industrial house of the Tatas. This episode is virtually unknown to Indian environmentalists – and in view of the remarkable parallels between the Mulshi *satyagraha* and ongoing protests against large dams, its history is worth recording at some length.

The Tatas had in fact planned an ambitious series of dams on the Sahyadri hills, chiefly to supply power to the rising industrial city of Bombay. When the first dam was built near the hill station of Lonavala, the farmers and herders whose lands were submerged were paid no compensation whatsoever. But when the Tatas came to Mulshi for the next phase of the project, they ran into trouble. At first, the company moved on to the farmers' lands and dug their test trenches without any legal formality. But Mulshi was very close to Pune (Poona), then an epicentre of the Indian freedom movement. So when a peasant objected to a trench being dug in his field and a British engineer threatened him with a pistol, there were strong protests in Pune.

The ensuing opposition to the dam was led by a young congressman, Senapati Bapat. Bapat and his followers succeeded in halting construction of the dam for a year. The Bombay government then promulgated an ordinance whereby the Tatas could acquire land on payment of compensation. Now the resistance to the dam split into two factions. Whereas the *Brahman* landlords of Pune, who owned much of the land in the Mulshi valley, were eager to accept compensation, the tenants and their leader, Senapati Bapat, were totally opposed to the dam project. With the landlords, the power company and the state all ranged against them, there was little the peasants could do, and the movement collapsed in its third year. Tragically, the compensation was pocketed by the landlords, and the actual tillers of the soil were left high and dry. However, the movement at least succeeded in forcing the Tatas to provide reasonable negotiated compensation for the submerged lands. In consequence, they did not proceed with the other hydroelectric projects they had intended for the Sahyadris (these were later taken up as state-sponsored projects on independence, with the displaced people still ending up greatly impoverished).

When the Mulshi *satyagraha* broke out, the British District Collector had toured the area, extolling the virtues of the dam. He remarked that the electricity produced by it would light up the latrines of the Bombay *chawls*, the dwelling homes of the city's industrial workers. This drew the sharp retort that the government and the Tatas sought to extinguish wick lamps in thousands of rural homes to light up the latrines of Bombay (Bhuskute 1968).

This exchange, apocryphal as it might be, could well have taken place in 1990 in either Ferkuva or New Delhi between proponents and opponents of the Sardar Sarovar dam. In fact, when the Narmada controversy was at its height, *The Times of India*, whether by accident or design, reproduced in its archive section a report, dated 2 May 1921, on the Mulshi *satyagraha*. Here the paper's correspondent had succinctly represented the main objections to the Tata project, as well as its most powerful justifications. The origins of the Mulshi *satyagraha*, he remarked, lie in:

1 A strong sense of wrong and deep feeling of resentment among the peasantry whose lands are affected by the project, against the government for sanctioning the scheme more than two years ago, without taking them in its confidence, i.e., without consent, knowledge or consultation of the peasant-owners of the land.

2 Suspicion and distrust in both the government and the company, due chiefly to the procedure of acquisition, as to the bonafides of their intentions to award full compensation, or equivalent . . . land somewhere else, and other facilities already enjoyed by them or necessary for fresh colonization.

3 Reluctance to part with the land on account of its extreme productivity, the natural facilities of irrigation and nominal amount of land revenue.

4 Reluctance to part with lands, ancestral homes, and traditional places of worship and see them submerged under water.

5 Natural reluctance in this class of peasantry to emigrate from one place to another.

On the other side the main claims of the project promoters were:

1 One and a half lakh [150,000] [of] electrical horse-power would be created by the Mulshi Peta dam.

2 It would save 525,000 tons [of] coal every year. This quantity of coal at the present rate costs Rs 18,300,000.

3 The saving of coal means a corresponding saving of Rs 10,750,000 worth of fuel to the mill industry of Bombay.

4 The quantity of coal saved on account of the scheme would require 26,250 wagons for transport. These would be saved and utilized for other public purposes.

5 Water once used can be directed for agricultural purposes after electrical power is created.

6 Electricity thus created would give work to 300,000 labourers. If it is utilized for cotton mills, every day 51 lakh [5,100,000] yards would be manufactured.

7 The projected electrification of the Bombay suburban railway lines would give to Bombay city much faster and more frequent trains, thus enabling the development of housing schemes in purer air and healthier circumstances.

Here lies an uncanny anticipation of the ideological roots of the conflicts over large dams that were to erupt half a century later. On the one side, the interests of subsistence-oriented peasants; on the other, the interests of urban centres and industry. When the major push towards river valley projects took place after Indian independence, it was easy to represent the former as static and backward, the latter as dynamic, forward-looking and coterminous with the national goals of progress and development. The villages to be submerged by the new projects were then expected to make way for the larger national interest, the more so as the new schemes (unlike Mulshi) were owned and executed not by private capitalists but by the state itself, the legatee of a broad-based, popular national movement.

Of course, displaced people were not entirely yielding. When the foundation stone of the Hirakud dam in Orissa was laid in March 1946, there were strong protests by the peasants of Sambalpur district who were to be ousted by the project. The dam was to inundate fertile rice tracts as well as substantial reserves of coal and other minerals. When the dam site was visited by the Public Works Department Minister, he was confronted with a demonstration of 4,000 people. Beginning in September 1946, several processions and strikes were organized in Sambalpur town. Section 144 of the criminal procedure code was served, forbidding large gatherings. Defying prohibitory orders, on 12 November a strong procession of 30,000, including many women, marched through the town shouting slogans against the dam. However, the agitation fizzled out shortly afterwards, in part due to the co-option of its leaders by the Congress and the administration (Pattanaik *et al.* 1987; Misra 1946).

A comprehensive if somewhat euphoric survey of the first wave of large dams built in independent India, by the political scientist Henry Hart, also noted the resentment of villagers confronted with the prospect of displacement. Thus in 1953, the residents of the town of Narayan Deva Keri, in present-day Karnataka, hoped desperately that the new reservoir on the Tungabhadra river would not fill up to capacity, thereby sparing their town. Disregarding the warnings of engineers, the townspeople stayed on until the last moment, having to be evacuated in a hurry when surrounded on three sides by water. Despite these signs, there was general agreement, at least among the votaries of dam building, that 'the suffering of the displaced people was for the good of the greatest number', as well as little doubt of the 'willingness of the Indian villager to make way for a nation building project, provided he is convinced that the sacrifice he is called upon to make is unavoidable' (Hart 1956: ch. 12).

It is true that the massive – one might, following Hart, call them heroic – river valley projects of the 1950s met with little opposition. These included the Bhakra-Nangal dam in Punjab, the Tungabhadra project on the Andhra Pradesh–Karnataka border and the Rihand dam in Uttar Pradesh, each displacing tens of thousands of people. And yet, over time the Indian villager was to develop a marked unwillingness to make way for 'nation-building' projects. A major reason for this growing hostility was the actual experience of communities displaced by earlier projects. The resettlement of dam evacuees has uniformly been inadequate: the rates of cash compensation have been low; the promise of land for land has very rarely been fulfilled (and where it has, the new lands are invariably of much poorer quality); not to speak of the difficulties of making a new home in unfamiliar, and often hostile, surroundings (see, for a review, Centre for Science and Environment 1985; Fernandes and Ganguly-Thukral 1988; Ganguly-Thukral 1992). Indeed, the experience of dam oustees in India validates the grim judgement of the anthropologist Thayer Scudder that 'next to killing a man, the worst you can do is to displace him' (quoted in Singh 1994: 259).

A significant acknowledgment of these failures has been the substitution, in recent years, of the term 'displacement' for the euphemistic 'resettlement' in public discussions of this process. Meanwhile, organized opposition to new projects gathered force in the early 1970s, with movements emerging independently in different parts. The most long-standing opposition has been to the Tehri dam, being built on the River Bhageerathi in the Garhwal Himalaya. For a decade and a half, the dam's construction has been opposed by the Tehri Baandh Virodhi Sangarsh Samiti (Committee for the Struggle against the Tehri Dam), a forum founded by the veteran freedom fighter Virendra Datt Saklani. The respected Chipko leader Sunderlal Bahuguna – whose own *ashram* is not far from Tehri town – has been very active in the movement, undertaking several fasts to pressurize the government to stop construction. The objections to the dam relate to the seismic sensitivity of the fragile mountain chain (hence the possibility of a dam burst), the submergence of large areas of forest, agricultural land and the historic town of Tehri, and the threat to the life of the reservoir owing to deforestation in the river catchment (D'Monte 1981; Valdiya 1992). These criticisms have gathered force after the massive earthquake in the upper Bhageerathi valley in October 1991, but the government appears resolved to go through with the dam.

Simultaneously, the other well-known Chipko leader, Chandi Prasad Bhatt, has been leading the resistance to a dam at Vishnuprayag, on the Alakananda river in eastern Garhwal. This construction is taking place very close to the famous Valley of Flowers, and fears that the ecology of the valley would be permanently disturbed are compounded by the geological features of the Vishnuprayag area, one peculiarly prone to landslides (Bhatt 1992). At the time of writing, the Vishnuprayag project, in part owing to such opposition, has been indefinitely put on the shelf.

The participation of Chipko activists in these protests is hardly accidental. Having largely lost their forests to commercial exploitation, Himalayan peasants now face further suffering owing to external pressures on the other resource their hills are abundant in, water. As in the case of forests, the benefits of intensive water exploitation have gone almost exclusively to the inhabitants of the plains. In a comparable fashion, the water-rich and heavily forested tribal areas of central India have also witnessed a surge of opposition to new hydroelectric projects. Three of the more notable movements have arisen in opposition to the Koel Karo and Subarnarekha dams in Bihar, and the Bhopalpatnam-Inchampalli project on the Maharashtra–Madhya Pradesh border. In all these cases, threatened tribal groups have responded spiritedly to defend their homes, by organizing demonstrations and work stoppages – all this in the face of police harassment, beating and other forms of repression by the state (Centre for Science and Environment 1985; Areeparampil 1987).

However, the groups affected by large dams have not always been tribal in origin. One successful movement was in fact led by prosperous orchard owners in the Uttara Kannada district of Karnataka. Here the Bedthi project was abandoned after it was opposed by influential spice garden farmers, largely *Brahman*, whose lands were to be submerged by the project. The Uttara Kannada farmers succeeded in organizing an alternative study and the country's first public appraisal of a development project. After hectic lobbying with political leaders, they managed to force the state government to abandon the dam (Sharma and Sharma 1981).

Another, and more striking, success was the abandonment of the Silent Valley hydroelectric project in the state of Kerala. No human community was to be displaced by this 120-kW dam, which nevertheless did intend to submerge one of the best surviving patches of rain forest in peninsular India. The opposition to the project was led by the Kerala Sastra Sahitya Parishad (KSSP), an organization dedicated to popular science education that has a wide reach and influence in Kerala. This left-leaning movement of school and college teachers here built up an unlikely collaboration with wildlife con-servationists (Zachariah and Sooryamurthy 1994). Each group has its own reasons for opposing the project. While the KSSP emphasized the techno-economic appraisal of energy-generating alternatives, their allies invoked the need for plant and animal conservation. Eventually, Mrs Indira Gandhi's desire to enhance her image among the international conservation community appears to have been critical in the central government's decision to shelve the project (see D'Monte 1985).

There is, then, a considerable history to the movement against dam construction on the Narmada river. The Narmada river valley project – which the writer Claude Alvares has termed the 'world's greatest planned en-vironmental disaster' – is a truly Gargantuan scheme, envisaging the con-struction of thirty major dams on the Narmada and its tributaries, not to speak of an additional 135 medium-sized and 3,000 minor dams (see Kalpavriksh 1988).

With two of the major dams already built, the focus of popular opposition has been the Sardar Sarovar reservoir, the largest of the project's individual schemes. Sardar Sarovar is unique in the history of dam building in India, in that the command area of major beneficiaries lies in one state, Gujarat, while the major displacement will occur in another state, Madhya Pradesh (of the 243 villages to be submerged by the dam, 193 lie in Madhya Pradesh). According to official estimates based on the already outdated 1981 census, over 100,000 people, of whom approximately 60 per cent are tribal, will be rendered homeless (Vinod Raina, personal communication).

As early as 1977, villagers in the Nimad region of Madhya Pradesh began protesting against the prospect of eviction due to Sardar Sarovar. Notably, social activists like Medha Patkar (now one of the Narmada Bachao Andolan's moving spirits) first began work on the proper rehabilitation of potential oustees. The government of Maharashtra had by then declared a very progressive 'land for land' policy for oustees of dams. This policy called for allotment of an equivalent amount of good land for the land being submerged.

But given India's population pressure, nearly all such land is already under cultivation or habitation. The most appropriate (and just) solution would, then, be to acquire surplus land from the larger landholders in the command areas of the dam. These landholders are the prime beneficiaries of investment by the state in enhancing the productive potential of their lands, investments to the tune of Rs 52,000 per ha. But these are people with political clout, and the state machinery is not motivated enough to acquire their surplus lands for distribution among displacees. It was therefore only on realizing that there was no land available in Madhya Pradesh, Maharashtra or Gujarat for the proclaimed 'land for land' policy that Medha Patkar and her colleagues turned to opposing the construction of the dam itself.

Although more than a decade old, the movement has really gathered force only since 1989. Over the past five years it has used a varied repertoire of protest to put forward its demands: *rasta rokos* (the blockade of roads and their traffic); public meetings (including some where oustees have pledged not to leave their homes even if the dam waters rise and drown them); fasts; and demonstrations, especially at state capitals. In one dramatic incident, villagers from the neighbourhood of Badwani town in Madhya Pradesh uprooted stone markers from the dam's submergence area, transported them several hundred kilometres to the state capital of Bhopal and flung them outside the Madhya Pradesh legislature (see *Narmada* 1989–90).

While localized protests have been occurring all along the Narmada valley, wider public attention has been drawn through spectacular events. Two of these have already been mentioned: the congregation outside the Prime Minister's house in New Delhi, and the protest march from Rajghat to Ferkuva. Yet the most successful of these public events was a great rally in the town of Harsud, held on 29 September 1989. Sixty thousand volunteers, mostly of tribal and peasant background, had gathered in Harsud, a town itself destined to come under 15 m of water. Representatives of citizens' groups from all over India also came to demonstrate their solidarity with the Narmada movement. A large public meeting, addressed by Amte, Patkar, Bahuguna and others, culminated in a collective oath to resist the pattern of 'destructive development' exemplified by the Sardar Sarovar dam (Alvares 1989).

Even though it lies in a path of continuity with them, there are several features which help distinguish the Narmada movement from earlier protests against large dams. Two of the most notable are its spread – it has activist groups working in three states and many supporting organizations elsewhere – and its tenacity in the face of government repression. Although the movement itself has been almost completely non-violent, its leaders and participants have been repeatedly harassed, and occasionally beaten and jailed. Again, unlike many other movements, the Narmada Bachao Andolan has been widely, and often sympathetically, covered in the print media, while it also has well-established links with environmental groups overseas. (Japanese environmentalists have persuaded their government not to advance money for the Narmada valley project, while US groups sympathetic to the movement tried hard to convince the World Bank to do likewise.) A final testimony of the movement's vigour is the active counter-movement of omnivores it has generated in support of the dam. Political leaders and social workers in Gujarat have strongly rallied behind the state's rich farmers – who, along with the building contractors, stand to gain most from the project – organizing demonstrations and press campaigns and mounting an ideological offensive wherein the Narmada movement's leaders are portrayed as 'anti-development' and 'anti-national'. The Narmada activists have even been accused of wanting to deny tribals the fruits of economic growth by keeping them in a perpetual state of nakedness, hunger and illiteracy (see Anklesaria Aiyar 1988; *Economic and Political Weekly* 1991).

One most disturbing aspect of omnivore opposition to the Narmada Bachao Andolan (NBA) has been growing state repression. Thus on 29 January 1993, two hundred policemen entered the tribal hamlet of Anjanvara, in the Jhabua district of Madhya Pradesh. Anjanvara was one of the villages which had been resisting government surveys conducted preparatory to submergence. The tribals had been organized by the Khedut Mazdoor Chetana Sangath, a local activist group active in the wider Narmada Bachao Andolan. The policemen who entered Anjanvara beat up some villagers, ransacked their houses and fired eight rounds. Three days later the police mounted another attack, following which they arrested twenty-one villagers.

Nine months later, in November 1993, tribal protesters in the Dhule district of Maharashtra found themselves at the receiving end of police brutality. In the village of Chinchkhedi, police fired forty-six rounds on a demonstration of tribals, killing a 15-year-old boy, Rehmal Punya Vasava. As in Jhabua, the tribals had themselves been militant but unarmed. In the most recent incident of this kind, the Gujarat police refused to intervene when political hoodlums broke into the NBA office in Baroda, beating up social workers and tearing up files.

For all this intense repression, at the time of writing (October 1994) the Narmada controversy is far from resolved. Thus in March 1993, the Indian government withdrew its request for a loan for the project from the World Bank – a pre-emptive move before the Bank itself was likely to have decided, on the basis of an adverse report by a review committee it had appointed, not to support the project (*The Times of India*, 31 March 1993). Some months later, in Manibeli village, Maharashtra, which was destined to be submerged by rising waters during the monsoon, a group of tribals and NBA activists resolved to drown rather than move out. They had to be forcibly removed from Manibeli by a police force. Then in June 1993, Medha Patkar went on a fifteen-day fast in Bombay, which attracted much attention and support. Following her fast, Patkar and her colleagues met with the union government's Water Resources Minister, who promised them a full review and reassessment of the project. Most recently, in March 1994 the new Congress Chief Minister of Madhya Pradesh, Digvijay Singh, proposed a reduction in the height of the dam, so as to lessen the burden of displacement. But his proposal has been dismissed by the government of Gujarat, and construction work continues on the dam.

In areas beyond the Narmada basin, the ecosystem people are losing their control over land in the context of other development and defence projects as well. Among the more notorious of these is the Baliapal missile range project in the Balasore district of Orissa. In the delta of the Subarnarekha River, the Ministry of Defence proposed to build a huge test range for missiles, a project that envisaged the acquisition of 190 sq km of land and would have displaced an estimated 70,000 people. But this is fertile land, with a highly developed agrarian economy based on betel leaves, coconut and fishing. In 1987 and 1988, a popular movement against the test range gathered force in the Baliapal-Bhograi region. The contempt of the peasants for the government was most dramatically manifested when they barricaded all entrances to the region, refusing to pay taxes or allow the entry of government officials. This movement forced a stalemate: the government has not yet been able to proceed with the project, and there has been talk of shifting it elsewhere. The general secretary of the People's Committee against the Test Range succinctly expressed the feelings of the people of Baliapal, and indeed of all those threatened with displacement by large projects, when he remarked: 'No land on earth can compensate for the land we have inherited from our forefathers' (*Indian Express*, 24 April 1988; *Deccan Herald*, 15 May 1988).

A Baliapal-type situation is currently being replicated in Orissa's neighbouring state of Bihar. Here, in the hilly, predominantly tribal and desperately poor district of Palamau, the Indian Army proposes to build a massive test firing range. Stretching over 40,000 ha in Palamau and the adjoining Gumla district, the construction of the range will affect a total of 190 villages. The firing range might also have a powerful negative impact on the biodiversity of the Betla National Park, a Project Tiger reserve. But here too the ecosystem people are resolved not to give up their lands and forests without a fight. Youth activists have formed a *jan sangharsh* (popular struggle) committee, and impressive protest demonstrations have been arranged in the towns of Ranchi, Gumla and Daltonganj. As one tribal told a visiting journalist, 'Hum jan denge lekin zamin nahin denge' ('We shall give up our lives, but not our land') (Sainath 1993).

## WATER AS SOURCE AND SINK

Water, along with land, is the most vital input for agricultural production. In a country in which the majority of people depend on agricultural production, access to water is of critical significance. But water is an equally important resource for many industries, at least as a sink for their waste products.

Drinking water is important for rural as well as urban settlements. Control over water is therefore a major source of social conflicts in India and inequitable control leading to mismanagement of water resources underlies many aspects of India's environmental crisis.

Water comes down as rain. It flows on the surface as well as under the ground. While access to surface water can be readily controlled and manipulated, ground water may be tapped at will by the landowner, so long as he or she has the resources to reach down and lift it. When the rate at which the ground water is withdrawn exceeds the rate at which it is being recharged, the level of the ground water table drops, so that more and more effort is required to get at it. Indians have been tapping ground water through open wells since the days of the Mohenjadaro civilization; but the rate at which this water could be extracted was previously limited by the capabilities of human and bullock power. This limitation has now been overcome with the development of diesel or electric power operated pumps, so that water below the surface can be pulled up at rates far exceeding those of recharge, leading to a rapid lowering and even local exhaustion of ground water. This opens up possibilities for conflict, with those capable of reaching down further and deploying more powerful pumps depriving others of access to ground water. With the state stepping in to subsidize digging of bore wells and purchase of pumps, and providing electrical power at nominal rates, over much of the country the better-off farmers have benefited at the cost of others.

This process is well illustrated in an exhaustive recent study of the ground water situation in rural Gujarat. Over a twenty-five-year period, there has been a nearly sixtyfold increase in the number of electrified pump sets and tube wells in the state. The interests of rich farmers have fuelled an indiscriminate expansion of water extraction devices and of water-intensive crops. These rural omnivores have been aided, in the tiresomely familiar pattern, by liberal financial assistance from the state and by greatly subsidized electricity. As a consequence, the water table has fallen alarmingly – in many parts of Gujarat, by several metres or more. This means that smaller peasants often no longer have access to water for their crops, for the ground water level is now well beyond the traditional dug wells, which are all that they can afford. Ironically, while the state as a whole has been plagued by water scarcity and drought, the rural omnivores are actually able to profit from drought conditions, owing to their near monopoly over water. Thus the ground water economy of Gujarat has been summed up in those two words, *overexploitation* and *inequity*, that so accurately capture the overall process of natural resource development in independent India (Bhatia 1992).

Another illustration from our own experience comes from the coastal Kumta *taluk* of the state of Karnataka. Here there are many villages where the richer, upper-caste *Haviks* own betelnut-cum-pepper orchards, while the poorer, lower-status *Halakki Vakkals* till small paddy fields. Prior to the 1960s, when water could be lifted only with the help of muscle power, the orchards were lightly irrigated and produced modest yields. The water table remained quite high, so that the *Halakkis* could grow two crops of rice a year. But then the infrastructure for lifting up water using electrical power became readily available and the better-educated, richer *Haviks* growing cash crops monopolized state subsidies. They began to irrigate their orchards more intensively, got higher yields, and as prices of betelnut and pepper rose came to accumulate more and more wealth and power. The rice-cultivating *Halakkis* found no entry into this system. Instead they were at the receiving end: unable to grow a second crop of rice after the monsoon as the water table plunged down, and further impoverished as the rice prices did not keep pace with the rate of inflation. The result has been growing tension between *Halakki* and *Havik*, a conflict tinged with a communal element.

Large-scale irrigation works are not new to India. Mauryan emperors were already constructing dams, and an extensive Yamuna irrigation canal was functioning in Mughal times. But larger-scale irrigation as well as hydro-electric projects really began in the British period, and took off as the most prominent ingredient of the development effort following independence. We have already recounted the struggles over land required for the construction of these projects. There are equally portentous struggles over water, especially acute because those who do get the water pay so little for it. The beneficiaries are in effect being enormously subsidized by the state exchequer and fight bitterly to corner these benefits.

Among the more celebrated of such conflicts is that over the sharing of the Kaveri waters between the omnivores of the states of Karnataka and Tamil Nadu – be they sugarcane farmers of Mysore, urban dwellers of Bangalore, industrialists of Mettur or rice growers of Tanjore. The Kaveri originates in Karnataka and Kerala and joins the sea in Tamil Nadu, being dammed several times on the way. The contention of Karnataka is that the formula for sharing of water arrived at under British rule was unfair, since the Karnataka stretch of the Kaveri was then under the native state of Mysore while the Tamil Nadu portion came under direct British rule. Tamil Nadu naturally insists on a continuation of this agreement. In Karnataka this led in December 1991 to *bandhs* (shutdowns) and counter-*bandhs* and mob violence while the state-controlled police watched idly. The chief victims of this violence were the poor migrant Tamil labourers and small cultivators, ecological refugees who had come to Karnataka to eke out a livelihood.

The state invests enormous amounts, often raised through World Bank loans, to bring water to the cities and distribute it practically free of cost to city dwellers. There are no firm computations of the extent of subsidies, but the citizens of Bangalore pay no more than 5 per cent of what it has cost the state to deliver water to them. The investment to Karnataka in bringing water to those 10 per cent of the citizens of the state concentrated in Bangalore is, in order of magnitude, greater than the investment for water supply to all the other villages, towns and cities of the state put together. In spite of this,

Bangalore, which is located on a high point on the Deccan plateau at an altitude of nearly 1000 m, and away from all the river courses, is chronically short of water. So while the five-star Ashoka Hotel at the highest point in Bangalore has running hot and cold water on tap for luxurious tub baths, and rich households can fill huge underground sumps, the poor must fight to fill a few pots at public taps which are dry except for a couple of hours every other day. The quarrels at these taps, along with the struggle to put up a hut. made out of old tar drums in crowded slums, symbolize the internecine strife that India's ecological refugees are tragically involved in today.

With water a free good, it makes perfect economic sense for a private entrepreneur to pollute his surroundings instead of investing in technology to properly treat and safely dispose of effluents. The state, as the repository in theory of the welfare of the public, then emerges as the agency most likely to pass legislation to check pollution and take punitive action against offenders. Indeed, in the industrialized world a major focus of the environmental movement has been on pressurizing the state to pass legislation and create enforcement agencies to check air and water pollution (see Hays 1987).

In India too, pollution control legislation is on the statute books, but because the gulf between omnivores who impose the cost and ecosystem people and ecological refugees who bear it is so large, industrial pollution has largely gone unchecked. In its executive functions, the Indian state apparatus alternates between being soft and predatory; in the first incarnation, laws are

*Plate 14* There is very inadequate control of industrial pollution in the country

not enforced, while the second allows offenders to buy official compliance. Early in 1993, for instance, the Ministry of Environment and Forests of the Government of India yet again relaxed its schedule for gross polluting industries to clean up or face closure. It extended its earlier deadline of 31 December 1992 by one year for new ventures and by two years for ventures set up before 1981. It also reversed its decision to periodically monitor pollution, so that industries are now required to obtain a permit just once, except when planning expansion (*Down to Earth*, 15 March 1993). Yet, in a democratic political system, citizens' actions can act as a partial corrective even when the state abdicates its role.

Among the most notorious of industrial polluters are paper and rayon factories. Three units of Gwalior Rayons – owned by India's largest industrial house, the Birlas – have been indicted by environmentalists for affecting the economic welfare of downstream villagers through pollution. The Gwalior Rayon factory on the Chaliyar river in Kerala was closed for seven years after a spirited movement, led by the Kerala Sastra Sahitya Parishat. In the adjoining state of Karnataka, Harihar Polyfibres – owned by the same parent company – has faced concerted opposition and a long-drawn-out court case for discharging untreated effluents into the Tungabhadra river. Villagers have complained of being hit by new diseases, declining fish yields and reduced availability of irrigation water (Hiremath 1987). A similar charge has been laid at the door of the massive Gwalior Rayons factory at Nagda in Madhya Pradesh, a state where the Birla-owned Orient Paper Mills, in the district of Shahdol, has also been criticized by social activists for its pollution of the Sone river (Roy *et al.* 1982).

Two other illustrations of the conflict between private profit and the public good come from Maharashtra, the western state with a highly developed industrial sector and a long tradition of social activism. In October 1987, farmers and fisherfolk from the Revadanda creek area of the Raigad district protested against the discharge of effluents from forty units operating in an industrial area owned by the Maharashtra State Industrial Development Corporation (MSIDC). Accusing the MSIDC of not treating effluents before discharging them into a nearby river, peasants inserted a wooden log into the discharge pipeline (*The Times of India*, 8 October 1987). Some months later, villagers in the Ahmednagar district of the state united to oppose the destruction of land and water by the discharge of south Asia's largest distillery. Despairing of remedial action, the villagers filed a suit in the Bombay high court, seeking Rs 10 million in damages from the offending unit, the Western Maharashtra Industrial Corporation, and the State Pollution Control Board (*Indian Post*, 8 August 1988).

Another example of citizen protest against pollution comes from the district of North Arcot in the southern state of Tamil Nadu. Here, effluents from a cluster of tanneries abruptly raised the chloride content of drinking water and contributed to declining crop yields by causing soil salinity. On

World Environment Day 1984, the town of Ambur, site of several tanneries, observed a total strike or *hartal*. Many women and children from the affected villages went in a procession through the town. Here women broke pitchers containing contaminated water, demanding that the authorities protect the health of their children. A huge effigy of an 'effluent monster' was also burnt on that day (*The Hindu*, 7 June 1984).

We have left until last what is perhaps the most tragic episode of the poisoning of the environment in human history. This took place at almost the exact geographical centre of India, the city of Bhopal, in December 1984. On the outskirts of the city lies the pesticide plant of the Union Carbide corporation, a US-based multinational. Early on the morning of 3 December there occurred an accident, allegedly caused by the introduction of water into the methyl isocyanate storage tank. This resulted in an uncontrollable reaction with the liberation of heat and the escape of methyl isocyanate in the form of a gas that killed, blinded or otherwise harmed humans. All around the Union Carbide plant was a shantytown full of ecological refugees. Within minutes of the gas beginning to escape, as many as 3,000 people were dead on the spot; altogether 50,000 were afflicted by chronic lung damage. But these numbers may err grossly on the low side, for there is a strong suspicion that the administration as well as the medical authorities greatly under-estimated the magnitude of the disaster (Dhara 1992).

## THERE'S TROUBLE IN FISHING

We now come to another category of ecosystem people whose dependence on living resources has also been undermined in recent decades. Distinct endogamous groups of specialist fisherfolk, both along the sea coast and on rivers, have long been a feature of the Indian landscape. These communities, which depend more or less exclusively on the catch and sale of fish, have recently been threatened by encroachments on their territory.

The problems of ocean-going fisherfolk have been well documented, particularly in the studies of the economist John Kurien. The clash between artisanal fisherfolk and modern trawlers, at its most intense in the southern state of Kerala, provides a chilling illustration of what can happen when one group's exclusive control over living resources is abruptly challenged by more powerful economic and political forces. For centuries, the coastal fish economy was controlled by artisanal fisherfolk operating small, un-mechanized craft, who supplied fish to inland markets. In the 1960s, big business began to enter the fisheries sector. The advent of large trawlers, catching fish primarily for export, led to major changes in the ecology and economy of fisheries in Kerala. A rapid increase in fish landings in the early years of trawling was followed by stagnation and relative decline. While some artisanal fisherfolk were able to make the transition to a more capital- and resource-intensive system, the majority faced the full weight of competition

*Plate 15* Rapid mechanization of fisheries has led to overexploitation of fish stocks and serious conflicts with artisanal fisherfolk

*Plate 16* Mechanized fisheries have become an important component of the economy of the Andaman and Nicobar Islands

from trawlers. This conflict gave rise to a widespread movement – comprising strikes, processions and violent clashes with trawler owners – in which small fisherfolk pressed for restrictions on the operations of trawlers. The movement also called for a ban on trawling during the monsoon, the breeding season for several important fish species. A partial ban finally imposed in 1988 and 1989 did in fact result in an increased harvest following the monsoon months (Kurien and Achari 1990; Kurien 1993).

So far as inland fisheries are concerned, there have been, as illustrated above, intermittent reports of localized opposition by fisherfolk affected by industrial pollution. In a class of its own is a unique movement to 'free the Ganga' that arose among fisherfolk in the Bhagalpur district of Bihar. Here, in a bizarre relic of feudalism, two families claim hereditary rights of control over a fifty-mile stretch of the Ganga. Claiming that these *panidari* rights originated in Mughal times, the water lords levy taxes on the 40,000-odd fisherfolk living along the river. A protracted court case has so far been unsuccessful in abolishing these rights, which by an anomaly escaped the provisions of the acts abolishing landlordism (*zamindari*) enacted after 1947. Since the early 1980s, the fisherfolk have been organized by young socialists in the Ganga Mukti Andolan ('free the Ganga' movement). With fish catch also declining owing to industrial pollution, the movement has been waged simultaneously on two fronts – against effluents and an anachronistic system of monopoly rights over water (Narain 1983).

As India's natural fisheries are being depleted by siltation and damming of rivers, and by pollution and overfishing, the culturing of carp in freshwater ponds and shrimps in brackish water fields is becoming a lucrative business. Establishing control over these water bodies, especially where they earlier served as common property resources, can lead to conflicts on many different scales. In the Kumta *taluk* of coastal Karnataka is the estuary of the River Aghanashini. The extensive tidal mud flats of this river support the cultivation of a salt-resistant rice variety by the *Halakki Vakkal* community in the monsoon and serve as rich fishing grounds for the fisherfolk community of *Ambigas* in the dry season. In the late 1960s, however, these mud flats were embanked at government expense. Around the same time were set up seafood canning factories, so that shrimp came to fetch much higher prices. The bunds have permanently demarcated the paddy fields, with the result that the *Halakki Vakkal* farmers now claim ownership rights over shrimp fishing as well. This was contested by the *Ambigas*, resulting in litigation. The courts have ruled that *Halakkis* do control part of the estuarine land, but that deeper channels must be left open to fishing by the *Ambigas*. But incomes from auctioning of rights to shrimp fishing are high and the farmers are refusing to abide by the court ruling, continuing to keep *Ambiga* fisherfolk – who are far less educated and much poorer than the *Halakki* farmers – out of their traditional fishing grounds.

A very similar conflict, albeit on a somewhat bigger scale, has now erupted over the proposal to take up shrimp culture in Chilika lake, India's largest body of brackish water. Spread over 11,000 sq km in the state of Orissa, this huge lagoon is connected by a narrow channel to the Bay of Bengal. An estimated 100,000 fisherfolk depend on this ecosystem for their livelihood. It is here that the large industrial house of the Tatas, in partnership with the state government, has proposed an 'Integrated Shrimp Farming Project' to augment productivity and exports. However, local fisherfolk anticipate a host of problems with the coming of the project. Aside from the declining availability of fish for them – which will be the first consequence of omnivore entry into Chilika – they argue that the construction of large embankments, which the project demands, will increase threats from floods and water-logging. The project will also pollute the ecosystem with its artificial protein feed, and keep away the great flocks of migratory birds that now visit the lake. A social movement, in which students have joined hands with ecosystem people, has arisen to try to stop the project. Aside from numerous petitions and a press campaign, the movement mobilized 8,000 supporters in a demonstration in the state capital, Bhubaneshwar, in September 1991 (Dogra 1992). This is a classic conflict between omnivores and ecosystem people, with the former using the state to capture resources previously under the control of the latter.

## FIGHTS IN THE FOREST

India's tropical soils are by and large poor in minerals and their cultivation demands substantial inputs of nutrients. The Indian cultivator has tradition-ally obtained these nutrients from non-cultivated lands surrounding the villages, either through loppings from trees, or by employing the livestock to convert grass into dung, which is then used as manure. Forest lands as grazing lands have therefore been an integral part of the subsistence base of India's ecosystem people. Not surprisingly, forest conflicts have been endemic ever since the British laid claim to India's vast forest tracts and attempted to wean the people away from their traditional use of these common lands.

Indeed, the origins of the Indian environmental movement can be fairly ascribed to that most celebrated of forest conflicts, the Chipko movement of the central Himalaya. Governmental officials, in their deferential way, tend to locate the origins of environmental concern in India to the love for nature frequently expressed by the long-time Prime Minister, Mrs Indira Gandhi. Yet as a 'grassroots' perspective the Chipko movement has an authenticity lacking in the interventions of politicians and scientists. As a powerful statement against the violation of customary rights to the forest by com-mercial timber operations, Chipko brought into sharp focus a wide range of issues concerning the country's forest policy which also impinged on the environment debate as a whole. (For a full account of the history and sociology of Chipko, see Guha 1989a.)

Owing to its novel techniques and Gandhian associations, the Chipko movement has rapidly acquired fame. Yet it is representative of a far wider spectrum of forest-based conflicts. In the tribal areas of central India, economic dependence on the forests is possibly even more acute than in the Himalayan foothills where Chipko originated. Here the 1970s witnessed escalating conflict between villagers and the forest administration in tribal districts of the states of Bihar, Orissa, Madhya Pradesh, Maharashtra and Andhra Pradesh. In tribal India, moreover, forest conflicts often have a sharper political edge. Thus in Bihar, they have been an integral element in the popular movement for a tribal homeland, while in the four other states mentioned above, the question of tribal forest rights has been actively taken up by revolutionary Maoist groups (see Sengupta 1982; Peoples Union for Democratic Rights 1982; Calman 1985).

Scholarly research inspired by the forest conflicts of the 1970s also revealed their long lineage. Indeed, local opposition to commercial forestry dates from the earliest days of state intervention. Before the inception of the Indian Forest Department in 1864, there was, by and large, little state intervention in the management of forest areas, which were left in the control of local communities. The takeover of large areas of forest by the colonial state constituted a fundamental political, social and ecological watershed: a political watershed, in that it represented an enormous expansion of the powers of the state, and a corresponding diminution of the rights of ecosystem people; a social watershed, in that by curbing local access it radically altered traditional patterns of resource use; and an ecological watershed, in so far as the emergence of timber as an important commodity was to fundamentally alter forest ecology (Gadgil and Guha 1992: Chs 5 and 6).

The imperatives of colonial forestry were largely commercial. From our point of view, its most significant consequence has been the intensification of social conflict between the state and its subjects. Almost everywhere, and for long periods of time, the takeover of the forest was bitterly resisted by local populations, for whom it represented an unacceptable infringement of their traditional rights of access and use. Hunter-gatherers, shifting cultivators, peasants, pastoral nomads, artisans – for all these social groups free access to forest produce was vital for economic survival, and they protested in various ways at the imposition of state control. Apart from forest laws, new restrictions on *shikar* for local populations (while on the other hand allowing freer hunting for sport by the British and the Indian elite) were another contributory factor in fuelling social conflict (Rangarajan 1992).

Throughout the colonial period, popular resistance to state forestry was remarkably sustained and widespread. In 1913, a government committee in the Madras Presidency was struck by the hostility towards the Forest Department, which was the most reviled government agency along with the Salt Department (likewise concerned with a commodity ostensibly low in value but of inestimable worth to every village household). Two thousand miles to the north, in the Garhwal Himalaya, a British official wrote at almost the same time that 'forest administration consists for the most part in a running fight with the villagers'. Popular resistance to state forestry both embraced forms of protest that minimized the element of confrontation with authority – e.g. covert breaches of the forest law – and organized rebellions that challenged the right of the state to own and manage forest areas. (For a comprehensive analysis of forest-based movements in modern India and a guide to sources see Gadgil and Guha (1992).)

Ironically, the post-independence period only witnessed an acceleration of this process. Economic development implied more intensive resource use which, in the prevailing technological and institutional framework, led inevitably to widespread environmental degradation. In the forestry sector, the industrial orientation became more marked, exemplified in the massive monocultural plantations begun in the early 1960s. Simultaneously, other development projects such as dams and mines exerted a largely negative influence on the forests.

Not surprisingly, the conflicts between the state and its citizens have persisted, and the Forest Department continues to be a largely unwelcome presence in the Indian countryside. However, forest conflicts in independent India have differed in one important respect from conflicts in the colonial period. Earlier, these conflicts emerged out of the contending claims of state and people over a relatively abundant resource; now these conflicts are played out against the backdrop of a rapidly dwindling forest resource base. In other words, a newer *ecological* dimension has been added to the moral, political, and economic dimensions of social conflicts over forests and wildlife.

Cumulatively, these processes have worked to further marginalize poor peasants and tribals, the social groups most heavily dependent on forest resources for their subsistence and survival. A long-time student of Indian tribals poignantly captures their frustration with state forestry thus:

> The reservation of vast tracts of forests, inevitable as it was, was . . . a very serious blow to the tribesman. He was forbidden to practise his traditional methods of [swidden] cultivation. He was ordered to remain in one village and not to wander from place to place. When he had cattle he was kept in a state of continual anxiety for fear they would stray over the boundary and render him liable to what were for him heavy fines. If he was a forest villager he became liable at any moment to be called to work for the forest department. If he lived elsewhere he was forced to obtain a license for almost every kind of forest produce. At every turn the forestry laws cut across his life, limiting, frustrating, destroying his self-confidence. During the year 1933–34 there were 27,000 forest offences registered in the Central Provinces and Berar and probably ten times as many unwhipped of justice. It is obvious that so great a number of offences would not occur unless the forest regulations ran counter

to the fundamental needs and sentiments of the tribesmen. A Forest Officer once said to me: 'Our laws are of such a kind that every villager breaks one forest law every day of his life.'

(Elwin 1964: 115)

Popular movements in defence of forest rights have raised two central questions regarding the direction of forest management. First, they have contended that the control of woodland must revert to communal hands, with the state gradually withdrawing from ownership and management. Second, the opposition to forest management has contrasted the subsistence orientation of villagers with the commercial orientation of the state. This latter contrast can be illustrated by two strikingly similar incidents, separated in time by a few months but occurring 3,000 km apart. The first took place in Kusnur village in the Dharwad district of the southern state of Karnataka. Protesting against the allotment by the state of village pasture land to a polyfibre industry which intended to grow eucalyptus on it, on 14 November 1987 the peasants of Kusnur and surrounding villages organized a 'Pluck-and-Plant' *satyagraha* wherein they symbolically plucked a hundred eucalyptus saplings and replaced them with useful local species. Less than a year later, and probably without knowledge of the Kusnur precedent, Chipko activists in the northern state of Himachal Pradesh were arrested on charges of causing 'damage to public property'. Their 'crime' had been to lead villagers in uprooting 7,000 eucalyptus saplings from a forest nursery in Chamba district to plant indigenous broad-leaved species in their stead. The Dharwad and Chamba episodes vividly illustrate the continuing cleavages between village interests and the commercial bias of state forestry (Kanvalli 1991; Modi 1988).

The clash between subsistence agriculturists and industry over the usufruct of state forests is only the most visible of forest conflicts. Localized opposition has also arisen among village artisans facing increasing difficulty in obtaining raw material from forest lands. Characteristically, the state has diverted to industrial enterprises resources earlier used for generations by artisans. But reed workers in Kerala, bamboo workers in Karnataka and rope makers using wild grass in the Siwalik hills of Uttar Pradesh have all resisted the Forest Department's desire to give preferential treatment to the paper industry in the supply of biomass from state lands.

However, in most areas forest-dependent artisans are yet to be politically organized. That is no longer the case with millions of tribals in central India for whom the collection and sale of 'minor' (i.e. non-wood) forest produce is vital to survival. For decades, tribals collecting non-wood forest produce have been severely exploited by merchants who control the trade. For these merchants, the most lucrative of all 'minor' forest produce is the tendu leaf, used in making the *bidi* or Indian cheroot. Over the past two decades, social activists have organized tendu leaf pluckers in a bid to increase their collection wages. On the eve of the 1991 plucking season, twenty-four organizations working among tribals in five contiguous states of central India unilaterally announced that they had fixed the price of tendu leaves at Rs 50 per 5,000 leaves (the merchants' acquisition rates varied from Rs 9 per 5,000 leaves in Bihar to Rs 25 in Maharashtra). In several areas, tribal forest labourers have been organized by left-wing revolutionaries, causing the alarmed traders to seek the protection of the state. Sadly but inevitably, violence has escalated in the tribal forest districts of Orissa, Madhya Pradesh, Maharashtra and Orissa (*The Statesman*, 17 March 1991; Peoples Union for Civil Liberties 1985).

## MINES AND MISERY

Conflicts over fish and forest resources have both arisen out of the competing claims of omnivores and ecosystem people, each coveting the same resource but for different reasons. By contrast, the conflicts we now highlight are a consequence of the negative externalities imposed by one kind of economic activity – open cast mining – upon another – subsistence agriculture.

The most celebrated of mining conflicts took place in the Doon valley in northwest India. Home to the Indian Military Academy as well as the country's most famous public school, this beautiful valley is a favourite watering hole of the Indian elite. Here, the intensification of limestone mining since 1947 has led to considerable environmental degradation: deforestation, drying up of water sources and the laying waste through erosion and debris of previously cultivated fields. Opposition to limestone quarrying has come from two distinct sources. On the one side, retired officials and executives formed the 'Friends of the Doon' and the 'Save Mussoorie' committees to safeguard the habitat of the valley. They were joined by hotel owners in Mussoorie, worried about the impact of environmental degradation on the tourist inflow into this well-known hill station. These groups may fairly be characterized as NIMBY (Not In My Backyard) environmentalists, pre-occupied above all with protecting a privileged landscape from overcrowding and defacement. On the other side, villagers more directly affected by mining were organized by local activists, many of whom had cut their teeth in the Chipko movement. Whereas the first group lobbied hard with politicians and senior bureaucrats, the latter resorted to sit-ins to stop quarrying. Finally, both wings collaborated in a public-interest litigation that resulted in a landmark judgment of the Supreme Court, which recommended the closure of all but six limestone mines in the Doon Valley (Dogra *et al.* 1983; Bandyopadhyay 1989).

At the height of the Dehradun limestone controversy, one of the valley's NIMBY environmentalists called – with characteristic disregard for the inhabitants of those areas – for mining to be shifted to the interior hills so that Dehradun and Mussoorie would be spared (Dalal 1983). Unknown to the writer, mining had already proceeded apace in the inner hills. Expectedly, it has met with resistance. In the Almora and Pithoragarh districts of Kumaun, soapstone and magnesite mining, by taking over or leading to the degradation of common forest and pasture land, has greatly reduced local access to fuel, fodder and water. With the onset of the monsoon, the debris accumulated through mining descends on the fields of adjacent villages. Meanwhile, tangible benefits to the village economy are few – and certainly inadequate in offsetting the losses due to declining agricultural productivity and biomass availability – with mining lessees preferring to bring in outside labour to act as a buffer between management and villagers.

Kumaun has a long heritage of social movements (Pathak 1987; Guha 1989a) and this has been invoked in the continuing struggles against un-regulated mining. Social activists have worked hard to form village-level *sangarsh samitis* (struggle committees) in the affected areas. The Laxmi *ashram* in Kausani, started by Gandhi's disciple Sarla Devi, has been quite successful in involving women in these movements. In other instances, villagers have acted independently to protest against the damage done by open cast mining. The forms of struggle have been varied: *padayatras* (walking tours), *dharnas*, fasts and efforts to persuade miners to go on strike. In many of these protests, women – whose own domain is most adversely hit by mining – have played a leading role. Several mines have been forced to close down, and villagers have then turned their energies to land reclamation through afforestation (Institute of Social Studies Trust 1991; Joshi 1983a, b).

Another movement with broadly similar contours has taken place against bauxite mining in the southeastern state of Orissa. In the Gandhamardan hills of the Sambalpur district of the state, the public-sector Bharat Aluminium Company (BALCO) has been granted permission to mine a heavily forested area of about 900 ha. The foundation stone for the project was laid by the Chief Minister of Orissa in May 1983, and mining commenced two years later; but by the end of 1986, BALCO operations had been forced to a halt.

As in the Himalaya, bauxite extraction in Gandhamardan led quickly to deforestation, erosion and the pollution of water sources. Blasting operations were also perceived as a threat to the region's ancient temples, visited by pilgrims from long distances. Characteristically, protest originated in a series of petitions sent to senior officials and politicians. When these had no effect, students and social activists began forming village committees. A three-day *dharna* in front of the block development office in October 1985 was followed two months later by a blockade that prevented BALCO vehicles from proceeding up the Gandhamardan plateau to the mines. Private vehicles carrying materials for BALCO operations were also blocked and unloaded. In January and February 1986, the movement shifted to the site of BALCO's proposed railway line, close to the Orissa–Madhya Pradesh border. According-ing to figures collected by a civil liberties group that visited the area, altogether 987 people were jailed in the course of the year-long struggle, including 479 women and 51 children (Peoples Union for Democratic Rights 1986; Concerned Scholars 1986).

*Plate 17* This site of an abandoned iron mine in Karnataka's Western Ghats once supported primeval rain forest

*Plate 18* South Indian granite is being exported in large volumes to Europe. Unregulated quarrying of granite is a significant cause of deforestation and disruption of watercourses

# THE WILD AND THE SACRED

The conflicts considered so far have related to economic interests of different actors in the use of natural resources. There is yet another category, where aesthetic, recreational, scientific and even religious interests of one group clash with the economic interests of another group. There is for instance the opposition to the submersion by the Upper Krishna Reservoir of Kudala Sangama, the site famous as the place where the Kannada saint Basavanna is believed to have attained salvation. But the more striking examples of such conflicts, considered below, relate to nature conservation.

Ecosystem people all over the world have viewed themselves as members of a community of beings, in coexistence with fellow creatures be they trees, birds, streams or rocks. Many of these are revered and protected as sacred objects. Sacred is the Himalayan peak of Nanda Devi, as too is Talakaveri from where originates the River Kaveri in the Western Ghats. Entire patches of forests, or pools in river courses, or ponds may be treated as sacred and accorded protection against human exploitation. All individuals of a species like the rhesus monkey might be treated as sacred objects not to be harmed. Such age-old traditions of nature conservation have played an important role in conserving India's heritage of biodiversity.

Sacred groves are an element of this system, older even than Gautama Buddha, who was born in a sal forest sacred to the goddess Lumbini. A rich network of sacred groves once covered all India, and Dietrich Brandis, the first Inspector-General of Forests, was already lamenting the destruction of much of this network under the British system of forest management by the 1880s (Brandis 1884). But remnants of this network still exist and are today the very last representatives of the climax vegetation in areas such as the Maharashtra Western Ghats. In 1972, one of us became interested in inventorying these sacred groves as exemplifying folk conservation practices of Maharashtra. When this survey was being conducted a letter arrived from a village called Gani in the Shrivardhan *taluk* of coastal Maharashtra. The letter stated that the village had a well-preserved sacred grove, of 10 ha, which had been marked for felling by the Forest Department, and asked for our help in saving it. We visited this remote village, after trekking for some three hours across totally devastated hills in a region that receives over 3,500 mm of rain, to find the sacred grove, which was one of the last remaining pieces of greenery. The villagers said that the only perennial stream near their village flowed out of this grove, and they desperately wanted the authorities not to cut it down. Nobody, however, cared for their protests so that when they learnt from a sympathetic forest ranger that we were circulating question-naires about sacred groves, they hoped we could help. And indeed we could, with the Forest Department agreeing to spare this grove, although one of the senior officials wondered why we were bothered about saving these stands of 'overmature timber'. Saving this grove was, however, an isolated success,

91

and all over India the last of the sacred groves are being felled, often against protests by the local population (Gadgil and Vartak 1975; Gadgil and Subash Chandran 1992).

While the protection of sacred groves is a traditional folk practice of ecosystem people, the omnivores have their own, primarily recreational, interest in nature conservation. Thus some 4 per cent of India's landmass has now been put under some form of protected area, as wildlife sanctuary, national park or biosphere reserve. The guiding philosophy behind the management of these protected areas has been one of keeping the local people out by force of arms, on the theory that they are the sole cause of the degradation of nature. There is little real evidence for this theory; indeed, there is accumulating evidence that pressures from outside omnivores, not those of local ecosystem people, are the main problem. Despite this evidence, attempts to deprive local people of access to resources of protected areas continue, and are leading to sharp conflicts. These conflicts have been reported from all over the country, from the Rajaji National Park in the north to the Nilgiri Biosphere Reserve in the south, and from the Betla Tiger Reserve in the east to the Nal Sarovar Sanctuary in the west. Sometimes they can have tragically destructive effects, as when a hostile local population is moved to burn large areas of a National Park which they perceive to be against their interests – acts of arson which they have resorted to in two important national parks, Kanha in Madhya Pradesh (in 1989) and Nagarhole in Karnataka (in 1992). Ironically, where ecosystem people living in and around parks have often been denied their traditional rights to forest produce, within the reserves commercial exploitation has continued with the connivance of officials. An enquiry by the Peoples Union for Democratic Rights into tribal discontent with the management of the Simlipal Tiger Reserve in Orissa came to this stark conclusion:

> Thus in Simlipal the choice is not between no poaching and poaching but between . . . poaching by tourists and organized smugglers and occasional tribal poaching. The choice is not betwen no cultivation and cultivation but between large scale illegal denudation of forests and cultivation by tribals ... The choice is not between [a] complete removal of human settlement and deforestation by tribals but between organized deforest- ation with the connivance of state agencies and limited de- forestation by tribals. In the end the choice is not between an ecosystem without human interference and that with human interference but it is between interference by tribals and interference by smugglers, traders and pleasure-seekers. It is a choice between two sets of human beings.
>
> (PUDR 1986)

– a choice, in our terms, between omnivores and ecosystem people.

A well-documented case of such a conflict is located at the Keoladeo National Park at Bharatpur in Rajasthan. Keoladeo is an artificially created

Plate 19 Women carrying fuelwood out of the Ranebennur wildlife sanctuary in Karnataka. There is perennial conflict between local communities and official guardians of protected areas

wetland, a shallow body of water of over 450 ha created by bunds constructed in the eighteenth century. This wetland attracts tens of thousands of wintering waterfowl as well as supporting an enormous number of herons, egrets, storks and ibises that breed during the monsoon. It was once a site of enormous shoots – tens of thousands of ducks and teals in a day – by British aristocrats coming down as guests of the maharaja of Bharatpur. The area had also always served as a grazing ground for cattle and buffaloes from local villages and as a source of irrigation in the post-monsoon period.

After independence the area was constituted as a national park, and taken over from the Bharatpur maharaja, who yielded it with great reluctance. Then began to operate the omnivore theory of the necessity of keeping ecosystem people at bay. Not only forest officials, but scientists, both Indian and American, claimed that a ban on grazing would be highly desirable – in the total absence of any evidence on this score. Such a ban was finally imposed in the early 1980s, but without any alternative provision for fodder supply. Peasants protested, there was police firing, killing some of them, and the ban was forcefully implemented. The results have been disastrous for Keoladeo as a bird habitat. In the absence of buffalo grazing, *Paspalum* grass has overgrown, choking out the shallow bodies of water, rendering this a far worse habitat, especially for wintering geese, ducks and teals, than it ever had been before (Vijayan 1987).

Wild animals protected in the conservation area, or protected species, outside it, can often affect the people adversely in various ways. Thus wild boar protected by the Bhimashankar Sanctuary in Pune district every year destroy extensive areas of standing crops in tribal hamlets which adjoin the reserve. In 1987 and 1993, a voluntary organization working with the tribals attempted to quantify the extent of crop damage by wild boars. In twenty-five hamlets surveyed in 1987, about 96,000 kg of grain had been destroyed by wild animals – a total monetary loss to the farmers of Rs 232,000. Six years later, the damage was computed at 90,820 kilograms, valued at approximately Rs 453,000 at current prices. The average loss per person was estimated at Rs 800 per year – in the circumstances, a considerable sum of money. To this must be added the difficulties faced by the tribals as a consequence of restricted access to forest produce, more freely available to them before the constitution of the sanctuary (Anil Kapur and Kusum Karnik, pers. comm.).

These enormous losses suffered by ecosystem people in the interior rarely attract the attention of wildlife lovers in the city. Ironically, these are the people whose energetic lobbying of government has contributed to the rapid and somewhat unplanned expansion of the protected area network. Curiously, urban wildlifers are prone to be more sensitive to animal depredations in their own vicinity. Thus bonnet macaque troupes still living in parts of the city of Bangalore damage papayas and guavas, roses and marigolds in the gardens of houses. We can recall an elite conservation group talking of the need to relocate these monkeys out of Bangalore, and simultaneously of the need to educate peasants to agree to bear crop damage caused by elephants in hilly regions to the southwest of the city. Not only do farmers in this area suffer extensive crop damage, but around twenty-five people are trampled to death every year by elephants. As is the case all over India, there is very inadequate provision for compensating for crop damage, while the family of the person killed has the solace of a princely sum of Rs 5,000 obtained after much difficulty and red tape. Farmers employ a variety of devices to keep elephants at bay, including killing them with country-made guns. If the elephant so killed has tusks, it can fetch a very handsome income, often equivalent to several years of earnings for a landless tribal or small farmer (Sukumar 1989, 1994). Thus elephant poaching coupled to sandalwood smuggling has become a thriving business, especially in the dry scrub to the southwest of Bangalore, where agriculture is quite unproductive and there are few other avenues of employment in forest areas. Gangsters have come up to organize this business – the most notorious of them is one Mr Veerappan. He and his gang have killed dozens of forest and police personnel, but for years they have remained at large. This is because while Veerappan has provided employment to the local people, government employees are viewed merely as aliens and exploiters. As a result, villagers provide no information on Veerappan's whereabouts, permitting him to escape capture (*Frontline*, 19 May 1995).

# THE ECOLOGICAL BASIS OF ETHNIC CONFLICT

In the ongoing inequitable process of development in India, not only are ecosystem people and ecological refugees everywhere denied access to resources, but certain regions are systematically drained to support others. The regions so impoverished usually have concentrations of the weakest segments of India's population, the tribals; the regions prospering at their cost have concentrations of those who wield economic and political clout, in metropolitan centres like Bombay and Delhi. The largest concentrations of India's tribal populations occur in the northeastern hills and in the so-called Jharkhand tracts extending over Bihar, West Bengal, Orissa and Madhya Pradesh in central India. Both these tribal areas are rich in forest resources; the northeast also has oil and the central Indian tracts have coal and iron ore. The forests are controlled by the state government, minerals by the central government. Although these areas generate enormous revenues, almost all of it flows out, leaving behind poorly paid wage labour and environments devastated by logging and mining. These inequities have fuelled numerous local separatist movements, be they the armed insurgencies of the Nagas, Mizos or Bodos, the Assam students' agitation or the Jharkhand movement. These movements demand more control over local resources and a greater share in the profits from their use – a demand vehemently opposed by omnivores sitting in Patna or Delhi.

The most long-standing of these movements is the Jharkhand agitation in the Chotanagpur region of Bihar. In an area rich in forest and mineral resources, the massive expansion of industrial and development projects has gone hand in hand with the impoverishment of the region's predominantly tribal population. Beginning in the late 1930s, there has been a consistent demand for a separate political entity ('Jharkhand') where the tribals would have effective control over their resources and their destiny. Foremost among the causes that lie behind the call for Jharkhand are the alienation of land and forest resources, the uncontrolled influx of outsiders who usually monopolize jobs and positions of power, and the grave neglect of infrastructural development by the government, so much so that a massive hydroelectric project which takes away power for industries would not even provide a light to local villages. Episodically, these demands have fuelled militant upsurges where tribals have surrounded or attacked government offices, or organized boycotts and blockades. Over the years, by systematically co-opting its leaders and through repression, the government of India has been able to thwart the creation of Jharkhand. All the same, the movement has not died out, and at regular intervals gathers renewed force (Devalle 1992; Ghosh 1991; Sengupta 1982). At last, in September 1994, the government of India has agreed to the creation of a Hill Council, where Jharkhand would remain within the state of Bihar but have a modicum of autonomy. It remains to be seen whether this concession will satisfy popular aspirations and diminish conflict.

In several of these regions, extremists have formed armed groups that virtually run parallel governments. These groups collect taxes, contending that this revenue must be collected and used locally, and not be permitted to flow out of the region to serve outside omnivores. Thus the Naxalites in the Bastar district of Madhya Pradesh demand Rs 5,000 from each lorry laden with bamboo going out to the paper mills. In the Chandrapur district of Maharashtra they have prevented operation of liquor shops that carried on a roaring business and were an important source of government revenue as well. Such movements have also precipitated major setbacks in the official nature conservation effort. Thus Naxalites have encouraged displaced tribals to encroach into and set fire to the Kanha Tiger Reserve, and Bodo extremist groups have permitted poaching in the Manas Tiger Reserve.

The inequities in contemporary India relate not only to control over land, water, fish, forest or minerals, but also to access to education, jobs in the bureaucracy, and the process of political decision making. There are growing social conflicts focused on each one of these concerns. Conflicts grow primarily because the gulf between omnivores and the dispossessed is continually widening. A most acute struggle therefore rages over entry into the omnivore class, whether through higher education, especially technical education, or more directly through a job in the government. The most recent manifestation of this conflict is the struggle over caste-based reservation in professional colleges and in government and public-sector employment. The principle of such reservation in favour of scheduled castes and scheduled tribes has long been accepted. It was long ago extended to other 'backward' castes in the southern states as a fallout of the anti-*Brahman* movement. In 1990, the Prime Minister accepted the Mandal commission report which would extend its coverage to employment in all central government undertakings. There was a surge of violent protests in the cities of northern India, marked by macabre incidents of self-immolation by upper-caste students.

The nation-wide debate sparked off by the Mandal controversy, however, failed to address itself to the deeper causes of why such a wide gulf exists in the first place between the omnivores and the dispossessed. This gulf has little to do with the fact that most members of backward castes can never hope to gain positions of power in government. Rather, its deeper causes are to be found in a pattern of development that has left the bulk of this class of people deprived of control over land, water, forest, minerals, education, employment and decision making. After more than four and a half decades of independence 38 per cent of males and 66 per cent of females in India remain illiterate. Notably, this lack of literacy is concentrated among the predominantly lower-caste ecosystem people. When, therefore, industries come to areas inhabited by them, as with the steel mills of Durgapur and Rourkela, all the well-paying jobs go to outsiders. A small number of the locals get ill-paid, unskilled jobs, while the majority merely suffer from the pollution and deforestation that inevitably follow the establishment of such industrial enclaves.

This chapter has provided an overview of the multitude of resource-based conflicts in present-day India. While these conflicts are endemic everywhere, some are more visible than others. They attract little attention when they concern only ecosystem people in the hinterlands, as with the landless encroaching on grazing lands valued by small peasants, or when they arise among ecological refugees, as with fights at public taps in city slums. The media pay far more attention to conflicts that pit omnivores either against each other as in urban land scams, or against ecosystem people as in the Chipko or Narmada Bachao *andolans*. In the colonial mode focusing on subjugation, these conflicts were essentially about the rapid drain of natural resources; here forest-related conflicts occupied centre stage. The conflicts have waned as forests have been exhausted over much of the subcontinent, while other sources of forest raw material such as imported Malaysian timber have provided a substitute. In the post-independence mode of governance as patronage, conflicts are sharpest where there are rich possibilities of cornering state-sponsored subsidies, as with river valley projects. Equally acute are conflicts over gaining a foothold in the omnivore class, as with the issue of caste-based reservations. But acute or subdued, endemic or suddenly erupting, resource-based conflicts have fissured Indian society along many, many axes. Their resolution is evidently an urgent national task. We offer our own perspectives in Part II of this book. But first let us locate the Indian environmental movement in the context of these conflicts.

# 4

# IDEOLOGIES OF ENVIRONMENTALISM

## THE MAJOR STRANDS

One may define an environmental movement as organized social activity consciously directed towards promoting sustainable use of natural resources, halting environmental degradation or bringing about environmental restoration. Viewed in this light India has a wide diversity of environmental movements involving members of one or more of our three categories of omnivores, ecosystem people and ecological refugees. In this multiplicity of movements one may discern seven major strands. Two of these are exclusively focused on nature conservation, one on aesthetic, recreational or scientific grounds and the other on the basis of cultural or religious traditions. The wildlife conservation movement, largely attracting urban omnivores, represents the first strand; the Bishnoi peasants of Rajasthan assiduously protecting *khejadi* trees and blackbuck, chinkara, nilgai and peafowl around their villages, the second. A third strand focuses on efficiency of resource use from a technocratic perspective. This has prompted the establishment of land use boards and integrated watershed development programmes, manned and run by omnivores.

However, the dominant strands in the Indian environmental movement are those that focus on the question of equity. These have largely arisen out of conflicts between omnivores who have gained disproportionately from economic development and ecosystem people whose livelihoods have been seriously undermined through a combination of resource fluxes biased against them and a growing degradation of the environment. Such movements most often tend to involve a small group of socially conscious omnivores working with larger numbers of ecosystem people or ecological refugees.

We might call these movements 'the environmentalism of the poor' to distinguish them from the environmentalism born out of affluence that is such a visible presence in the advanced capitalist societies of the West (Martinez-Alier 1990). There are four broad strands within these movements. The first emphasizes the moral imperative of checking overuse and doing justice to the poor, and largely includes Gandhians. The second emphasizes the need to dismantle the unjust social order through struggle, and primarily attracts Marxists. The third and fourth strands emphasize reconstruction, employing technologies appropriate to the context and the times. These might arise either out of the concerns of scientists or, more significantly, through the revival of community-based management systems. The latter include spontaneous village-level efforts to protect and sustainably use local wood lots or ponds, or to pursue environmentally friendly agricultural practices.

## ENVIRONMENTAL STRUGGLES

In analysing environmental movements, one may distinguish between their material, political and ideological expressions. The material context, discussed in preceding chapters, is provided by the wide-ranging shortages of, threats to and struggles over natural resources. Against this backdrop, the political expression of Indian environmentalism has been the organization by social action groups of the victims of environmental degradation. Even the urban well-to-do, increasingly subject to noise and air pollution, and deprived of exposure to nature, might be viewed as victims of environmental degradation, and their organization into societies like the World Wide Fund for Nature (WWF-India) is an environmental movement of sorts. Indeed, some Western scholars such as Thurow (1980) have contended that environmentalism is predominantly an interest of the upper middle class of the rich countries, and that poor countries and poor individuals are simply not interested in environmentalism. It is abundantly clear, however, that Indian environmental movements very much involve the poor and the disadvantaged victims of environmental degradation. In the rest of our discussion, therefore, we shall almost exclusively focus on environmental movements involving struggles of ecosystem people.

Environmental action groups working with such people have embarked upon three distinct, if interrelated, sets of initiatives. First, through a process of organization and struggle they have tried, with varying degrees of success, to prevent ecologically destructive economic practices. Second, they have promoted the environmental message through the skilful use of the media, and, more innovatively, via informal means such as walking tours and eco-development camps. Finally, social action groups have taken up programmes of environmental rehabilitation (e.g. afforestation and soil conservation), restoring degraded village ecosystems and thereby enhancing the quality of life of its inhabitants.

Although these myriad initiatives may be construed, in the broad sense, as being political in nature, they have been almost wholly undertaken by groups falling outside the sphere of formal party politics. Across the ideological spectrum of party politics in India – from the Bharatiya Janata Party on the right to the Communist Party of India (Marxist) on the left – the established parties, whose higher-level leadership has become an integral part of the omnivore establishment, have turned a blind eye to the continuing impoverishment of India's natural resource base, and the threat this poses to the lives and livelihoods of vulnerable populations. In the circumstances, it has been left primarily to social action groups not owing allegiance to any political party – what the political scientist Rajni Kothari (1984) has termed 'non-party political formations' – to focus public attention on the linkages between ecological degradation and rural poverty.

Through the process of struggle, the spreading of consciousness and constructive work, action groups in the environmental field have come to develop an incisive critique of the development process itself. Environmental activists, and intellectuals sympathetic to their work, have raised major questions about the orientation of economic planning in India, its inbuilt biases in favour of the commercial–industrial sector, and its neglect of ecological considerations. More hesitantly, they have tried to outline an alternative framework for development which they argue would be both ecologically sustainable and socially just. Although perspectives within the movement are themselves quite varied, in its totality this fostering of a public debate on development options constitutes the ideological expression of the environmental movement.

## CREATING AWARENESS

In most of the conflicts over natural resources, collective protests – against the agencies of the state or against private firms – have been closely accompanied by coverage in the print media. Sometimes, leading environmental activists – Sunderlal Bahuguna and Baba Amte come immediately to mind – themselves write signed articles in newspapers drawing attention to the struggle they are engaged in. More frequently, though, sympathetic journalists write on these struggles and their wider implications. Since the mid-1970s there has been a virtual explosion of environmental writing in English- and Indian-language newspapers and magazines. Among the most notable of such publications have been the citizens' reports on the state of the Indian environment and the magazine *Down to Earth* published by the Delhi-based Centre for Science and Environment, and the books and magazines brought out by the Kerala Sastra Sahitya Parishat, Kerala's popular science movement. With radio and television controlled by the state, the print media have played an important role in reporting, interpreting and publicizing nature-based conflicts in modern India.

And yet, in understanding the spread of environmental consciousness, one must not underestimate oral means of communication. For example, to increase popular awareness of deforestation and pollution the Kerala Sastra Sahitya Parishat has performed plays and folk songs in all parts of Kerala. In the neighbouring state of Karnataka, themes of environmental abuse and renewal have figured in the traditional dance-drama of the west coast,

Yakshagana. An activity that combines discussion and practical action is the 'eco-development' camp, widely used by action groups to promote afforestation and other forms of environmental restoration (see Bhaskaran 1990).

But in the sphere of communication, too, the most innovative technique of the environmental movement recalls its acknowledged patron saint, Mahatma Gandhi. This is the *padayatra* or walking tour. Used by Gandhi to spread the message of communal harmony and by his disciple Vinoba Bhave to persuade landlords to donate land to the landless, the *padayatra* has been enthusiastically revived by environmental activists. The first environmental *padayatra* was in fact undertaken by one of Bhave's own disciples. This was the Kashmir to Kohima trans-Himalayan march, covering 4,000 km, accomplished by Sunderlal Bahuguna and his associates in 1982-3.

The most notable *padayatra* of this ilk was the Save the Western Ghats March of 1987-8. Following seven months of preparation involving over 150 voluntary organizations (from the states of Kerala, Tamil Nadu, Karnataka, Goa and Maharashtra), the march commenced on 1 November 1987 simultaneously from the two extremities of this 1,600 km long mountain chain – Kanyakumari in Tamil Nadu and Navapur in the Dhulia district of Maharashtra. Three months later, marchers from the north and south converged at Ponda in Goa for the meeting that marked the march's conclusion. By then they had collectively covered 4,000 km of hill terrain, making contact with over 600 villages *en route*. The predominantly urban marchers themselves came from a variety of backgrounds and age groups. Their aim was threefold: to study at first hand environmental degradation and its consequences for communities living along the Ghats; to try and activate local groups to play a watchdog role in preventing further ecological deterioration; and to canvass public opinion in general (Vijaypurkar 1988; Hiremath 1988).

One of the objectives of the Western Ghats march, in which it largely succeeded, was to draw attention to threatened mountain ecosystems other than the Himalaya, whose plight had hitherto dominated the Indian environment debate. As a haven of biological diversity (1,400 endemic species of flowering plants alone) and the source of many rivers, the Ghats are as crucial to the ecological stability of peninsular India as indeed the Himalaya are to the Indo-Gangetic plain. Notably, the Western Ghats march inspired *padayatras* across other vulnerable mountain systems. In the winter following the Western Ghats campaign, a 'Save the Nilgiris' march was organized. Covering villages in four *taluks*, this march culminated in a public meeting at the hill station of Ootacamund on Christmas Eve 1988 (*The Times of India*, 15 December 1988). Again, a 'Save the Sivaliks' march was undertaken across 200 km of the Sivalik range in Jammu and Kashmir the winter following the Western Ghats enterprise, while in early 1991 a fifty-day march was carried out through the Eastern Ghats of Andhra Pradesh and Orissa. The latter effort, termed the 'Vanya Prant Chaitanya Yatra' (forest areas awareness march), focused on the interconnections between environmental degradation and tribal poverty, as exemplified by deforestation, pollution, land alienation and displacement (see Saraf 1989; Vinayak 1990). Most recently, a group of social activists, predominantly Gandhian in orientation, organized a two-month-long 'Aravalli Chetna Yatra' in late 1993, traversing over 600 km on foot through a mountain chain that extends over the states of Gujarat, Rajasthan and Haryana apart from the capital city of Delhi. The marchers drew particular attention to the damage caused by illegal mining and logging in the Aravallis (Kishore Saint, pers. comm.).

Our final illustration of an environmental *padayatra* highlights not a region but a threatened resource – water. This was the Kanyakumari march, organized by the National Fisherfolk Forum in April 1989 under the slogan 'Protect waters, protect life!' As in the Western Ghats, two teams started independently – one in a fishing village in Bengal on the east coast, the other near Bombay on the west coast. Making their way on foot and by van, the marchers organized a variety of meetings and seminars in villages along the way. Although initiated by organizations working among fisherfolk, the march had a wider ambit. Apart from declining fish yields, the marchers studied the pollution of coastal waters by industry and urban sewage, and the destruction of key ecosystems such as mangrove swamps and estuaries.

The objectives of the march, as enumerated by its organizers, were:

(a) To widen people's awareness of the link between water and life and to encourage popular initiatives to protect water;
(b) to form a network of all those concerned with these issues;
(c) to pressurize the government in evolving a sustainable water utilization policy, and to democratize and strengthen the existing water management agencies;
(d) to assess the damage already done, identify problem areas for detailed study, and evolve practices for rejuvenating water resources; and
(e) to revive and propagate traditional water conservation practices and regenerative fishing technologies.

(National Fisherfolk Forum 1989)

The marchers on both coasts converged in Kanyakumari, on the southern-most tip of India, on May Day 1989 (this culminating date reflecting the trade union connections of the organizers). An exhibition on water pollution and conservation, held at a local high school, was followed by a march to the sea. Here the participants, led by 100 women, took a pledge to 'protect waters, protect life'. Finally, a crowd of nearly 10,000, at least half of whom were women, wound their way to the public meeting that was to mark the culmination of the march. Sadly, an incident provoked by a government bus disrupting the marchers led to a police firing in which several people were killed, and the rally was called off. Despite its aborted ending, the Kanya-kumari march had fulfilled its aim of highlighting the threats to a liquid resource which, in the Indian context, must be reckoned to be as important as oil (see Dietrich 1989; Kumar 1989).

As tactics of struggle and consciousness raising, the *satyagraha* and *padayatra* have received generous media coverage. Less visible, but equally significant, are the programmes of ecological restoration that various social action groups have undertaken. With the state's manifest inability to restore degraded ecosystems, many voluntary organizations – some exclusively. involving local people, and others relying on outside catalysts – have taken it upon themselves to organize villagers in programmes of afforestation, soil and water conservation, and the adoption of environmentally sound technologies.

## ENVIRONMENTAL REHABILITATION

In focusing on environmental rehabilitation in preference to struggle or publicity, some groups are merely reviving indigenous traditions of community control, while others have been variously influenced by the Gandhian tradition of constructive work, by religious reform movements or by the example of international relief organizations. Often, voluntary groups with a background of work in health care, education or women's uplift have turned in recent years to promoting sound natural resource management. Of a wide range of groups we have chosen here to highlight two initiatives involving contact with the outside world, and two others that exclusively involve local people.

We start with the Dasholi Gram Swarajya Mandal (DGSM), the group that pioneered the Chipko movement, under the leadership of Chandi Prasad Bhatt. One wing of Chipko, identified with Sunderlal Bahuguna, has preferred to connect Himalayan deforestation with national and global environmental concerns. The DGSM, however, has turned from struggle to reconstruction work at the grassroots. Over the past decade, the DGSM has concentrated chiefly on afforestation work in the villages of the upper Alakananda valley. Two notable features of these plantations have been the lead taken by women, and the high survival rate of saplings – an average of 75 per cent in contrast to the 14 per cent in Forest Department plantations. In addition, in heavily eroded landscapes volunteers have taken up appropriate soil conservation measures such as the plugging of gullies, the construction of small check-dams and the planting of fast-growing grass species. Finally, the DGSM has enthusiastically promoted energy-saving devices such as fuel-efficient cooking stoves and biogas plants (Centre for Science and Environment 1985; S.N. Prasad, pers. comm.; Mukul 1993).

A recent investigation by the Satellite Applications Centre at Ahmedabad underscores the efficacy of this approach. Cultivated lands constitute only 4 per cent of the total landscape in this mountainous terrain – 1 per cent less than land under permanent snow cover. By 1972, when the first satellite pictures came, over 9 per cent of the land, mostly close to roads, had come to be covered by landslides or degraded scrub. In the 1970s the efforts of DSGM focused on checking the pace of deforestation. The satellite imagery shows that despite these efforts another 2 per cent of the land came to be covered by degraded scrub and landslides between 1972 and 1982. But plantation efforts were beginning to pick up in this decade, and an equivalent amount of old wasteland was nursed back to tree cover in this period. The tide was fully turned around in the 1980s. In this period only 0.5 per cent of land was newly converted to wasteland. At the same time, over 6 per cent of land was successfully brought under newly planted tree cover between 1982 and 1992 (Space Applications Centre 1993).

Our next case study originated not in a movement but in a remarkable individual, Anna Saheb Hazare of the village of Ralegaon Shindi in the Ahmednagar district of Maharashtra. Ahmednagar is in a drought-prone region; speaking of the scarcity of water there, the *Bombay Chronicle* of 2 March 1913 called it 'the most unfortunate and heavily tried district in India'. Thus when Anna Hazare returned to the village on retirement from the army in the mid-1970s, food production was barely 30 per cent of its requirements. Quickly locating the problem as insufficient retention of rain-water, he organized villagers in building a series of storage ponds and embankments (*nallah bandhs*) along the low hills surrounding the village. Very soon, runoff was reduced and aquifers recharged, and the ground water table rose considerably. There is now sufficient water for household use and irrigation, and crop yields have increased dramatically (the village has even started exporting food). Alongside, Hazare has mobilized villagers to plant 400,000 saplings. With his village now acknowledged as a model of eco-restoration through self-help, Hazare is training volunteers to work in other villages. He has simultaneously launched a movement against corruption in government-run forestry and drinking water programmes. Awarded the Padma Shri, a high national honour, Hazare returned the award to the Government of India in April 1994, following its failure to take effective action against forest officials accused of corruption (Rai *et al.* 1991; *Indian Express*, 18 April 1994).

Chandi Prasad Bhatt and Anna Hazare are among India's most celebrated environmentalists. Their own exposure to the wider society, through the Sarvodaya movement in the one case and army employment in the other, undoubtedly helped crystallize the ideas and strategies of action which they then applied in their own localities. However, there are many other initiatives, often totally unknown to the outside world, in which a group of local people have spontaneously organized efforts at eco-restoration and the sustainable use of natural resources. We report here two examples, previously unrecorded in the literature, that we are personally acquainted with.

The first of these is from two villages, Hosdurga and Rampura, in the semi-arid *taluk* of Pavgada on the Karnataka–Andhra Pradesh border. In this undulating terrain, the hillocks were once wooded with many hardy species.

But gradually they have been shorn of all tree cover, often through the sale of timber by local villagers to charcoal merchants. Some ten years ago a group of forty youths of Hosdurga, who had organized a small mutual fund for their own purposes, decided to reforest the hillock near their village. They sought and obtained the co-operation of the representative political body, the village *panchayat*. Investing some of the money from their mutual fund, they employed a watchman and strictly protected the emerging regeneration on the hillock. The vegetation is now coming back, and the group makes a little money by permitting harvests at a moderate level. The group has subsequently assumed an active role in other village development activities as well. Witnessing this success, a group from Rampura, a hamlet of the village of Hosdurga, has similarly taken to protecting the forest on a hillock near their habitation.

Our last example comes from the state of Manipur on the India–Myanmar border. Hill areas of this state are inhabited by shifting cultivators who led almost completely isolated lives until about 1910. Their traditional system of shifting cultivation involved leaving intact substantial patches of forest areas, abodes of nature spirits where cutting was taboo. These tribals embraced Christianity between 1910 and 1960, and on conversion cut down almost all these sacred forests. The results were disastrous, with fire from plots being brought under cultivation entering the villages and reducing them to ash. In many of the now Christian villages, such as the Gangtes village of Saichang in the Churchandapur district of Manipur, villagers have re-established a so-called 'safety forest' fringing the entire habitation. This safety forest is given strict protection, including a ban on harvest of canes, which have a lucrative market. While no longer believed to be an abode of nature spirits, this safety forest receives community protection, including the traditional punishment to any offender of having to sacrifice a pig and give a feast to the entire village. These tribals have also re-established protected bamboo forests from which no shoots are harvested as food, while bamboos may be harvested only for use in the construction of one's own house.

As these examples show, reconstruction work may proceed hand in hand with struggle. Yet in many other instances, groups temperamentally unsuited to confrontation have done estimable work in promoting environmentally benign technologies and in rehabilitating degraded lands. All in all, re-construction work constitutes a valuable third front of the environmental movement, complementing the activities of consciousness building and popular resistance to state policies.

## GREEN DEVELOPMENT

Individual groups working in the environmental field are typically confined to a small area. In the past decade, some attempts have been made to develop a macro-level organization to co-ordinate these varied groups and activities.

*Plate 20* A procession through Harsud, a town that was about to be submerged under the rising waters of the Narmada

This process received a considerable boost with the rally against 'destructive development' held in Harsud in September 1989. In a follow-up meeting held in Bhopal in December – to coincide with the fifth anniversary of the gas tragedy in that city – groups that participated in the Harsud rally initiated the formation of the Jan Vikas Andolan ('People's Development Movement') or JVA, a loosely knit national level organization to co-ordinate local struggles, chiefly of ecosystem people.

Over the past four years, the JVA has had meetings in different parts of the country, involving a wide range of movements and individuals. In defining itself as a movement against the existing pattern of development, the JVA's own objectives are fourfold:

(a) to co-ordinate collective action against environmentally destructive policies and practices;
(b) to provide national solidarity to these struggles;
(c) to mobilize wider public opinion on the need for a new development path; and
(d) to work towards an alternative vision, ecologically sustainable and socially just, for India's future.

(Jan Vikas Andolan 1990)

To this end, it has joined groups representing construction workers, fisherfolk and other non-party formations in a national alliance of people's movements (*The Times of India*, Sunday edition, 3 April 1994).

Social action in the three generic modes outlined above constitutes the bedrock of the Indian environmental movement. While such activism has characteristically been localized, with most groups working within one district, the links between the micro and macro spheres have been made most explicit (recent initiatives like the JVA and the national alliance excepted) through the environmentalists' critique of the ruling ideology of Indian democracy, that of imitative industrialization. Environmentalists have insistently claimed that the intensification of natural resources conflict is a direct consequence of the resource-intensive, capital-intensive pattern of economic development, modelled on the Western experience, followed since independence. The resource illiteracy of development planning, they claim, is directly responsible for the impoverishment of the resource base and of the millions of rural people who depend on it (see also Jan Vikas Andolan 1990).

While there is widespread agreement within the environmental movement as regards the failures of the present development model, there is little consensus on, indeed inadequate effort at working out, plausible alternatives. Here we might identify three distinct ideological perspectives within the movement. It is of course entirely possible that none of the ideologies so identified is present in a particular struggle, or that adherents of all three viewpoints might participate unitedly in a specific initiative. However, interaction over many years with groups spread all over India does suggest that the three strands analysed below are the dominant ideologies of Indian environmentalism.

## CRUSADING GANDHIANS

The first strand, which we may call 'crusading Gandhian', relies heavily on a moral/religious viewpoint in its rejection of the modern way of life. Here, environmental degradation is viewed above all as a moral problem, its origins lying in the wider acceptance of the ideology of materialism and consumerism, which draws humans away from nature even as it encourages wasteful lifestyles. Crusading Gandhians argue that the essence of 'Eastern' cultures is their indifference, even hostility, to economic gain. Thus, if India were to abandon its pursuit of Western models of economic development, it would only be returning to its cultural roots. These environmentalists call, therefore, for a return to pre-colonial (and pre-capitalist) village society, which they uphold as the exemplar of social and ecological harmony. Gandhi's own invocation of Ram Rajya (the mythical but benign rule of King Rama) is here being taken literally, rather than metaphorically. In this regard crusading Gandhians frequently cite Hindu scriptures as exemplifying a 'traditional' reverence for nature and lifeforms.

Crusading Gandhians have worked hard in carrying their message of moral regeneration across the country and indeed across the globe. They have sharply attacked the stranglehold of modernist philosophies – particularly those upholding rationalism and economic growth – on the Indian intelligentsia. Through the written and spoken word, they propagate an alternative, non-modern philosophy whose roots lie in Indian tradition (for a statement by the leading crusading Gandhian, the Chipko activist Sunderlal Bahuguna, see Bahuguna (1983); for an argument by a feminist follower of Bahuguna see Shiva (1988); and for more sophisticated intellectual treatments in the same vein, see Nandy (1987 and 1989)).

## ECOLOGICAL MARXISTS

The second trend, in many ways the polar opposite of the first, is Marxist in inspiration. Marxists see the problem in political and economic terms, arguing that it is unequal access to resources, rather than the question of values, which better explains the patterns and processes of environmental degradation in India. In this sharply stratified society, the rich destroy nature in the pursuit of profit, while the poor do so simply to survive (the crusading Gandhians would tend to deny altogether that the poor also contribute to environmental degradation). For ecological Marxists, therefore, the creation of an economically just society is a logical precondition of social and ecological harmony. In their practical emphasis, socialist activists concentrate on organizing the poor for collective action, working towards their larger goal of the redistribution of economic and political power.

While including various Naxalite and radical Christian groupings, in the Indian context ecological Marxists are perhaps most closely identified with the People's Science Movements (PSMs), whose initial concern with taking 'science to the people' has been widened to include environmental protection. Ecological Marxists can be distinguished from Gandhians in two significant respects: their unremitting hostility to tradition (and corresponding faith in modernity and modern science) and their relatively greater emphasis on confrontational movements (for representative statements of this viewpoint, see Kerala Sastra Sahitya Parishat 1984; Raghunandan 1987).

## APPROPRIATE TECHNOLOGISTS

Crusading Gandhians and ecological Marxists can be seen as being, respectively, the 'ideological' and 'political' extremists of the Indian environmental movement. Because of their ideological purity and consistency, their arguments are often compelling, albeit to different sets of people. In between these two extremes, and occupying the vast middle ground, lies a third tendency, which may be termed (less controversially) as 'appropriate technology'. This strand of the environmental movement strives for a working synthesis of agriculture and industry, big and small units, and Western and Eastern (or modern and traditional) technological traditions. Both in its ambivalence about religion and in its criticism of traditional social hierarchies it is markedly influenced by Western socialism. Yet in its practical emphasis on constructive work, it taps another vein in the Gandhians' tradition. Thus appropriate technologists have done pioneering work in the generation and diffusion of resource-conserving, labour-intensive and socially liberating technologies. Their emphasis is not so much the Marxists' challenging of the 'system', or the Gandhians' the system's ideological underpinnings, as it is demonstrating in practice a set of socio-technical alternatives to the central-izing and environmentally degrading technologies presently in operation (see Reddy 1982; Agarwal 1986; Bhatt 1984).

All three tendencies are represented in that most celebrated of environmental initiatives, the Chipko movement (Guha 1989a). The Gandhian trend, associated above all with the figure of Sunderlal Bahuguna, is best known outside the Himalaya. The Marxist trend within Chipko has been repre-sented by the Uttarakhand Sangarsh Vahini, a youth organization that has organized popular movements against commercial forestry, unregulated mining and the illegal liquor trade. Finally, the appropriate technologists are represented by the organization under whose auspices the movement began, the Dasholi Gram Swarajya Mandal, whose fine work in ecological restora-tion has already been alluded to.

These contrasting perspectives may be further clarified by examining each strand's attitudes towards equity and science, as well as their style and scale of activism. Most crusading Gandhians reject socialism as a Western concept: this leads some among them to gloss over inequalities in traditional Indian society, and yet others even to justify them. Clearly the Marxists have been most forthright in their denunciations of inequality across the triple axes of class, caste and gender. The appropriate technologists have been sufficiently influenced by Marxism to acknowledge the presence and pervasiveness of inequality, but have rarely shown the will to challenge social hierarchies in practice. Attitudes towards modern science and technology also vary widely. The Gandhians consider science to be a brick in the edifice of industrial society, and responsible for some of its worst excesses. Marxists yield to no one in their admiration, even worship, of modern science and technology, viewing science and the 'scientific temper' as an indispensable ally in the construction of a new social order. Here, the appropriate technologists are the most judicious, calling for a pragmatic reconciliation between modern and traditional knowledge and technique, to fulfil the needs of social equity, local self-reliance and environmental sustainability.

Appropriate technologists prefer to work on a micro scale, a group of contiguous villages at best, in demonstrating the viability of an alternative model of economic development. The Gandhians have the largest attempted reach, carrying their crusade on world-wide lecture tours. They have often tended to think globally and act globally, even as the appropriate technologists have acted locally and occasionally thought locally too. The Marxist groupings work in the intermediate range, at the level of a district perhaps, or (as in the case of the Kerala Sastra Sahitya Parishat) the level of a state. Finally, the three strands also differ in their preferred sectors of activism. Their rural romanticism has led the Gandhians to emphasize agrarian environmental problems exclusively , a preference reinforced by their well-known hostility to modern industry. While appropriate technologists do recognize that some degree of industrialization (though not of the present resource-intensive kind) is inevitable, in practice they too have worked largely on technologies aimed at relieving the drudgery of work in the village. Here it is the ecological Marxists, with their natural constituency among miners and workers, who have been most alert to questions of industrial pollution and workplace safety.

Crusading Gandhians, appropriate technologists and ecological Marxists represent the three most forceful strands in the Indian environmental debate. But two other points of view also have a significant measure of support, especially among the omnivores. First, we have the Indian variant of that vibrant strand in global environmentalism, the wilderness movement. Indian naturalists have provided massive documentation of the decline of natural forests and their plant and animal species, urging the government to take remedial action (see Krishnan 1975). Although their earlier efforts were directed almost exclusively towards the protection of large mammals, more recently wildlife preservationists have used the scientific rhetoric of biological diversity and the moral arguments in favour of 'species equality' in pursuit of a more extensive system of parks and sanctuaries and a total ban on human activity in protected areas (Guha 1989b).

## SCIENTIFIC CONSERVATION

We come, finally, to an influential strand of thinking within the state and state agencies. Focusing on efficiency, this strand might be termed 'scientific conservation' (Hays 1957). Pre-eminent here is the work of B.B. Vohra, a senior bureaucrat who was one of the first to draw public attention to land and water degradation. In a pioneering and impressively thorough paper (Vohra 1973), he documented the extent of erosion, waterlogging and other forms of land degradation. There was, he noted, no country-wide organization or policy to deal with these problems; nor was there co-ordination between concerned government departments. For Vohra, as for the early scientific conservationists (see Hays 1957), the solution lies in the creation of new ministries and departments to deal with problems of environmental degradation. The central government, he has written, 'has no option but to obtain a commanding position for itself in the field of land and soil

Table 2 Ideological preferences of the various strands of the Indian environmental movement

| | Crusading Gandhians | Ecological Marxists | Appropriate Technologists | Scientific Conservation | Wilderness Enthusiasts |
|---|---|---|---|---|---|
| Polity | Highly decentralized democracy, 'village republics' | Dictatorship of the proletariat | Decentralized democracy, with women, low-caste participation | No firm view | No firm view |
| Decision making | Highly dispersed power of decision making | Centralized planning | Decentralized planning | Centralized planning | Strongly centralized administration |
| Society | No firm view | Economically equitable, but centralized political power | Economic and political equity | No firm view | No firm view |
| Economy | Mixed economy | State occupying 'commanding heights' | Mixed economy | Mixed economy | No firm view |
| Scale of economic enterprises | Predominantly small, village level | Predominantly large | Focus on small, complemented by large | No firm view | No firm view |
| Appetite for consumption | Limited through moral choice | Limited only on grounds of equity | Limited on grounds of both equity and ecology | Unlimited | Unlimited |
| Linkages to global economy | Weak | Weak | Weak | Weak | No firm view |
| Rate of technological change | Exceedingly low | High | Moderate | No firm view | No firm view |
| Commitment to military expenditure | Very weak | Strong | Weak | No firm view | No firm view |

management through financial and administrative measures' (Vohra 1973: A12; but-see also Vohra 1980, 1982).

Neither wilderness protection nor scientific conservation commands a popular following, yet each has had a considerable influence on government policy. Both tendencies look upon the state as the ultimate guarantor of environmental protection, and their energetic lobbying has informed stringent legislation in pursuance of this ideal – as for example the Wildlife Protection Act of 1972 (modified in 1991), the Forest Conservation Act of 1980, and the Environment Protection Act of 1986. However, in so far as neither group is cognizant of the social roots of environmental use and abuse, they tend to be dismissed as 'elite' conservationists by environmentalists owing allegiance to Gandhian or Marxist traditions.

So much for a thumbnail sketch of the main ideological strands of the Indian environmental movement. Table 2 summarizes their respective positions on a series of choices relevant to a new developmental paradigm. It is useful to construct such a table in order to bring out, first, that only Gandhians and Marxists have an overall, largely consistent philosophy of development, and, second, that there is very little agreement on any of the pertinent issues. Indeed, the ideological debate has been marked by a level of acrimony and abuse perhaps only to be expected in a youthful, radical movement – but distressing nevertheless. Little wonder then that the environmental movement has been quite unable to articulate a coherent alternative to correct the many shortcomings it has been so persistently fighting against. This has allowed the proponents of 'business as usual' (the ideologists of the omnivores) to dub environmentalists as being 'anti-progress', or even agents of foreign powers out to sabotage India's forward march. Such criticism must properly be met on its own ground, by articulating a coherent alternative path of development that accepts the fact that an overwhelming majority of human beings are engaged in the pursuit of their own self-interest. Part II of the book sketches out the elements of such an alternative. Chapter 5 sets out the broad philosophical underpinnings of the alternative paradigm. The following chapters then investigate three critical sectors, those of science and information, forestry, and population respectively. The conclusion highlights the resources we can draw upon in this effort.

# Part II

# THE INDIA THAT MIGHT BE

# 5

# CONSERVATIVE–LIBERAL–SOCIALISM

## A SYSTEM IN TROUBLE

India is a country full of contradictions. On getting up in the morning, we are expected to beg forgiveness from Mother Earth for stepping on her:

Samudravasane prithvi, parvatastan mandale, Vishnupatni namastubhyam, padasparsham kshamaswa me!

(O earth, consort of Vishnu, the Lord of creations, with mountains for thy breasts, and oceans for thy garments, forgive me for stepping on you.)

But not only do we not mind stepping on the earth, we blithely tolerate disasters like the Bhopal gas leak. India gave birth to Gautama Buddha, in a sacred grove of sal trees dedicated to the goddess Lumbini; Buddha achieved enlightenment under a peepul tree and preached a doctrine of compassion towards all creatures on earth. Today we are cutting down peepul and banyan trees, protected over centuries, to bake bricks to build our cities and to crate mangoes sent to the Middle East. We respect Mahatma Gandhi as the Father of the Nation; above all he wanted independent India to be rejuvenated as a land of village republics. But over the past forty-eight years, we have systematically sabotaged attempts to empower village people to control and manage their own destiny. We acknowledge a debt of gratitude to Jawaharlal Nehru, who wholeheartedly supported the development of modern science and technology in India in the belief that a scientific temper was the key to the rational development of the country's resource potential. But we continue to import almost all the technology we need, while using the prestige of science to push through development programmes without the open, fearless scrutiny that is so central to the scientific approach. B.R. Ambedkar, born into an untouchable caste, is called the father of the Indian constitution. But while we continue to extend caste-based reservation in educational institutions and government jobs well beyond the time limit Ambedkar himself had set, masses of our lower-caste peasants remain assetless in the absence of genuine land reform over most of the country. We have had Marxist parties

115

in power in one or more states; one of these went out of its way to lure a private rayon mill by promising virtually free supply of forest raw material, and no checks or controls on polluting effluents.

For four decades after independence, India adopted a model of centralized socialist planning for all-round development. Yet no integrated view of the development process ever emerged. Planning degenerated into a process of merely allocating the state funds to meet various sectoral, often disparate, demands, depending on the relative clout of the interests being served. India, as the economic historian Dharma Kumar has observed (pers. comm.), is now a 'nation of grievance collectors' – the state is happy to submit to grievances in turn, even if they be contradictory. Thus we simultaneously have money sanctioned for giving loans to buy goats, and to undertake afforestation on the presumption that goats will be totally banned.

The previous chapters abundantly document the conflicts that have resulted from these manifold contradictions. The solutions sought for these problems too have been piecemeal, more in the nature of fire fighting than a systematic strategy tackling the root causes. This is as true of political problems of separatism and communalism as it is of environmental degradation.

Environmental activists have demanded that 'mega' projects like the Narmada or Tehri dams or the Gandhamardan mine be abandoned. But they have given little thought to how they can succeed in stopping them when the projects permit a narrow elite to corner large amounts of resources at state expense; and when the total lack of accountability in the way projects are planned and executed permits an alliance of politicians, bureaucrats and contractors to misappropriate a significant fraction of the resources deployed. There has been public agitation against pollution of water as in the Chazhiyar river case, or against air pollution as with the Bhopal gas leak. But can polluting industries really become motivated to mend their ways so long as their victims have no clout and when there is no open, publicly transparent process of the monitoring of pollution levels? There have been hesitant moves to create a stake for villagers in the health of nearby forest areas. But the activists who support these moves have not adequately examined the role of the high level of subsidies that forest-based industry enjoys, or the ways in which the forest bureaucracy can undermine popular participation. Wilderness lovers have lobbied for the total protection of some of the country's few remaining natural areas, without thinking clearly of how this is feasible when the iron triangle can profit so handsomely from exploiting these tracts, while the local tribals and peasants are being increasingly impoverished.

Some Indian environmentalists have gone so far as to contend that India must altogether eschew industrial development. But they have not squarely addressed the question of how this is possible in an increasingly unified world dominated by industrial countries, and where military might depends heavily on technological and industrial capabilities. More recently a demand was raised that India should decide not to pay its international debt and refuse to honour the conditions imposed by the World Bank for being given further loans. But again no thought has been given to how India can actually do this and still continue to function, given the foreign exchange needs of an economy that depends on large-scale import of petroleum products and a defence system which continually demands sophisticated hardware that India itself cannot manufacture.

Fighting against huge odds and the awesome power commanded by omnivores, the environmental movement in India has tended to be excessively defensive, taking on projects one by one. But surely it is time to go beyond the fire-fighting approach. We need to look at the process of development in a holistic fashion, while proposing a broad-based strategy for tackling the enormous difficulties we face as a nation. In our understanding, the system-wide difficulties arise out of six root causes:

(a) Ecosystem people have been suffering from an increasingly circumscribed access to natural capital, the resource base on which they still depend to fulfil many of their basic needs. This is because of the shrinkage of this resource base, as grazing lands are encroached or overgrazed, or natural forests give way to eucalyptus or *Acacia auriculiformis* plantations; and because the ever-growing state apparatus increasingly hinders them from using resources, as when 'open-access' revenue wastelands are taken over as strictly controlled reserved forest lands.

(b) Ecosystem people have very limited access to human-made capital, i.e. the resources of the organized industry–services sector. This is because employment in this sector has grown at a much slower pace than the population, and because education has failed to reach the large masses of ecosystem people, who must therefore eke out a living through unskilled labour on farms or in the informal sector.

(c) The process of building of human-made capital has been highly inefficient and greatly destructive of natural capital. This is because it has been conducted as a monopoly of a state apparatus without any public accountability, because the beneficiaries of these state-mediated interventions are given access to resources at highly subsidized rates and therefore do not care if the process is grossly inefficient, because the state apparatus has failed to force private enterprise to internalize environmental costs and finally because the cost of destruction of natural capital is passed on to the masses of people who are largely assetless, illiterate and, despite the democratic system, have no role in deciding on the direction in which the development process is moving.

(d) Omnivores are establishing, with the help of state power, an ever stronger hold over natural capital, as witness the displacement of Narmada refugees against their wishes, without any appropriate plans for their resettlement. Since the omnivores can pass on to others the costs of degradation of natural capital, their stranglehold promotes patterns of inefficient, non-sustainable use.

(e) Omnivores have a strong hold over human-made capital to the exclusion of both ecosystem people and ecological refugees. This means that the masses must subsist primarily through unskilled labour. In consequence they have no incentive to invest in quality of offspring, but instead produce large numbers of them, contributing to continuing population growth and adding to the resource crunch.

(f) There are large-scale outflows adversely affecting natural capital, whether this be iron and manganese mining silting up estuaries, overfishing in the sea or overgrazing to support the export of leather goods. These pressures are a result of the country's heavy dependence on import of technology and petroleum products, rooted in excessive concentration of economic development in a few islands of prosperity, as well as of high levels of demands for imported military hardware.

## THE GANDHIAN WAY

Three among the existing political ideologies in India are, at first sight, comprehensive enough to address the whole range of issues pertinent to the development debate. These are the Gandhian, Marxist and liberal-capitalist philosophies. Gandhism, which is very much the dominant strand in India's environmental movement, is grounded above all in a moral imperative. With respect to the six issues raised above it proposes that:

(a) Ecosystem people should be given far greater access to and control over the natural resource base of their own localities. Ecosystem people should also be given an important role in a new, largely decentralized system of governance.

(b) Ecosystem people should remain content with their requirements of subsistence, without aspiring to greater access to material goods.

(c) The process of building up human-made capital at considerable cost to natural capital should be halted, by simply giving up the endeavour to step up resource use, to industrialize, or to intensify cultivation.

(d) and (e) Omnivores should not aspire to enhance their own material consumption, and in consequence give up their attempts to establish a stronger hold over the nation's natural and human-made capital.

(f) India should check the drain of its natural capital to the outside world by doing away with the need for foreign exchange through acceptance of a way of life with very low material demands, and a foreign policy based on non-violence leading to low military demands.

The serious flaw in the Gandhian philosophy is its emphasis on voluntary restraint on material consumption, and on the voluntary surrender of power.

While small numbers of people may do this, most people appear voluntarily to accept restraint on consumption or social power only if they are convinced that to do so is in their self-interest. Now restraint over resource consumption will be in the self-interest of an individual or a group of people only if such a group itself bears the cost of excessive levels of consumption. For instance, Manipur hill tribals have agreed among themselves not to harvest shoots or market bamboos from village forests because they are convinced that such restraints are essential to ensure an adequate supply of bamboo for their own requirements of house construction. The only route to a proper regime of restraint in resource use is therefore to pass on resource control to social groups who would themselves reap the benefits of prudent use. Today these are by and large India's ecosystem people; hence the Gandhian prescription of empowering them is quite sensible.

But the accompanying prescription that such a transfer of power should be a voluntary act on the part of the omnivores is obviously impractical. This, for instance, is the lesson of Vinoba Bhave's largely fruitless drive to bring about land reform through the voluntary surrender of land by larger land holders. For even where land was so gifted, Gandhian workers could never organize peasant recipients to put it to effective use. Indeed, it has been reported that when urged to experiment with new forms of social organization in villages where all lands had been surrendered, the Gandhians were afraid that such attempts would release forces that would have widespread violent repercussions and therefore withdrew from the scene. This and all other historical experience suggests that moral exhortations, Gandhian or otherwise, are unlikely to work widely.

India's omnivores are increasingly being swept into the global frenzy of consumption, which is daily fuelled by glossy magazines and satellite television. In the country's relatively open society the masses too are quite aware of and strongly attracted towards this consumer culture. While it appears materially impossible for India's 150 million families to own refrigerators and automobiles, they are all bound to aspire to do so. If meeting these aspirations is going to exact unacceptable environmental costs, then ways should be found of compelling all of India's citizens to share these costs and therefore accept restraints on consumption. But Gandhian moral exhortations are unlikely to bring this about.

Present patterns of resource consumption by India's omnivores cannot be sustained without heavy dependence on imports. Given the failure of India's industrial development to generate innovative products that can compete on the world market, these foreign exchange needs are met by draining the nation's natural resources. Again, Gandhian prescriptions seem incapable of getting us out of this bind. For what is required is a larger transformation of the system of resource use existing today.

Thus the key Gandhian prescriptions that make perfect sense are that ecosystem people must be empowered, and that material consumption should be maintained within limits compatible with a reasonably equitable sharing of the products of nature and the economy. However, the means suggested, that this should be brought about by voluntary acceptance by all people, in and out of power, as a moral imperative, seems impracticable.

## THE MARXIST UTOPIA

While Gandhians view environmental problems as being caused primarily by materialistic greed, Marxists lay the blame at the door of capitalist exploitation simultaneously of nature and the working classes. By and large, they wholeheartedly approve of a concerted, nation-wide effort at stepping up natural resource use, so long as this is controlled and guided by a state acting on behalf of the people. With respect to the six key issues outlined above therefore, the Marxist approach may be summarized as being:

(a) Ecosystem people should be given far greater access to and control over the natural resource base. The leftist governments of West Bengal and Kerala indeed lead the country in efforts at land reform, involving local people in forest management, and in the establishment of decentralized political institutions at village and district-levels.

(b) Ecosystem people should have better access to human-made capital. Again the left-oriented government of Kerala leads the country in taking literacy to all and in providing fuller access to employment in the modern industries–services sectors.

(c) Leftist state governments have, however, been no more successful in tackling the great waste and inefficiency in the process of conversion of natural into human-made capital. This problem arises in part from the naive Marxist faith in an all-powerful state apparatus which in practice behaves as irresponsibly (in an environmental sense) in India as it did in the erstwhile communist countries of eastern Europe.

(d) and (e) Indian Marxists, while in power, have indeed taken some steps to break the monopolistic access of omnivores to the capital of natural and human-made resources, but have not done enough to curb the state apparatus, which has become a significant component of the omnivore complex.

(f) Marxists wish to reduce the drain of natural resources to capitalist countries, and are active today in opposing what they view as the US-led conspiracy to lay India open to further exploitation through GATT and the series of economic reform measures demanded by the World Bank. But in the past they were happy enough with hefty exports of natural-resource-based goods such as tea and leather to the Soviet Union; nor have they any clear analysis of how to tackle India's compulsion to earn foreign exchange.

In fact, the emerging evidence of environmental degradation in eastern

Europe and the former Soviet Union abundantly demonstrates that a monolithic state apparatus behaves in a terribly wasteful and environmentally insensitive fashion. This is in part because most state employees have inadequate motivation to function efficiently, but more importantly because under the communist dictatorships citizens could not organize to protest against environmental degradation and were unable to move the state machinery to take effective action. There is little doubt that the Western capitalistic democracies have a far better record in this context. In this system, capitalist enterprises, often in league with the state, as under the Reagan administration in the United States of America (1980-8), do try to externalize the cost of environmental degradation as much as possible. However, where this personally hurts large numbers of people, as with air and water pollution, citizens can force the system to take effective remedial action (see Hays 1987). Equally importantly, market forces do tend to promote efficient resource use by private enterprises. Thus while Indian and Russian industries have remained among the least efficient in the use of energy, Scandinavian and Japanese industries are by far the most efficient.

## THE CAPITALIST DREAM

Clean and efficient technologies have been favoured in some capitalist countries, thereby promoting more effective conversion of natural to human-made capital. The need to reduce adverse environmental impacts has also induced substitution of information for material and energy; thus carefully crafted electronic devices perform a variety of functions far more effectively than heavier, unwieldy mechanical devices which consume more materials and energy for their fabrication.

While the capitalist system thus promotes more efficient use of resources it also continually encourages higher levels of consumption of material, energy and informational resources. Private enterprise thrives on ever-greater consumption of the goods and services it produces. In all capitalist societies people are sucked into a rat race, consuming more and more of the resources of the earth. However, this has by no means led to more satisfactory lives. As Ivan Illich points out, the availability of private automobiles in the United States has not reduced, but rather increased, the time an average citizen spends in daily commuting to work. The frenzy of earning and spending in the capitalist system is thus not at all compatible with moderating the impact of people on the environment (Durning 1992; Illich 1978).

But then many in the First World question the need for moderating the impact of people on the resources of the earth (Simon 1981; Simon and Kahn 1984). Global omnivores too see few if any signals of environmental degradation hurting them at a personal level. This is for the same reason that Indian omnivores have little concern with environmental degradation, namely, that they have successfully passed on these costs to the ecosystem people and ecological refugees of the Third World. Europeans and North Americans drive automobiles fabricated in Japan made of iron mined in India. The devastation of the Indian hinterland stripped of ore from open pits is not a signal that reaches them in any form. Nor is the pesticide pollution and resultant extinction of several species of river fish endemic to the eastern Himalaya at all evident to European consumers sipping cups of Darjeeling tea. The forest cover of Japan is among the most extensive in the world, around 60 per cent of the land surface, only because the Japanese can import and lavishly use and throw away wood coming from southeast Asia. The suffering inflicted on the ecosystem people of these forest tracts is emphatically not a signal reaching an average citizen of Japan. It is primarily through the operations of capitalism that the global sisterhood and brotherhood of omnivores can effectively shield itself from the larger environmental consequences of its actions.

With Marxism on the retreat and Gandhism never having acquired any real popular following, the philosophy of capitalism is today in the ascendant. In India too, economic liberalization is at the centre stage of the national debate on development. The proposals for liberalization call for a loosening of bureaucratic control and opening up of the economy to the outside world by the abolition of tariff barriers and restrictions on the operation of foreign capital. What implications does this programme have for the six root causes we have identified as lying behind the degradation of India's environment?

(a) The world capitalist system thrives on passing on the costs of environmental degradation to the ecosystem people of the Third World; the Indian version of economic liberalization is therefore quite unlikely to enhance the access of India's ecosystem people to natural resources.

(b) Economic theory claims that, under ideal conditions, market forces result in an efficient pattern of resource use, which leads in turn to the maximization of the level of satisfaction of the parties concerned. So in theory ecosystem people and ecological refugees too should, under a system of privatization, enjoy greater access to the human-made capital of India. The reality is far from such an ideal. In India's grossly inequitable society the market assigns a far lower weighting to demands by weaker segments of the society, so that resources flow towards production of commodities in demand by the omnivores. Moreover, the market is so manipulated that labour or commodities which the weaker segments have to offer are obtained from them at a very low value, with organized trade and industry usurping the larger share of profits accruing from economic transactions.

(c) The philosophy of economic liberalization advocates pruning the bureaucratic apparatus, doing away with state subsidies, permitting market-driven competition to operate more freely. These are all measures that ought to enhance the efficiency of conversion of natural capital into human-made capital. But piecemeal application is unlikely to achieve this to any significant degree. Today, the size and powers of the Indian bureaucracy are hardly being pruned. Subsidies are being only selectively pared down; thus, cutting down on subsidies for the supply of water to industry, to city dwellers or to rich farmers is a subject conspicuously absent from policy debates. Nor are private enterprises being compelled to act in an environmentally responsible fashion. This does not augur well for long-term improvements in the efficiency of resource use in India.

(d) Economic liberalization is not likely to cut down the power of omnivores to capture the country's capital of natural resources; rather, this power may increase as indigenous omnivores ally more strongly with those of the First World.

(e) Economic liberalization will not put a brake on the omnivore's appetite for, and access to, human-made capital either, especially as omnivores are likely to hold on to their ability to pass on the costs of production to the ecosystem people and ecological refugees of the country.

(f) Economic liberalization may further accelerate the drain of the country's natural resources abroad, unless it sufficiently strengthens capabilities of exporting products of manufacture or high-technology services.

The philosophy of economic liberalization does have at its core three significant themes: namely, the need to do away with state-sponsored subsidies; to prune the powers and size of the state apparatus; and to create an open, democratic society that should in theory imply a better deal for India's environment and people. Implemented in the context of a highly inequitable society, however, this philosophy is unlikely to lead to any genuine progress.

## A WORKING SYNTHESIS

Each of the three contending philosophies – Gandhism, Marxism and liberal capitalism – thus has components which are very desirable when viewed from an environmental perspective. But each philosophy also has components that deserve to be decisively rejected. What is evidently needed is a synthesis of the several positive elements: decentralization and empowerment of village communities along with a moderation of appetite for resource consumption from Gandhism; equity and empowerment of the weaker sections from Marxism; and an encouragement of private enterprise coupled to public accountability in an open, democratic system from liberal capitalism. In so far as Gandhism seeks to conserve all that is best in our traditions, it might be called the Indian variant of conservatism, with this significant caveat: that it seeks to conserve not the hierarchy of aristocratic privilege, but the repository of wisdom and meaning vested in ecosystem people. Marxism is of course the best-known strand within the socialist movement, while democratic capitalism is the ideology of the liberal. If our arguments are correct, then the environmental philosophy most appropriate for our times is in fact, nothing but conservative-liberal-socialism.

The alternative development paradigm flowing out of conservative-liberal-socialism would have the following elements with respect to the key issues identified by us (see Table 2, p. 111):

1 India should move towards a genuinely participatory democracy where the political leadership as well as the bureaucracy is made accountable to the masses of people. This requires the strengthening of grassroots democracy, conferring substantial powers on *mandal panchayats* and *zilla parishats*. It would also be desirable to reconstitute the existing states of the Indian union into smaller, more homogeneous units, breaking up huge provinces like Uttar Pradesh and Bihar. Following Gandhi's vision the country might come to be made up of self-governing village republics with no powers available to higher-level political authorities arbitrarily to dissolve or suspend elections to their *panchayats*. The *zilla parishats* and state assemblies should be similarly protected against arbitrary interference from higher levels. This nurturing of widely participatory democratic institutions should be complemented by an opportunity for people to decide directly on a wide range of developmental issues. In this context the system operating in the US state of California might be appropriately adapted to Indian conditions. Under this system any issue may be put on the ballot at the time of election provided that a minimum number of constituents so demand it, by attaching their signature to a petition. Thus the voters in California have decided against installing any more nuclear power stations through a public referendum at the time of state elections, a decision then binding on the state government. Analogously, an appropriate constituency of Indian districts might be empowered to decide on whether they want high dams in the Himalaya after a debate has exposed them to the merits and demerits of various alternatives.

2 The process of control, planning, implementation and monitoring of natural resource use should be radically restructured to render it an open, democratic process with full public accountability, and with substantial powers of controlling the resources of each locality being devolved to the local population. This calls for a pruning of bureaucratic authority and a transfer of most of its powers to local grassroots-level democratic institutions. Thus forests, grazing lands and irrigation tanks should revert to management by local communities, on their own or (as in the case of joint forest management, which is discussed in Chapter 7) in partnership with the state. The government should be relieved of its draconian powers to acquire land and water resources without the consent of the local people, without paying due compensation, or without appropriate arrangements for resettlement.

Instead, the local people should be empowered to work out plans for developing natural resources and managing local environmental affairs in a manner fine-tuned to the specific local situation and in accordance with their aspirations. Such a system would permit a much more fruitful utilization of the traditional knowledge and wisdom possessed by ecosystem people. The programmes of natural resource development so formulated should also be implemented largely by local people, closely supervised by them to ensure proper public accountability. In all this process there should be full freedom of access to and sharing of information. The local educational institutions should become publicly accessible repositories of environmental information pertinent to their own localities and should play a key role in planning natural resource use working with local communities. Higher-level political authorities should not drain away all the profits arising from use of local natural resources; a substantial fraction of such profits must be ploughed back to motivate local people to husband local resources prudently .

This is not to advocate that natural resource management should become an affair exclusively grounded in small-scale village-level programmes totally controlled by small communities. Larger-scale enterprises would undoubtedly continue and require proper co-ordination at larger scales. What is suggested, however, is that such programmes should be designed not by riding roughshod over local communities, as happens today, but through their involvement and after winning their consent. This would again require full freedom of information to the public, as well as public appraisals of the social, economic and environmental consequences of all development projects, whether they be on the scale of a small irrigation tank, or a series of large dams on the River Narmada. The public should also be involved in developing a detailed environmental management plan for each locality, ensuring that no undue damage is inflicted, even on a smaller scale, in rushing through large projects. Monitoring of project performance should be made mandatory and involve the public, again to ensure efficient execution sensitive to the local environment and the welfare of the local people.

3 Decentralization and wider public involvement can also put a stop to the widespread undervaluing of natural resources. In numerous localities of the southern states of Karnataka and Andhra Pradesh, for instance, the quarrying of granite has emerged as a lucrative business. Leases are granted by the state government to private operators, without taking into account the wishes of the villagers in whose vicinity these lands lie. But granite quarrying has contributed greatly to environmental degradation, through deforestation and the blockage of streams used for both drinking water and irrigation. At present, the concessionaires of stone and sand quarries pay a very small royalty, and that too to the state. So they have little interest in careful, efficient use of the resources. Rather they would blast or dig away as rapidly as possible, quickly carting away whatever they can. If the resource was

properly priced, and a goodly share of the profit were to come to the local communities, they would ensure much more prudent use – as indeed happened in the case of the Baliraja dam mentioned in Chapter 2. There the villagers claimed the rights over the royalty from quarrying sand in the river bed. They then ensured that the sand was quarried in such a way as to help in digging the foundation of the dam they wanted built, rather than haphazardly disrupting the river course as had been the practice of the contractors. Thus an important reform that environmental movements and grassroots-level politicians must press for is the proper valuation of natural resources, with a substantial share accruing to local communities from the profits realized from the utilization of these resources, whether timber, granite, coal or oil.

A proper price tag must also be attached to degradation of the environment, charged directly to the agent responsible. Thus villages suffering crop loss due to pollution from a cement factory must have the power to levy an appropriately scaled tax on the industry to compensate it for the loss. Only then would the 'polluter pays' principle be properly implemented and lead to effective action. In addition, the government should collect an environment management cess on a range of economic activities that have environmental consequences. This cess should not go into the kitty of the central or state governments, but directly, without any attenuation, to the village *panchayats* covering the length and breadth of the country. The *panchayats* could then use these funds to organize appropriate monitoring of environmental degradation, through a combination of local ecological knowledge, basic information collected through schools and colleges, or by other technical non-government organizations and research institutions hired at the discretion of the *panchayat*. This could create an entirely new support base for generating detailed, locality-specific scientific information that would be of great value in then deciding on the pricing of natural resources within *panchayat* territory, or on the level of the environment tax to be imposed on offending industries. This information (and revenue) could additionally be put to use in designing a development strategy that would at once benefit local communities and be gentle on the environment.

4 We are quite clear that such a development process could succeed only in a far more equitable society than India is presently. Halting the pace of environmental degradation, then, depends on progress towards a more equitable access to resources, to information and to decision-making power for ecosystem people and for ecological refugees. In India's predominantly agricultural society, a key reform to promote equity is radical land reform. At the same time we need to reverse the current inequitable pattern of flow of state-sponsored subsidies – a flow which is at present biased towards the better-off omnivores, at the cost of the weaker sections. Industry, chemicalized agriculture and urban islands of prosperity must all pay a fair price for the resources they utilize and must bear the full cost of treating the pollutants

they discharge into the environment. The utterly inefficient bureaucratic apparatus that presides over this function of mobilizing resource fluxes in favour of the omnivores should be dismantled. This in itself would cut down significantly on the ranks of omnivores. The state should then pass on the finances thus saved to the decentralized political institutions at village and district levels, for the local inhabitants to use in accordance with their own priorities. Evidence suggests that both health care and education would be a high priority were people given a free choice (in the state of Kerala, where popular participation in the political process is more intense and rewarding than in other parts of India, universal literacy and excellent health care facilities have indeed been its consequence (Jeffrey 1992)). However, the content of education should be broadened to emphasize first-hand observations of people's own surroundings. Investments in health care and in meaningful education linked to the development process, and creation of decentralized political institutions along with land reform would move Indian society towards a far more equitable condition. Since the interests of eco-system people as well as ecological refugees are more compatible with the prudent use of natural resources, such moves in the direction of economic and political equity would promote a people-oriented as well as environmentally sensitive process of development.

5 The collapse of communist regimes in eastern Europe and the recent shifts towards private enterprise in China are striking proof of the failures of state socialism. Here the Indian experience, with its corrupt, inefficient and wasteful bureaucratic apparatus, is entirely in consonance. It is clear therefore that we should continue the shift towards encouraging private enterprise on all fronts, for producing goods and services which people would pay for on the market, while increasingly providing social services such as health and education through the voluntary sector. While freeing the private sector from undue restrictions such as licensing, we must also withdraw all state-sponsored subsidies, adopting instead pricing policies to promote environment-friendly behaviour. State subsidies have so far merely encouraged exhaustive, wasteful resource use and focused the attention of entrepreneurs on manipulation and bribery, rather than on delivering goods and services in an efficient fashion. Pollution regulation should be strongly enforced, but should not remain a business carried out in secret by official agencies, since such an arrangement promotes corruption. A culture of efficient resource use and control of pollution by industrial enterprises would not only serve the cause of environment, but would also enhance the competitive ability of these enterprises in the international marketplace.

Private enterprise should also be encouraged in delivering services such as education, health care or watershed-based soil conservation for which communities might pay through public funds. India's experience of delivering such services through the state apparatus has been very disappointing indeed.

Once guaranteed a salary, many state-employed teachers or doctors simply do not go and work in remote villages. By contrast, voluntary agencies have often done an excellent job of delivering education and health care in rural areas. This is primarily because the voluntary-sector workers are not guaranteed permanent jobs regardless of how good their performance is. They have to compete with each other and to demonstrate continually to funding agencies that they are indeed fulfilling their tasks. At the present time, voluntary agencies play a minor role limited mostly by the availability of foreign funding. Instead, internally generated funds must devolve to local communities, which should have the option to pay for such services only if they are delivered efficiently.

6 The appropriate scale of economic enterprises, especially of development projects, has been a major subject of controversy. Proponents of 'small is beautiful', Gandhians and appropriate technologists have challenged the overwhelming bias in development projects towards the large scale. This preference has undoubtedly favoured the concentration of benefits in the hands of a small number of people. However, the opposition to big projects, for their own sake, is not always productive. It is not so much the scale of enterprise as the way it is conducted that is at the heart of the deprivation of ecosystem people and the degradation of their environment. Carefully treating individual watersheds, and building a series of small dams, may indeed be, in a technical sense, a more optimal solution than building one huge dam. However, watershed treatments may well be carried out in an even more inefficient fashion than the construction and management of larger dams, in which case little is gained by merely choosing projects on a smaller scale. On the other hand, if local communities can get together, experiment with alternative soil and water conservation techniques, cropping patterns and water requirements, and have an opportunity to consider a whole range of technological options before making an informed choice, a proper scale is far more likely to be selected.

It is very likely that today the choice of economic enterprises is unduly biased towards the large scale. But the solution lies not in merely correcting the choice in terms of scale, but in putting in place new mechanisms of deciding upon what projects should be chosen, while ensuring that they are implemented in an efficient and properly accountable fashion.

7 Technological advances thus far have supported a continual increase in the scale of economic enterprises: bigger dams, giant power projects and larger ships. Technological advances have also permitted larger resource fluxes from more remote areas. These developments have permitted a small proportion of the world's population to capture resources at the cost of the rest of the people. Technocrats have often been in league with the small number of omnivores, turning a blind eye both to social deprivation and to environmental degradation. Many environmentalists, especially those of a

Gandhian persuasion, have therefore come to reject technology altogether, dreaming of a return to an idyllic pastoral-agrarian system (Nandy 1989). But pre-colonial, agrarian society in India was beset with its own set of evils, including untouchability and the oppression of women. It is not at all clear if it was indeed in the idyllic state of Ram Rajya as some Gandhians believe it to have been. But that apart, control over technology is a potent force of domination in the world today. Any society that turns its back on advances in technology will quickly find itself exploited and subjugated by others.

Thus, the opting out of scientific and technological advance, which after all is a thrilling adventure of the human spirit, is not a route that India could possibly follow. What is needed instead is to look for ways in which this advance could be directed away from the current path of speeding up the drain of the country's natural resources, producing more polluting substances and concentrating power in the hands of an ever narrower elite. We believe that this can be achieved by taking advantage of the tremendous possibilities of rapid communication opened up by modern technologies to put in place a genuinely participatory democracy with decentralized political institutions endowed with decision-making power. Indeed, we are now very close to a situation in which the role of elected representatives can be severely pared down, with more decisions taken through broader-based referenda. Modern technology also favours greater access to information on the part of a wider public. For instance, with satellite pictures beamed to the earth and readily available for a modest sum of money, the forest cover or patterns of siltation of river beds in any part of the country can easily be assessed by an interested citizen. Such information can also be rapidly analysed and transmitted to thousands of other interested parties. Our emphasis should be on adopting a more open, people-oriented pattern of development that would counter the other, negative implications of technological advance.

This is a crucial theme that is elaborated in the next chapter. But we may note here that the country-wide 'people's science' movements, pioneered by the 25-year-old Kerala Sastra Sahitya Parishat, are attempting just this. They are involving people in assessing the implications of ongoing development processes, in what is happening to their environment and health, and in campaigning for positive alternatives. Demystifying science and technology, making it accessible to all people and involving them in deciding how it should be fruitfully employed is obviously the direction in which we should proceed; to turn our back on modern science and technology (as some radical Greens would have us do) is neither desirable, nor within the realm of practical possibility.

8 Environmental change is clearly related to human demands on the resources of the earth; these demands have been escalating both because of increase in human numbers and because of rapidly increasing per capita demands. The increase in numbers may be largely attributed to the ecosystem

people of the world; while the increase in per capita resource demands has largely occurred because of the omnivores, whose own numbers have been growing at a far slower rate. Now the transition to smaller families will take place only when parents can fruitfully invest substantially in enhancing the quality of their offspring, in equipping them to compete in the market for skilled labour. Ecosystem people would thus become motivated to rear a small number of offspring only when a more equitable development process gradually draws them into the ambit of modern industry, services and intensive agriculture. But with India's huge population the resource demands of such a large number of people in these modern sectors may simply become too large a drain on the country's resources. Industrial nations today get away with their own huge demands by passing on the consequences to Third World countries. This option can be open to only a small minority of the global community; it is certainly not open to India as a whole. The massive expansion of the resource demands of Indian omnivores has been enabled only through a type of internal colonialism. If India is to move towards a sustainable pattern of development, then the omnivores will have to accept restraints on their own consumption levels.

The Gandhian philosophy proposes that this limitation can be brought about through voluntary restraint. But this is impossible to put into practice. What one must urge instead is that omnivores should not be permitted to capture resources at the cost of ecosystem people and ecological refugees; they must bear such costs themselves. This calls for a discontinuation of state-sponsored subsidies benefiting the rich, a compulsion to take care of pollutants, and a democratic dispersion of powers to decide on the use of the country's resources. At the same time, however, the level of resource consumption of India's population need not be brought down or remain stagnant, for the present pattern of resource use is incredibly wasteful. Enhancing its efficiency can greatly enhance availability of resources and services that the resources provide for all of the country's citizens. There are also enormous possibilities here (as we show in the next chapter) of bringing this about through an information-intensive process of development: that is, the substitution of information for energy and materials that modern science is swiftly making possible.

9 During the colonial period India suffered grievously from a drain of its natural resources, while serving as a market for goods of British manufacture at adverse terms of trade. The Second World War greatly improved India's position *vis-à-vis* Britain, and on independence concerted attempts were made to protect India from the drain of its resources by creating tariff barriers against imported goods. This created a sellers' market, actively helped by state-imposed restrictions on production through licensing. On top of this Indian industry was pampered by state subsidies in the supply of natural resources. As a result, industrialists have concentrated on making huge profits through

130

obtaining licences and government subsidies by manipulation and bribery, while using imported, if often obsolete, technologies. They have completely neglected to use resources efficiently, or advance technologically. This lopsided industrial development has created a high-cost, low-quality economy unable to hold its own in the global marketplace. Meanwhile, India's demands for foreign exchange have gone on soaring with the increasing consumption of petroleum products and military hardware. That has led India into a debt trap, from which the only escape route is to beg for further loans by agreeing to demolish the barriers keeping foreign business out of India.

Where should India go from here? The solution favoured by Marxists, to renege on the international debt and to continue behind closed doors, is quite unworkable, unless the country is able drastically to cut down on its import bill. That would call for a new strategy of far more dispersed development of agriculture and industry requiring much lower levels of energy inputs, as well as a matrix of international relationships that would permit a deceleration of investment in the defence sector. Both of these developments would be highly desirable. But if they were to come about, then little would be gained by keeping out of global trade, provided only that India does take good care that this trade does not impose undue pressures on its resource base.

We have outlined above the broad parameters of a new development model for India, which we have called (only partly in jest) conservative-liberal-socialism. Subsequent chapters more closely investigate the implications of this model for scientific advance, for the management of forest resources and for the urgent task of stabilizing India's population. While we deal only with three sectors in this book, on account of both space and our own spheres of competence, our broader framework might be fruitfully applied to other key sectors of Indian development, such as energy, water utilization, transport and housing. But to conclude the present discussion, let us reaffirm that our alternative path of development would effectively address the six root causes of environmental degradation in India. Thus:

(a) It would confer on ecosystem people a substantial degree of control over the country's natural resource base. With the quality of their own life linked to the well being of the local environment, this is likely to lead to a far more prudent use of nature.

(b) It would create conditions under which ecosystem people would gradually come to share equitably in the country's capital of human-made resources. India cannot indefinitely maintain a dual society with the masses of people at subsistence level; only when these people enter the modernized sectors of industry–services–intensive agriculture will they become motivated to invest in the quality of children and produce a small number of offspring, thereby moving India towards a stable population.

131

(c) By drastically restructuring the management of natural resources, ensuring accountability and empowering segments of population with a stake in sustainable use, this development strategy would greatly enhance efficiency in the generation of human-made capital from natural capital.

(d) This strategy would restrict the level of access of omnivores to energy and natural resources. This is essential to put brakes to the current process of exhaustive use by omnivores, where the costs of degradation are passed on to ecosystem people and ecological refugees.

(e) This strategy would also temper the appetite of omnivores for consumption by ensuring that they have to bear the true costs of resource degradation.

(f) By reducing excessive dependence on external energy inputs, this strategy would put India in a far stronger position to organize trade with the outside world without putting excessive pressure on its natural environment.

An environmental strategy informed by conservative-liberal-socialism would be in the spirit of tolerance and assimilation of a diversity of strands of thought that is so characteristic of Indian culture. It would keep India alive as a country of vigorous, powerful communities controlling their own destinies to a significant degree. It would create a more open, democratic and egalitarian society which would also nurture socially responsible private enterprise. It would participate in the adventure of modern science and technology, putting it to use in the protection of the environment and in enhancing the quality of life of its people. It would permit India to integrate with the international community from a position of strength.

This is of course a strategy that would be opposed resolutely by the omnivores, themselves happy to liquidate the country's resource base, let it fall prey to a debt trap while continuing their own wasteful, inefficient ways. But it is a strategy very much in the interests of a vast majority of India's people. Moving in its direction would be a long and arduous, but by no means hopeless, struggle.

# 6

# KNOWLEDGE OF THE
# PEOPLE, BY THE PEOPLE,
# FOR THE PEOPLE

## THE DINOSAUR GOES UNDER

India has been and remains a biomass-based civilization. By this we mean that a majority of Indians depend on biomass gathered by their own labour, or produced through low-input agriculture to meet most of their subsistence needs. They also exchange such biomass, at best processed through simple manual labour, to acquire other materials and services they consume on the market. According to the 1991 census, fully 74.3 per cent of the Indian population is rural based, and most of these people depend on cultivating their own lands or working as labour on other people's lands for a significant fraction of their earnings. The Anthropological Survey of India (ASI) recognizes 4,635 communities as making up the Indian population. For each community they record as current occupations those in which a notable proportion of community members are engaged. According to the ASI, 5.1 per cent of communities are still involved in hunting-gathering, 2.4 per cent in trapping birds, 8.3 per cent in fishing, 3.7 per cent in shifting cultivation, 4.7 per cent in terrace cultivation in hills, 54 per cent in settled cultivation, 4.3 per cent in horticulture, 21.5 per cent in maintaining domestic livestock, 0.8 per cent in nomadic herding, 3.5 per cent in weaving mats or baskets, and 3.9 per cent in woodworking. Of these 4,635 communities at least some members of as many as 4,285 communities still collect fuelwood or crop residues for domestic cooking, while members of 2,514 communities collect dung for this purpose. In another study of eighty-two villages from the drier parts of India, it was found that 14–23 per cent of the earnings of all households was derived from collection of biomass from common lands around villages (Jodha 1990; K.S. Singh 1992; Gadgil 1993).

Of course, the subsistence of most of humanity was grounded in biomass before the beginning of the industrial revolution two centuries ago. But the industrial revolution brought to the fore the effective use of a range of new energy sources: coal, petroleum, hydroelectric power and most recently nuclear power and solar power. The availability of such substantial additional sources of energy has permitted the fabrication of an enormous body of material artefacts that now spread across the earth. To begin with, such artefacts were rather crude, guzzling energy and belching smoke like steam locomotives, or leaving ugly scars on the earth like open-pit mines. But with time they have become more efficient in resource use, and less polluting, as witness modern-day electric locomotives or mines that are carefully filled up and revegetated afterwards. Not that people are once again treading lightly on the earth. Far from it. But we are surely moving in the direction of economic processes that are much more efficient in energy and material terms, and we are doing this primarily by using more, and better, information. There is no doubt that humanity as a whole, having passed from a civilization grounded on biomass through one grounded on material artefacts, is now on course to erect a civilization grounded on information.

Thus far at least, Indian society has barely been touched by this transformation. Indeed, in a technological sense India is still a land of dinosaurs. Its heroic dams are among the largest of human-made objects, the web of high-tension power lines and irrigation canals fanning out from these dams among the most extensive of such networks. But the gigantic no longer commands the centre stage of world technology; rather, it is the diminutive that holds sway. It is as if the age of the ponderous dinosaurs were giving way to that of itsy-bitsy shrews. These tiny mammals may be seen, metaphorically at least, as having outstripped the behemoths on the strength of their brainpower. They could absorb, process and deploy information far better than could the cumbrous reptiles. Somewhat analogously, the minuscule artefacts taking over the world by storm are far more intricately constructed than the Titanics of yesteryear. These artefacts are, above all, devices capable of processing information very, very effectively. We are entering the age of hand-held video cameras, miniature computers, automated manufacture; of ways of absorbing, processing, deploying information to ever greater purpose.

The societies that dominate the world today – indeed, that have come to dominate the world over the past three centuries – have been those capable of handling information well. At the core of the modern scientific and technological revolution is a method of continually augmenting the social store of knowledge. This method involves an open sharing of information, continually relating it to the hard, verifiable facts of the world out there. No progress is possible unless the information is thus openly available, at least to the concerned social group.

It is thus no accident that democratic values have flourished hand in hand with the development of scientific knowledge. For science and technology can flourish only in an open society that encourages scepticism. The Soviet Union did for a while make some significant advances in fields such as space technology, but these were limited achievements. The same society fell far behind in other important disciplines such as genetics precisely because open discussion was suppressed, and because appeals to state authority or Marxist dogma, rather than empirical facts, came to dominate intellectual life. The closed Soviet society, in the tight grip of a political-bureaucratic elite, also fell back on two other crucial fronts: the electronic revolution that ushered in the modern information age, and adequate control of industrial (including radioactive) pollution. Focusing on heavy industry, Stalin ushered in the age of ponderous, polluting artefacts, an age of dinosaurs. This despotic society collapsed when the information age dawned in the more flexible, democratic societies of the West. It collapsed in good part because the Soviet military machine was handicapped by being so far behind in automatic control and guidance systems. Just as dinosaurs inevitably made room for the smarter mammals, so has European communism given way to a democratic, capitalist system.

Shaking off the colonial yoke with its suppression of information, India elected to follow the Soviet model of a centrally planned society, albeit in a democratic political order. Here it took from colonialism the state monopoly of information, and from the Soviet model a centralized, supposedly all-knowing bureaucratic-technocratic apparatus. This apparatus has since made all major decisions in secrecy, with no open discussions, no sharing of information with the people at large. As in the Soviet Union, this has very quickly degenerated into a system where a narrow alliance of omnivores has come to enjoy enormous power, including the power to suppress scrutiny to serve their own limited ends. At least under communist states universal education has been vigorously promoted, so that China has today over 62 per cent female and 84 per cent male literacy. Even this has not happened in India, so that it remains a country where half the population is illiterate and assetless, where all development decisions are made in a closed fashion, where disinformation to further the interests of those in power rules the roost.

This pattern of development has inevitably favoured the dinosaurs. When energy policy is being formulated, for example, the discussion always turns on the need to build more large power generation plants, be they nuclear, superthermal or hydroelectric, and on the amount of money to be sunk in these unwieldy projects with their long gestation periods. The discussion always bypasses a critical assessment of the tremendous wastage of power, of the large losses that state electricity boards have been incurring. There is no good basic information on simple parameters such as who uses how much of the electric power generated. Thus agricultural pump sets are not metered, and the pump set owners merely pay a small flat rate based on the horsepower of the pump. This makes it easy to arbitrarily assign some large power consumption to the pump sets to cover up for gross inefficiencies and irregularities of power use elsewhere. The electricity boards are widely suspected of routinely using this device to lower the figures of transmission and distribution losses. Even then these losses are officially of the order of 23 per cent; it is speculated that they are actually of the order of 30 per cent. This may be compared with losses in other Asian countries such as South Korea and Thailand, which are of the order of 12–15 per cent. In this manner, good information is simply not available, not just to the public, but with the electricity boards themselves. Bad management of information and indeed disinformation thus perpetuates a system in which enhancing the efficiency of energy use, which ought to receive a very high priority in a desperately poor country, is sidelined in favour of the building of more and more, larger and larger, less and less efficient power plants.

Small may not be inherently beautiful, nor large by definition ugly. But there are definite disadvantages associated with the large, given the long gestation between conception and actual commissioning. These long delays did not matter so much at a time when technology was changing at a snail's pace. It may have taken decades to construct the magnificent old Meenakshi temple in Madurai in south India, but building technology very likely stood still over the whole period. However, between the time we now start constructing a large hydroelectric project and its completion, twenty or thirty years later, solar power generation cells may well advance enough to make that a far more economic and environmentally desirable option. Keeping all options open and quickly taking advantage of new possibilities is therefore becoming more significant with every passing day in the modern world. A preoccupation with the execution of large, cumbersome projects is simply not in tune with the times. What is in tune is a spirit of remaining constantly alert, absorbing new information as it comes in, and putting it to good advantage in projects that can yield handsome returns before the technology they employ becomes thoroughly outdated.

## PRACTICAL KNOWLEDGE

India has but one course open to it: to move forward towards an information-based society that generates for its large population the many services it needs, while sparingly using material and energy resources. This could permit us to retain intact, even replenish, the biomass that continues to be a basis of Indian civilization. Unfortunately the course on which we are currently launched at once abuses both biomass and information. That is why our forest managers massacre prime rain forests with claims of enhancing their productivity through raising eucalyptus plantations without any proper trials as to whether eucalyptus will succeed. That is why they find the eucalyptus plantations falling prey to the pink disease that subsequently wipes them out, converting into desert hill slopes once clothed by lush green vegetation (Sharma *et al.* 1984).

What would an alternative development strategy respectful of both biomass and information imply? It would involve putting into practice the approach of dealing gently with complex natural systems, of trying to enhance the services these systems provide for the people with many, small, timely inputs. It would focus on what people get, not on what they – or the state apparatus – spend. It would try to maximize the ratio of output of useful services to input of investment in terms of labour, money and so on. It would properly discount for any incidental loss of useful services as a consequence of the intervention. Since interventions in complex systems inevitably have un-expected, often undesired, consequences such a strategy would call for continual monitoring and feedback. Since our knowledge of the functioning of complex natural systems is limited, such a strategy would emphasize flexible, adaptive management. It would not call for rigid, inflexible, cen-tralized plans, but locality-specific plans that are continually adjusted in the light of experience.

For India's biomass-based society, it is vital to ask the question as to who has the motivation, the knowledge and the competence to enhance the vital services the country's lands, forests and waters provide. Would it be eco-system people, the local communities, who have served as stewards of these natural resources for centuries? Or would it be the bureaucracies, that have taken over as regulators over the last century and a half; or the manipulators, be they contractors involved in dam construction, or manufacturers of sugar or paper? And what of ecological refugees, such as peasants from the plains of Kerala encroaching on Western Ghats forests who fit into none of these categories of stewards, regulators or manipulators? It is, then, of interest to examine how these four categories rate with respect to motivation, knowledge and competence to manage the land and water resources of India, recognizing of course that none of the categories are homogeneous; that local communities are especially apt to be divided into many interest groups pulling in different directions.

Consider first the motivation for careful, sustainable use. Such motivation would depend on whether imprudent use today adversely affects the well-being in the future of the party indulging in misuse. One of the factors that could decide whether there is such a link is the size of the resource catchment: that is, the area over which the concerned parties garner resources. A social group with access to resources from an extensive area is unlikely to be motivated to use the resources of any particular locality in a sustainable fashion, for it would always perceive the option of moving on to another locality on exhaustion of the resources in any given locality. Thus manipu-lators like the commercial users of biomass resources of non-cultivated lands, be they forest-based industry, manufacturers of herbal medicine or suppliers of milk to metropolitan cities, deal with large resource catchments. They therefore tend to focus at any time on the resource elements that bring them the greatest profit, switching to other localities, or other kinds of resources, as these are depleted. Thus India's paper and plywood industries have always concentrated initial harvests in localities close to the mills; as these become depleted, they move on to localities further and further away. They also have options of switching to a different type of resource – for instance, to eucalyptus if bamboo is depleted. Manipulators with their vast resource catchments are therefore unlikely to be motivated to use the natural resources of any part of the country in a sustainable fashion (see also Gadgil and Guha 1992).

So far as the bureaucratic regulators and their political masters are concerned, their future wellbeing has no relation whatsoever to the prudent use of resources of any locality under their control. Given the present lack of accountability, with no rewards for honest performance as custodians, and no punishment for misappropriation of the resource base, the regulators stand only to gain from profligacy – except, occasionally, when a major misdemeanour comes to light and they are exposed to adverse publicity. One example is the case of the Coorg forests. Lying near the centre of the Karnataka Western Ghats, Coorg is an undulating plateau at an altitude of 1,000–1,500 m, famous for its coffee and cardamom estates and its martial traditions. The estates have extensive areas especially in the catchments of springs, traditionally maintained under natural forest cover. The planted areas also have large numbers of shade trees. Over the years the ownership of the land and the trees of Coorg has remained in flux with large tracts being neither under clearcut private ownership, nor under full control of the Forest Department as reserved forest. This has created a situation of uncertainty where constant shifts in government policy have ensured that the estate owners are tempted to dispose of the trees illegally and smuggle them across the state border into Kerala. A serious attempt was made to bring this difficult situation under control in 1991 at the personal behest of the Karnataka chief minister through a government order regulating tree felling. The order was apparently deliberately flouted for over six months by a group of government officials who issued licences for tree cutting in contravention of the order. This misconduct was brought to light in 1994 because a former Conservator of Forests of Coorg who temporarily had to leave the Forest Department for his protests against the working of the department was commissioned by the state legislature to investigate the whole matter. His report recommended that the Corps of Detectives be called in to investigate the malpractices indulged in by over fifty officials. It remains to be seen whether any of them are actually punished. But unfortunately few such cases are ever exposed. With little accountability, the regulators as a class have indeed no motivation to strive for responsible resource use, although there are undoubtedly a minority of public-spirited officials who do attempt to enforce prudent use in the broader national interest.

What of the local communities, then, divided along many lines of class and caste? With increasing penetration of the market, the sizes of resource catchments of local communities are rapidly expanding; more and more of what they use is being brought to them from further and further away. Thus, throughout India, a large number of poorer rural people depend on weaving baskets and mats to supplement their earnings outside the agricultural season. But as bamboo, reeds and other materials have been locally exhausted, such activities increasingly depend on import of raw material from considerable distance and marketing of the produce in distant towns and cities. However, the near-total dependence of large numbers of poorer rural people on locally collected plant material for fuel and fodder continues to this day. Since women are especially involved in gathering of such biomass, they are particularly concerned with maintenance of sources of such material locally, while men, more strongly involved with the market economy, are less concerned. This, for instance, has been the experience of the Dasholi Gram Swarajya Sangh in the Garhwal Himalaya. This organization, which pioneered the Chipko movement, primarily operates through conducting week-long eco-development camps in village after village. During such camps people of surrounding villages come to help work on building protective stone walls and to undertake plantation activities. The local villagers collectively cook and feed the entire volunteer corps. Over the years women have taken an increasingly active role in these camps, as they did in the early protests against deforestation. As recounted in Chapter 4, a recent study of satellite pictures showed that these camps have played an important role not only in halting the pace of deforestation, but in helping rebuild the tree cover of the Alakananda valley.

But over much of rural India, and increasingly even in the tribal areas of the northeast, there are members of local communities whether from old large landholding families or the newly prosperous political bosses who depend little on the natural resources of their own localities. Rather, they are happy enough to see these liquidated if it means a fast buck for themselves. These contradictions are reflected in some of the experiences of social forestry programmes, for instance in Ranebennur *taluk* in the semiarid parts of Karnataka state. In several villages of this region the poorer segments of the local communities wanted the produce of these plantations distributed among residents at concessional rates, while the village council president, coming from a richer family, wanted these auctioned on the open market.

So, while a significant fraction of India's rural and tribal communities does favour long-term, prudent use of local natural resources, there are elements among them with no such motivation. However, the actual behaviour of local communities pretty much uniformly throughout the country tends to short-sighted, exhaustive use. This is because these communities today have almost no control over their local resources. Except in parts of northeastern India, such control was taken over by the state, through its Forest, Revenue or Irrigation Departments. With this state takeover local communities have no rights to control use of the common property resources, either by their own members or outsiders, although they may enjoy certain privileges of limited use. Consider, for example, the recent failure of the villagers of Halakar in the Uttara Kannada district of Karnataka to stake a claim on the timber harvested from their village forest. As mentioned in Chapter 2, Halakar was one of the three villages specially earmarked for praise by Collins, the British official who was appointed to look into the settlement of forest lands of the district in 1921. As a result of Collins's recommendations a village forest committee for Halakar was formally established in 1928, and has continued to function effectively to this date. It has overcome the arbitrary dissolution of the committee by the state Forest Department in 1968, winning a court case against this order. So there is no doubt that the villagers of Halakar deserve full credit for the current good state of forest cover in their village forest. Now, part of this land was acquired by the government for the West Coast railway line in 1992, with the Forest Department stepping in to organize clear-felling of this area. The villagers then claimed rights over the timber, but failed to sustain their claims; the timber was taken over by the Forest Department.

All over the country such usurpation of rights over resources by the state at the cost of local communities ensures that people fear that fruits of their own prudence will go to somebody else, and therefore have little motivation to exercise restraint. The situation is even more precarious with the ecological refugees, with no roots in the localities they migrate into, and with a very insecure control over the resource base. Under the current regime, then, no segment of Indian society is motivated to use the common-property natural resources of the country in a prudent fashion, least of all the state apparatus that has monopolistic control over it. The only exception is in parts of northeastern India, where local communities do retain a measure of resource control and occasionally do exhibit prudent behaviour, as for instance in the case of the establishment of safety forests by the Gangtes of the Churchandapur district of Manipur discussed in Chapter 4 above.

Prudent management of natural resources also calls for knowledge and capability along with motivation. The technocracy, be it foresters, agricultural scientists or civil engineers, claims monopoly over this knowledge. But these are dubious claims, because these resource managers are dealing with highly complex natural systems. The behaviour of a complex natural system depends on many subtle interactions among the diverse components making up the system. Thus bamboos are among the most prominent and useful constituents of India's tropical forests. They have been used by rural people for centuries, as house construction material, for weaving baskets, for fabricating fishing or agricultural implements. In recent years bamboo has come to serve as an important raw material for the paper and polyfibre industry as well. As noted in Chapter 2, the state has made bamboo resources available to the industry at throwaway prices, and bamboo resources have declined rapidly following industrial exploitation. The question is the role of many different factors impinging on bamboo stocks in their decline. One of us was involved in an investigation of this problem between 1975 and 1980. We worked in the district of Uttara Kannada, in forest areas leased out to the West Coast Paper Mill. The paper mill's forest department, headed by a very experienced retired forest official, prescribed the clearing of the covering of thorny branches that forms at the base of any bamboo clump. This was in the belief that this facilitated good growth of the bamboo culms. It turned out that the practice exposed the new bamboo shoots to grazing by langur monkeys, wild pigs and cattle. Our field investigations over three years conclusively demonstrated that this was a major factor responsible for the failure of new culms to develop, resulting in the depletion of bamboo stocks. The local people had also been harvesting bamboos for their own use for centuries. They never cleared this thorny protective covering of young shoots. Indeed, we later discovered that they were aware of the problems caused by removal of the thorns. But the forest department of the mill had never consulted them – nor for that matter did we until well after the investigations were launched. In consequence the mill was paying its labourers to carry out operations that actually hurt the resources they valued.

Ecosystem people have for centuries depended on the natural resources from a limited resource catchment to provide them with manifold services. They have therefore discovered a spectrum of uses for the local natural resources, be they soils or rocks, plants or animals. Omnivores, on the other hand, have been able to draw resources from vast areas, and to process them to provide many different services. They have therefore tended to force any given locality into exporting one particular resource, in the process writing off a whole range of other possibilities. Thus the British attempted to convert much of the forest of peninsular India into single-species stands of teak (*Tectona grandis*), a tree that was valued first for shipbuilding, and later for furniture and house construction. In this single-minded pursuit of teak plantations they ring barked and killed a wealth of tropical rain forest trees. A century later, the forest managers of independent India decided that the tropical forests of peninsular India must now produce eucalyptus as pulpwood, and in pursuit of this goal clear-felled vast stretches of rain forests. Notably, in both cases monocultural plantations often failed because of lack of attention to locality-specific factors (Gadgil and Guha 1992).

Ecosystem people know of a vast number of uses of the many plants of tropical rain forests; not just trees, but also herbs, shrubs, climbers, the epiphytes. One such useful plant is a herb, *Rauwolfia serpentina*, which produces an alkaloid reserpine useful in the treatment of high blood pressure. This herb, endemic to the Western Ghats, has been a part of the tribal medical repertoire for a long time; this information was picked up by practitioners of modern medicine some twenty years ago. Soon a commercial drug extracted from *Rauwolfia* came on the market, triggering off rapid exploitation and near extinction of the herb before its export was banned.

Ecosystem people have been in the business of extracting services from nature without large inputs (primarily because they had no access to them) for a very long time. Their practices have therefore been moulded to working closely with nature. This repertoire includes a great variety of land races of cultivated plants and domesticated animals adapted to particular environments which often are reservoirs of valuable genes conferring resistance to diseases, permitting salt or drought tolerance and so on. These land races are currently being rapidly lost, not necessarily because more productive varieties have been introduced, but often simply because of government pressures for acceptance of a package of modern practices carrying with it inducements of subsidies for supply of fertilizers, pesticides and so on. Such subsidized packages have tended to destroy many desirable practices, for example the desilting of irrigation tanks and the use of silt as a manure, or the use of plants for insect control, for example leaves of *Strychnos nuxvomica* in betelnut orchards of Uttara Kannada district.

## A SCIENCE FOR SUSTENANCE

The practical knowledge and wisdom of India's ecosystem people must therefore once again come to assume an important role in enhancing the services being provided by natural systems, a role that is today being completely denied. We are, however, by no means proposing that modern science and technology should therefore withdraw. Folk knowledge is particularly relevant for processes that are evident to the eye and that take place on time scales of less than a few years. It does not extend to microbes that may be polluting drinking water sources, or to processes of soil erosion that are manifest only over decades. Enhancing ecosystem services beyond traditional levels must therefore depend on wise use of newer understanding and technologies. Indeed, in coming decades the startling new capabilities of moving genetic material at will from one organism into another are bound to revolutionize the whole process of biological production for human use. It would be folly not to take full advantage of such developments.

As important as these technical developments is the application of the scientific methodology in the endeavour to enhance ecosystem services. At the heart of the scientific methodology is the careful recording of empirical observations, the development of a model of how any natural system functions on the basis of such observations, the predictions of how it may behave in response to specific interventions, the verification through further observations of whether indeed changes take place as predicted, the incorporation of appropriate changes in the model of the system behaviour, and so on. This is an open-ended process through which the understanding of natural processes and human abilities to intervene effectively progresses step by step. Of course, from time to time such a process yields a spectacular new understanding, and often such understanding permits entirely new and highly effective forms of human intervention. Such was the appreciation that matter and energy are interconvertible, an understanding that permitted the release of atomic energy. Such also is the appreciation of the chemical nature of genetic material, which is now permitting the creation of genetically engineered life forms. These landmark developments are of course important; but every one of these major advances in understanding, as well as the development of techniques based on them, has taken place outside India. Importing these techniques could bring considerable benefits to India; but that should not be the sole context in which India employs modern science and technology. Rather India must get down to applying the scientific methodology across the length and breadth of the country in enhancing ecosystem services, not necessarily in spectacular leaps and bounds but in slow, steady increments.

This the current system utterly fails to accomplish. The irrigation engineers take care neither of the catchment nor of the distributories, so that dams silt up at far greater rates than projected, while large parts of lower command areas fail to receive any irrigation water. Soil conservation measures are prescribed without considering the nature of the substrate, and are so sloppily executed that they often enhance, not reduce, rates of soil erosion. The agricultural strategy promoted by the agricultural universities and state departments prescribes standard, rather heavy, doses of fertilizers and pesticides with no reference to the field-to-field variation in many relevant factors such as levels of soil nutrients, or factors that vary from day to day such as levels of pest populations.

While natural resources are everywhere treated in this sloppy fashion, the technocratic agencies make bogus claims of scientific management. These have no substance, for these agencies do not maintain any careful data on the state of the system, they do not monitor whether their interventions have indeed had the projected consequences, and they do not adjust their operations to correct for any deviations between the projected and realized outcomes. Rather, state agencies employ the jargon and prestige of science to cover up for the stark inefficiency and wastefulness of the technocratic management of India's natural resources (see Gadgil and Guha 1992; Singh 1994).

An alternative strategy of providing good information inputs evidently needs to be put in place. This should be a system attuned to expose rather than obfuscate what is happening to the resource base; that would look for ways to minimize external intervention, rather than hunt for excuses to spend state funds; that would try to work out a programme of timely, small, appropriate interventions to make the most of the potentialities of variable natural systems. Such a programme of injecting scientific inputs would not be glamorous and its results would have very limited international recognition. Nor would the prescriptions generated create large opportunities of profit for omnivore enterprises. In consequence the sophisticated, largely American-trained, leadership of the Indian scientific community has little interest in such an endeavour. Neither, of course, has the state technocracy, which has unfortunately developed vested interests in wasteful resource use. It is instead the 'people's science' movement groups that have so far promoted such exercises, albeit in small ways. Perhaps the most notable example of such an exercise comes from the Baliraja dam movement of the peasants of the village of Tandulwadi in the drought-prone Sangli district of Maharashtra, to

143

which we have already alluded in Chapter 2. To recapitulate, the peasants of this village decided that the standard package of using irrigation water to grow a limited repertoire of water-hungry crops on a small area by a few farmers was inappropriate. They then worked closely with a voluntary group called Mukti Sangarsh and a highly qualified irrigation engineer, K.R. Datye, to work out a strategy that would not simply maximize commercial profits per hectare of land under unlimited irrigation, but rather would maximize the total quantity of additional biomass produced as a result of irrigation. Such a strategy favours deployment of irrigation water in more limited quantities in any given area, but its more uniform use over a much wider region. To work out such a strategy calls for careful experiments with different crop mixes and different schedules of irrigation. The strategy has to be flexible to permit effective use of small quantities of water in years of drought, with a shift to more water-demanding crops to take advantage of abundant rains in good years. Furthermore, the crop mix is to be so chosen as to produce varied biomass components to meet the manifold requirements; onions to be marketed as a cash crop, sorghum for consumption as food, subabul to provide green manure, eucalyptus to generate wood for local landless artisans, and so on. Obviously this is a challenging scientific problem, albeit of specific application to a particular locality. It is rather like the Japanese way of doing things – a process of deliberate, patient, continual improvement in productivity; every year making the product a little better, the process a little more resource efficient. It is time Indians applied this Japanese philosophy to the way the natural resources of the country are managed and tried slowly, patiently, continually to enhance their productivity. That is surely the key to success in the long run.

This would call for a radical restructuring of the scientific effort pertaining to the management of the country's natural resources. Today this is a narrow-based, bureaucratic effort pursued through centralized institutions. Even where there has been an attempt to set up dispersed research centres, as with the agricultural universities, there is little genuine understanding of conditions on farmers' land. True, some trials are carried out in farmers' fields, but the farmers themselves are scarcely involved in deciding on the contents of the experiment. The agricultural scientists by and large have little respect for people's knowledge; they are also bent upon pushing a package of uniform practices involving heavy, standardized applications of water and agrochemicals. At the extreme is forestry science, dominated totally by bureaucrats who take an occasional year or two to take on research assignments. The research is conducted in a thoroughly non-academic atmosphere with results published in house journals where acceptance is related to bureaucratic status and not scientific merit. Nor is the research usually related to what is happening in real forests 'out there'. Thus a hundred years after the Forest Research Institute was founded, the National Commission on Agriculture had to remark that it was quite unable to decide on whether the area controlled by the Forest Department was 69 or 75 million ha (National Commission on

Agriculture 1976). And after continually pointing for decades to rural biomass needs as the main cause of the many failures of forest management, the Forest Department has not conducted a single careful empirical study of this problem.

A bottom-up research strategy involving the wider masses of people would look to them to pose research problems of relevance. This would follow if the local communities are permitted to reassert control over the resource base. Then the communities would begin to ask for scientific inputs that would benefit them in a sustainable fashion, creating a genuine demand for environment-friendly science and technology. This demand could be effectively met by a much more decentralized network of institutions of learning. The undergraduate colleges, some 6,000 of them scattered throughout the country, could come to play a critical role in this endeavour. Today these colleges are merely involved in teaching routine, unimaginative courses with little relevance to real life. But they have a large body of students and teachers who could be stimulated to take up the challenge of investigating the status of the environment, to create a publicly accessible database on local natural resources, to continually monitor how various developments are impinging on it and to help work out ways of carefully, incrementally enhancing resource productivity through timely, appropriate inputs. A consortium of local community institutions, of high schools, colleges and research bodies would have to be organized to provide the necessary scientific inputs for this enterprise. That would be the organization of a science for sustenance, indeed a science for continuous progress.

Good management of natural resources would also call for the capability of continually monitoring and regulating such use. This is an enormous task given the vastness of the country and the complexity of its natural and social systems. It is no doubt true that modern technological advances, such as in remote sensing and computer-based databases, permit efficient collection and handling of large masses of information. But in spite of these advances, centralized planning of locality-specific interventions and monitoring of their consequences is an impossible task, especially when one is dealing with complex natural systems that vary greatly from location to location. The consequences of adding a certain amount of irrigation water to a field depend on the type of the soil, its slope, its aspect, the crops under cultivation, the weeds, insect pests, fungal diseases affecting the crops, the amount of manure or fertilizer used, the amount of money, labour available with the farming family, the human diseases prevalent, and so on. Equally variable would be the consequences of releasing carp seed in a freshwater pond, dependent in turn on the seasonal fluctuations in water level, the temperature regime, weeds, other fish present, eutrophication and chemical pollution, as well as on who owns the pond, who have been traditionally fishing in it and so on. Such detailed information would only be available in each locality at any point in time. Moreover, the values of relevant parameters can go on changing with time. For instance, a new insect pest may arrive; people may learn to like carp although they had looked down on them before; or there may be an exceptionally heavy downpour that breaks down the bund. Tackling all the changes in myriad localities in a centralized fashion is not a practical proposition, despite rapid improvements in capabilities of handling information. Instead, a far more cost-effective and efficient process of planning and management would be to carry out the component operations in a parallel fashion in each of the many different localities, restricting centralized functioning only to necessary co-ordination.

There are two very different ways of dealing with complex natural systems that are highly variable in space and time. The first is to try to fine-tune the inputs to the specific context, to gently nudge the system through many small, but timely and appropriate, inputs towards enhancing the services people want. The other approach is to ride roughshod over the complexity and variation, to homogenize the system through heavy external inputs and then manage it, employing some standard interventions uniformly prescribed for large areas. Consider, for instance, the problem of drinking water becoming ever more scarce in many Indian villages. One approach is to depend on local precipitation, however scanty, by properly regulating its flow, ensuring that it gets collected in some appropriate means of storage, protecting the stored water against pollution or bacterial contamination, and then using it in appropriate quantities at appropriate seasons. The other is to permit local water storages to get silted up or polluted, and then supply water to the village by bringing it over large distances in tankers, a solution that is indeed in vogue over large portions of the state of Maharashtra. The first approach calls for very careful local-level planning that can have some central scientific and technical inputs, but ultimately must depend on local motivation, knowledge and community-based management. The second approach needs little local planning and management, but can be run as a centrally planned, controlled operation. It uses a lot of external inputs – diesel for the tankers, for one. But that only creates a strong vested interest in the omnivore community in favour of the second, and far less desirable, approach.

Evidently, then, on all criteria – motivation, knowledge, management capabilities – the local communities are the most appropriate agency to look after and organize fine-tuned, prudent use of India's natural resource base, whether of forests, grazing grounds, irrigation tanks or agricultural lands. The state and the technically more sophisticated people of course have an important role to play, but that role should be of facilitation, co-ordination, of adding information not available to the local communities. Instead the state and its technocracy have assumed a very different role: that of agents of an extractive economy with no interest in prudent, sustainable resource use, controlling the resource base, claiming monopoly over all pertinent information and making all decisions on how to deal with the natural resources of the country.

How might we then move in the direction of strengthening the motivation, the knowledge base, the capabilities of the local communities towards prudent management of the country's natural resources? How might we transform the present predatory role of omnivores into one of mutualism with the ecosystem people of the country, paying a fair price for access to the natural resources, and contributing to their sustainable utilization through their stock of rapidly growing scientific knowledge and entrepreneurial abilities? The next chapter outlines the steps that might facilitate this transition in the crucial sector of forests and biodiversity conservation.

# 7

# WHAT ARE FORESTS FOR?

## THE FORESTRY DEBATE

In the global history of natural resource management, there are few institutions as significant as the Indian Forest Department. Set up in 1864, it now controls over one-fifth of the country's land area, backed by an impressive administrative and legal infrastructure. Not only is the Forest Department India's biggest landlord, it has the power to affect the lives of virtually every inhabitant of the countryside. In what is still dominantly a biomass-based economy, all segments of Indian society – peasants, tribals, pastoralists, slum dwellers and industry – have a heavy dependence on the produce of the forests, as the source of fuel, fodder, construction timber or industrial raw material.

And yet, in the century and a quarter of its existence, the Forest Department has been a most widely reviled arm of the Indian state. As narrated in Chapter 3, popular opposition to the workings of the Forest Department has been both sustained and widespread. One might say that underlying these varied protests have been two central ideas: that state control over woodland (as opposed to local community control) is illegitimate; and that the Forest Department's programmes of commercial timber harvesting have seriously undermined local subsistence economies. Since the early 1970s, grassroots organizations active in forest areas have called repeatedly for a complete overhaul of forest management. State policies, they contend, have excluded the dominant majority of the Indian population from the benefits of forest working while favouring the interests of a select group of industries and urban consumers. At the same time, they have asked for a reorientation of forest policy, towards more directly serving rural subsistence interests (see Fernandes and Kulkarni 1983). Nor are tribal and peasant groups the Forest Department's only critics. Thus conservationists have argued that commercial forestry has contributed significantly to the decimation of biological diversity and to an increase in soil erosion and floods. More recently, industrialists, who have hitherto been the prime beneficiaries of forest policy, have hit out at the department's partial withdrawal of subsidies and at being at last denied preferential supply of raw material. Faced with sharp criticism from several quarters, senior forest officials have claimed that they are unjustly being made scapegoats while the real causes of forest destruction escape identification (Misra 1984).

Despite nearly two decades of vigorous debate, however, policy changes in the forestry sector have been slow. Where earlier policies had narrowly focused on commercial exploitation for industry and the market, there is now greater talk of 'people's participation' and 'ecological security'. All the same, there is a lack of conceptual clarity – both within the government and outside – of how best to manage the country's forests in a sustainable fashion, while at the same time minimizing conflicts among the varied and often competing demands on its produce.

This chapter outlines the elements of what we believe to be an appropriate forest policy for India, one consistent with the alternative development agenda outlined in Chapter 5. Forestry is a resource sector to which we have devoted much of our own work. It is thus that we use forestry, rather than water or energy, to illustrate how the principles of conservative-liberal-socialism might be put into practice in one vital area of India's development efforts. Specifically, we show how the cardinal weaknesses of forest management to date lie in the failure to identify clearly the competing demands on India's forests. We then go on to prescribe ways of meeting these demands in a manner consistent with the imperatives of ecology, equity and efficiency – with clearly specified roles for the state, the market and the local communities respectively.

Let us first present the arguments of the key actors in the ongoing debate on Indian forest management. There are four important groups in this debate, whom we might characterize as wildlife conservationists, timber harvesters (i.e. industrialists), rural social activists and scientific foresters respectively. These groups are indeed 'interest groups' in the classical sense of the term: in other words, each has a specific claim on the resource under contention, and lobbies actively to defend and promote this interest. What is noteworthy is that in each case, the management proposals advanced by the group seek wider support from a sophisticated theory of resource use in which their own specific interests are presented as being congruent with the general interests of society as a whole.

## Wilderness conservationists

We begin with wilderness conservationists, an interest group small in number but with a major influence on policy. While their practical emphasis concerns the preservation of unspoilt nature, defenders of the wilderness are prone to advance moral, scientific and philosophical arguments to advance their cause. Although the initial and possibly still the dominant impulse is the aesthetic

*Plate 21* Tropical rain forests still cover extensive areas of the Western Ghats hill chain, identified as one of the world's eighteen biodiversity hot spots

*Plate 22* The pristine rain forest of the Andaman and Nicobar Islands is a treasure house of biodiversity

value of wilderness and wild species, this sentiment has found strong support from recent biological and philosophical debates. The theme of biological diversity as an essential component of direct and indirect, known and yet to be discovered, survival value for humanity and an emphasis upon the intrinsic 'rights' of non-human species have been prominent in recent debates on wilderness preservation. The quite specific interests of nature lovers in the preservation of wilderness are thus submerged in the philosophy of 'bio-centricism', which validates strong action on behalf of the rights of non-human nature (Guha 1989b).

These philosophical claims notwithstanding, in India a select group of ex-hunters and naturalists has been in the forefront of wilderness conservation. Their concern has been overwhelmingly with the protection of endangered species of large mammals such as the tiger, rhinoceros and elephant. Their influence is manifest in the massive network of parks and sanctuaries, many of which are oriented around the protection of a single species, notably the two dozen or so parks under the celebrated Project Tiger. They share with senior bureaucrats in particular a similar educational and cultural background, and this proximity has in no small way influenced the designation and management of wildland. It is thus quite fair to characterize them, as we have done in Chapter 4, as being environmentalists with an 'omnivore' background.

### Timber harvesters

The second important group in the Indian forestry debate consists of those who view the forest as a source of industrial raw material. In terms of their management preferences, timber harvesters are the polar opposite of the wilderness purists. While the latter stand for a 'hands off' management style that involves the minimum possible interference with natural processes, industrial demands on the forest often involve substantial and even ir-reversible modifications of natural ecosystems.

The industrial view of nature is simply instrumental. The forest is a source of raw material for processing factories, and the pursuit of profit dictates a pragmatic and flexible attitude towards its management. In the past, timber harvesters had been content with letting the state manage forests, so long as they were assured abundant raw material at rock-bottom prices; now, with increasing deforestation and the withdrawal of subsidies, wood-based indus-try has been lobbying hard for the release of degraded forest lands as captive plantations. Yet, in response to the environmentalist challenge of the past two decades, some industrialists have been quick to develop their own general theory of resource use. On the one hand, to justify their claims on public land they continually invoke the equation in conventional develop-ment thinking of industrialization with 'progress' and prosperity. At the same time, they argue that captive plantations will significantly lessen the

*Plate 23* Having economically exhausted the timber resources of mainland India, forest-based industry is now increasingly dependent on the as yet little-exploited forests of the Andaman and Nicobar Islands

pressure on natural forests, whose destruction, largely at their own hands, they had hitherto been indifferent to.

## Rural social activists

By 'rural social activists' we refer to those individuals and groups who work among ecosystem people dependent on the forest for a variety of economic needs, both subsistence and commercial. In India this would include groups of hunter-gatherers, shifting cultivators, pastoralists, artisans, landless labourers, and small, medium-sized and big farmers. A large proportion of the rural population lives close to a biological subsistence margin, and access to fuel, fodder, small timber and non-timber forest produce is critical to their ways of making a livelihood. Moreover, most Indian villages have stayed on one site for centuries, and the collective consciousness of their inhabitants stretches far back into the past; state usurpation of the forest is from this perspective a comparatively recent phenomenon, and thus resolutely opposed by tribals and peasants, who cling tenaciously to traditional conceptions of ownership and use.

These deep-rooted animosities are invoked by rural social activists in their polemic against state forest management. Some among them call for a radical reorientation of forest policy, towards more directly serving the interests of

152

subsistence peasants, tribals, nomads and artisans; others go further in asking for a total state withdrawal from forest areas, which can then revert to the control of village communities which, they claim, have the wherewithal to manage these areas sustainably and without friction. These specific recommendations draw sustenance from a powerful philosophy of agrarian localism, namely Gandhism. Where Gandhi, the 'Father of the Nation', always gave theoretical and practical primacy to rural interests, the policies of independent India are indicted – often rightly – as being heavily biased in favour of urban and industrial interests.

## Scientific foresters

We come finally to the group in actual territorial control of forests and wildlands. The brief of the Forest Department is the adjudication of the competing claims of the three interest groups dealt with above, and this in a 'scientific' and 'objective' manner. Historically, scientific foresters saw themselves as heralding the transition from *laissez-faire* to state-directed capitalism, in which they were, along with other professional groups, the leading edge of economic development. Conservation is for them the 'gospel of efficiency', and scientific expertise and state control its prerequisites. These ideological justifications aside, in practice foresters have often tended to act on the basis of narrow self-interest, tenaciously clinging to control over all forest areas and the discretionary power that goes with this control (see Lal 1989).

In this manner, through the skilful submergence of specific interests in a general theory of the human and natural good, these competing groups have legitimized their actual claims on forests and woodland. The territorial aspirations of foresters are advanced by claims to a monopoly over scientific expertise; the aesthetic longings of nature lovers are legitimized by talk of biological diversity and environmental ethics; the profit motive of capital masquerades as a philosophy of progress and development; and the demands of agrarian populations are juxtaposed to an ideology of the rights of the 'little man'.

Finally, let us note the varying positions on state control over forests and wildland. Two groups are unambiguous in supporting state control over the commons, even if they insist that the state enforce only their definitions of forest use. Wilderness conservationists see an interventionist and powerful state as indispensable in both designating wildlands and keeping out intruders, while for scientific foresters state control is virtually a *sine qua non*, in that it allows scientific experts the room to plan rationally and at a nation-wide level. Agricultural interests, for their part, while clearly expecting the state to take their side against other interest groups, are by and large opposed to state forestry, arguing instead for community ownership and management.

*Plate 24* Hills in dry parts of peninsular India have been laid totally bare by charcoal production to meet urban demands followed by overgrazing and hacking for fuelwood by villagers

Lastly, industrialists are characteristically opportunistic about the question of forest ownership, calling for privatization of forest land when it suits them, and for state control and subsidized raw material when it does not.

These four contending groups apart, a fifth category of resource users has also exercised a major influence on the direction of Indian forestry. This consists of the urban middle and upper classes, who constitute a substantial and growing market for a variety of forest produce. Unlike the other interest groups in the forestry debate, urban consumers, while numbering in the tens of millions, do not have a co-ordinated perspective on forest policy. None the less, their demands for paper, plywood, quality furniture and processed non-wood forest produce have powerfully stimulated processes of forest destruction in many parts of the country. Ironically, the upper class also constitutes the constituency from which wilderness conservationists spring and to which their arguments are largely addressed. Where on the one hand the consumption ethic of the urban elite contributes to forest exploitation, on the other hand their aesthetic and recreational preferences help determine the priorities of park management. Squeezed between these twin processes of destruction and conservation are the vast bulk of the rural population, who have little stake either in commercial forestry or in wilderness areas as presently managed.

## HARMONIZING CONFLICTS

The functions that forests perform can be classified under five major heads:

(a) maintenance of soil and water regimes;
(b) conservation of biological and genetic diversity;
(c) production of biomass for subsistence (i.e. as fuel, fodder, agricultural implements, building materials, etc.);
(d) production of woody biomass for commercial purposes (i.e. as raw material for industry); and
(e) production of non-timber biomass (e.g. cane, sal seeds, tendu leaves) for commerce.

The fulfilment of functions (a) and (b) may be said to constitute the *ecological* objective of forest management; of (c) the *subsistence* objective; and of (d) the *developmental* function. As for (e), in so far as it generates substantial employment among local (especially tribal) communities it constitutes both a subsistence and developmental objective.

In terms of our interest group framework, one can see immediately that wilderness conservationists have been concerned overwhelmingly with function (b) (and, to a much lesser extent, with (a)); rural social activists primarily with (c) and to a limited extent with (e); and timber harvesters almost exclusively with (d). These varied, often conflicting, demands are then made on the state agency physically in command of forest land. Indeed, since its inception the Indian Forest Department has had primary, and often sole, responsibility for assuring the ecological, subsistence as well as developmental objectives of forest management. Moreover, under its management system of working plans, these competing demands are often placed on the same territorial area or patch of forest.

In practice, however, the developmental objective has come to assume overwhelming importance in forest management. As we have documented in some detail elsewhere (Gadgil and Guha 1992: Chapter 6), the industrial thrust of state forestry has severely undermined subsistence options by restricting access of villagers to forest areas, and contributed greatly to deforestation and allied ecological degradation (see also Centre for Science and Environment 1985; Fernandes and Kulkarni 1983). Clearly, an alternative forest policy must separate out these three objectives – in a conceptual, territorial and management sense. Following from this premise, the following discussion offers distinct policy options and institutional frameworks for fulfilling the ecological, subsistence and developmental objectives of forest management.

## THE ECOLOGICAL IMPERATIVE

India has a great diversity of environmental regimes. In the northeast it can boast of areas with the highest levels of annual precipitation in the world; at the same time parts of the Thar Desert may see no rain for years on end with the annual average being below 300 mm. Its northern plains may experience temperatures of 50°C in the summers, while the higher reaches of the Himalayas remain perpetually snow-bound. India contains some of the highest mountains of the world as well as tiny coral islands in the Indian Ocean. Its river systems include the mighty Ganga and Brahmaputra with their huge flood plains and the short and swift west-flowing rivers discharging from the Western Ghats into the Arabian Sea. In consequence, India's natural vegetation ranges widely over tropical evergreen and mangroves to dry deciduous and desert scrub.

Indeed, India qualifies as one of the top twelve countries in the world in terms of biodiversity because of its tropical and subtropical climate, its great variety of environmental regimes, and its position at the junction of the African, Palaearctic and Oriental regions. It is also a part of a secondary, diffuse centre of domestication of plants and animals. Conserving the Indian heritage of biodiversity is a major challenge, especially given the large human population with its subsistence needs, and the growing resource demands of the urban–industrial sector.

This effort must consider biodiversity along many dimensions and on many scales of organization, ranging over the genetic, species, ecosystem, landscape and regional levels. At any of these levels, we must consider the

*Plate 25* Earthen mound inside which an unusual bird, the Nicobar megapode, lays its eggs. India, one of the world's top twelve countries for megadiversity, harbours a rich variety of plant, animal and microbial species

*Plate 26* This fig tree, possibly new to science, is one of the many species of trees endemic to the Andaman and Nicobar Islands

relative value of the different elements; after all, we set a higher conservation value on elephants and tigers than on the smallpox virus. In general, the rarer elements would be valued higher, as would elements of greater economic promise such as the wild relatives of cultivated plants. Other aesthetic, scientific and moral considerations would also enter into setting conservation priorities. In particular, if a major justification for conservation is the 'transformative value' of wilderness (Norton 1987) – that is, its value in moulding human attitudes towards the environment as a habitat for humanity then we would decide in favour of the biodiversity being widely dispersed rather than concentrated in a few large reserves.

Thus far, efforts at protecting diversity have focused almost exclusively on large nature reserves. These cover about 4 per cent of the country's surface; not all of this constitutes 'core areas' under strict protection. That is, it also includes areas from which timber harvests may be continuing (see Kothari *et al.* 1989). These large reserves have tended to be managed in the interests of

157

the more spectacular larger mammals such as deer and tiger, although now there are reserves aimed at conserving other groups such as orchids. The philosophy underlying the management of nature reserves has focused on the banning of hunting by all segments of the population, and exclusion of subsistence demands by the local rural and tribal populations. Thus long-settled villages have been shifted out of nature reserves, sometimes leading to damaging conflicts (see Chapter 3). Cultivators living on the periphery of these reserves are inadequately protected against crop damage and man-slaughter by wild animals, and very rarely compensated (see Sukumar 1989). The formal state-sponsored conservation effort has also been indifferent or inimical to traditions such as the belief in sacred groves and the protection of monkey populations.

Little thought or effort has gone into the conservation of biodiversity outside the nature reserve system. Indeed, the management regime prevalent outside the category of reserve forests has been an open-access regime in which no segment of society has a long-term interest in sustainable resource use. Development policies in general have also tended to promote exhaustive use of resources, both in reserved forests and outside them. Two elements that seem to be particularly significant in this context, as we showed in Chapters 1 and 2, are high levels of subsidies to the resource users and the lack of accountability on the part of resource managers. The pros and cons of doing away with resource subsidies, injecting responsibility into the actions of resource managers, including possibilities of participation in management by groups outside the state bureaucracies, and finally trans-forming the open-access regimes currently responsible for the tragedy of the commons are key issues that must be debated widely.

The current emphasis on nature reserves as the tool of biodiversity conservation, to the exclusion of all else, flows out of the overall development strategy. As we have shown, this strategy has equated development with the intensification of subsidized flows of resources to urban–industrial–irrigated agriculture sectors. These sectors have come therefore to acquire an in-creasingly larger share of economic and political power to the exclusion of the masses of peasants dependent on rain-fed agriculture and the small tribal population dependent on forest resources. It is these latter ecosystem people who relate directly to natural environments and depend on a diversity of biological resources for their personal wellbeing. On the other hand, those in power are well shielded from any direct negative consequences of all forms of environmental degradation including the loss of biodiversity. At the same time, they successfully monopolize most of the benefits flowing out of developmental activities. For them, therefore, environment and development often appear in contradiction, with environmental interests being naturally the ones to be sacrificed. Biodiversity conservation tends to be low on the priorities of the development strategy moulded by them, usually being restricted to nature reserves with a recreational appeal for the elite.

This development strategy explicitly attributes all environmental problems to the population pressure of the large masses of the ecosystem people directly dependent on the natural resources of their immediate environment. Therefore the exclusion of these people from nature reserves by the official machinery of the state has become a major focus of the state-sponsored conservation effort.

An examination of the adequacy of this approach is an appropriate starting-point for our discussion. This may be looked at from two perspectives. From the perspective of biological processes, the focus on nature reserves means that eventually all land or water areas outside this system are likely to be reduced either to highly degraded desert-like vegetation or to monocultures of just a few species such as eucalyptus, *Acacia auriculiformis* or a small number of high-yielding varieties of crops such as rice, wheat or tobacco. There may additionally be a few bodies of water dedicated to the production of carp or shrimp. The net result would be the persistence of a few species-rich fragments of a restricted range of habitats in the matrix of a biologically poor landscape and waterscape. Ecological theory tells us that such fragments are bound to lose a large proportion of their species in the long run. Such loss is all the more likely in the event of the global warming that may materialize over the next few decades.

It may be important, in this context, to strive to maintain, or better still re-create, a biologically richer overall matrix in which the species-rich fragments of nature reserves will be embedded. This would enhance the possibilities of long-term persistence of biodiversity in two ways. First, this matrix, which may include up to 40 per cent of the country's land surface under natural vegetation in various stages of degradation through human intervention, could on its own support a fair diversity of species of either commercial or subsistence value; this could easily run into several hundred species of trees, climbers, shrubs, herbs and grasses. These economically valued species would in turn support a large number of other species, especially invertebrates and smaller vertebrates. In addition, a small proportion of land in such a matrix, perhaps 5 per cent, could be devoted to biodiversity conservation in the form of species-rich islands such as individual fig trees or groves ranging from a few hundred square metres to a few hectares in size, to constitute a system of highly dispersed patchy refugia. This could end up adding another 2 per cent of the country's land surface to the nature reserve system. Such a matrix would also be of value in facilitating the dispersal from one reserve to another of many species at present restricted to the larger nature reserves. Ecological theory tells us that long-term persistence of species in any given locality is related to the rate of immigration into that locality of individuals of the different species present in the overall species pool. Such immigration would become all the more critical if global warming sharply increases rates of local extinctions. All in all, the maintenance of a biologically rich matrix with a system of small refuges serving as stepping-stones for dispersers would be a highly desirable complement to the system of nature reserves.

The cultivated lands and human habitations could also play a useful role in the maintenance of biodiversity through offering very small refuges such as the traditional banyan and peepul trees that dot India's villages and towns. Farm bunds, stream and river sides, canal sides and roadside avenues could all contribute towards the maintenance of a biologically friendly matrix of conservation areas.

The second important issue to be explored is how society could organize the maintenance of biodiversity in the system of nature reserves, as well as in the matrix of non-cultivated vegetation in which the nature reserves would be embedded. The current approach is one in which technical decisions as well as policing are in the hands of the state apparatus. One may, in this context, ask three crucial questions:

(a) Does this apparatus have at its disposal the information needed to make appropriate decisions?
(b) Is this apparatus adequately motivated to maintain biodiversity?
(c) Is this apparatus competent to carry out effectively the task of regulating human interventions in the interest of the maintenance of biodiversity?

There is room for questioning the adequacy of the state apparatus on all three counts. The personnel of this apparatus primarily derives from the Forest Service. The bureaucracy in general, including the forest bureaucracy, has monopolized the collection and interpretation of information on natural resource management in India. However, this has been done in a most casual and unscientific fashion, so that the scientific competence of this apparatus is in grave doubt. Moreover, ecological systems are exceedingly complex systems. The behaviour of such systems can be predicted only to a very limited extent on the basis of general principles. At the same time, historical observations of the system behaviour are a valuable input into predicting the outcomes of human interventions in specific systems. This implies that knowledge of folk ecology available with the local people may compare to and even exceed in value the knowledge base available with the official apparatus (see Chapter 6).

Second, no member of the official apparatus has a personal stake in the maintenance of biodiversity in any given locality. Of course, the local people may or may not have such a stake depending on how different elements of biodiversity affect their personal wellbeing. But if they receive direct benefits from maintenance of biodiversity for themselves (for instance, through supply of wild fruit or herbal medicine or protection of stream flow) or if they are rewarded by the wider society for protecting biodiversity, they are far more likely to be motivated to maintain the biodiversity of the localities to which they are intimately attached.

Finally, the state apparatus is quite ineffective in discharging a policing function unless it has full local co-operation, as has been strikingly brought out by the experience of the failure to apprehend the notorious sandalwood and ivory smuggler of Karnataka, Veerappan. There is, therefore, every reason to believe that the system of biodiversity conservation could be made much more effective by utilizing local folk ecological knowledge, and by creating a stake for the local population in conservation of biodiversity, if necessary by a system of payment of monetary reward as service charges (Gadgil and Rao 1994).

A decentralized system of biodiversity conservation, which might come to cover all of rural India, provides a great opportunity for developing a symbiotic relationship between systems of folk knowledge, traditional knowledge systems such as Ayurveda, Siddha or Unani medicine and modern scientific knowledge. A great deal of locality-specific knowledge of biodiversity elements significant to their own lifestyles resides with ecosystem people, many of them illiterate. This, for instance, is the case with specialist fisherfolk, who have an intimate knowledge of water bodies and ongoing changes in them and their snail, bivalve, shrimp, crab, fish fauna; or nomadic shepherds, who know a great deal about large tracts of scrub savannas and grasslands and their vegetation. But this knowledge can and should be tapped to feed into a wider process of biodiversity conservation. Some pioneering attempts at documenting such knowledge have already been initiated through an organization called SRISHTI under the leadership of Anil Kumar Gupta at the Indian Institute of Management, Ahmedabad. SRISHTI has organized a series of biodiversity contests with the help of primary schools in different states, which have discovered everywhere exceptional individuals, children as well as adults, who know hundreds of local species. SRISHTI also runs a network called HONEYBEE for giving due credit for and sharing such knowledge of distribution of biodiversity, as also its uses, often newly discovered by various innovators.

As we visualize it, rewards for effective custodianship of biodiversity and knowledge of its use would flow primarily to a geographically defined community; though they may also go to individuals, to caste or tribal groups or to clusters of village communities. It is important for the rewards to be in the form of assertion of community rights over public lands and waters within their defined territory. With the communities standing to gain in the long run they are likely to organize sustainable use patterns for these lands and waters, and to manage them in such a way as to also enhance their biodiversity value. However, it is absolutely essential that they should have adequate authority to exclude outsiders, and to regulate the harvests by group members, as well as an assurance of long-term returns from restrained use for such a system to operate effectively.

Such additional rights of access to publicly held resources would serve as a positive incentive for making prudent use of public lands and waters to meet local biomass needs. But the decentralization of natural resource management may by itself be inadequate to promote maintenance of high levels of distinctive elements of biodiversity within the community, since there may be better economic returns from monocultures, be they of high-yielding crop varieties on private lands or eucalyptus on public lands. Specific incentives, which should be viewed as service charges, are therefore necessary to maintain diversity, whether of cultivars on farm lands, of indigenous livestock breeds, of fruit trees in homesteads, of medicinal plants, wild relatives of crop plants, or troops of primates or crocodiles on public lands and waters. Individuals or communities participating in such efforts must therefore be paid certain rewards linked to the levels and value of biodiversity within their territory. Such rewards could be untied funds coming to the community to be devoted either to community works such as educational or health facilities, or to be shared among all community members. The rewards could also take the form of building community capacity for maintaining enhanced value of bio-diversity within their territory, or for setting up biodiversity-based enter-prises, such as chemical prospecting or extraction of active ingredients. Similar rewards may also flow for making available knowledge pertinent to uses of biodiversity, for instance in pest control.

Apart from these rewards, which might provide sustained positive incen-tives to custodians of biodiversity, there could be one-time rewards such as fees for collecting some genetic resource from the territory, or fees for sharing some piece of knowledge relating to use of biodiversity. There may also be shorter-term rewards such as royalties from commercial application of some element of biodiversity or some piece of knowledge relating to its use. It will, however, be very difficult to channel royalties of this nature properly to particular individuals or communities, since this would require every such element of biodiversity or knowledge to be traceable to a particular set of localities, communities or persons. It might therefore be better to pool such royalties in a national biodiversity fund and use this for rewarding com-munities for the ongoing maintenance of biodiversity within their territories (see, for details, Gadgil and Rao 1994). In this manner, community efforts at biodiversity conservation would innovatively supplement and perhaps in the long term largely replace more conventional schemes for the protection of wild areas and wild species through national parks and sanctuaries.

## THE IMPERATIVE OF EFFICIENCY

The industrial orientation of state forestry in India has passed through three stages (see Gadgil and Guha 1992). In the first phase, c. 1870–1960, traditional selection felling methods of timber harvesting were relied upon in the belief that industrial demands could thus be met in perpetuity. However, for a variety of reasons, including a poor understanding of the ecology of species-rich tropical forests, an inadequate database, no proper planning for local subsistence demands and corruption, the objectives of 'sustained yield' forestry have been honoured mostly in the breach (see also Food and Agricultural Organization 1984). In the second phase, c. 1960–80, selection

*Plate 27* Raft after raft of timber float down many streams of the Andaman and Nicobar Islands

felling was supplemented by large-scale clear-cutting of natural forests, which were then replaced by monocultural plantations, chiefly of exotic species such as eucalyptus and Caribbean pine. In the third phase, *c.* 1980 to date, wood-based industry has turned increasingly to purchasing raw materials from tree farmers or through imports.

As numerous studies have shown, commercial tree felling and the raising of species-poor plantations have proved to be most incompatible with the ecological functions of natural biological communities. At the same time, by largely excluding village communities from its fruits, industrial forestry under state auspices has also seriously violated the imperative of social equity. Not surprisingly, wood-based industry and the forest departments that tend to serve its interests have been the prime target of environmentalists.

As the proposals we are about to offer are likely to be controversial among those very environmentalists, a few clarifying remarks are in order. In their often justifiable attacks on the insatiable appetite and ecological insensitivity

163

*Plate 28* India's forest-based industry is now tapping the rain forests of the Andaman and Nicobar Islands

of the forest industries sector, environmentalists, particularly those of a Gandhian persuasion, have sometimes tended to argue that India should turn its back on industrialization altogether. Undeniably, luxury furniture and consumer goods (such as chocolate and shampoo) based on the processing of select non-timber forest produce are hardly central to the developmental goals of the country. But the same cannot be said of paper, rayon, packaging materials, tendu leaves, etc., materials indispensable to the welfare of hundreds of millions of Indians.

If we are to reject, as we must, the extreme option of closing down all industrial units, the question remains – how best can we assure the raw material requirements of forest-based industry consistent with the imperatives of ecology, equity and efficiency? There are four alternative paths to choose from:

(a) continue to vest this responsibility with the Forest Department and its methods of timber harvesting from public land, with the caveat that subsidies be withdrawn and industry be made to pay a fair ecological price for forest produce;
(b) rely increasingly on the import of wood and wood pulp, thereby shifting the burden of deforestation on to other countries;
(c) turn over state forest land directly to industrial units as 'captive' plantations; or

164

(d) consolidate and promote the efforts of individual tree farmers, and effectively link them to regional markets and industrial units.

Given India's ever precarious balance of payments, the second alternative is hardly desirable. There is little doubt that industrialists would themselves prefer alternative (c), and have thus for several years been lobbying vigorously with state and central governments to hand over land on long lease (see Jain 1989). As for the Forest Department, notwithstanding the serious failures of 'sustained yield' forestry in the past, it would most likely prefer to continue these methods, under its control, of supplying raw materials to processing units.

Ecological considerations apart, from the viewpoint of efficiency too the state machinery is quite inappropriate for undertaking economic productive functions (indeed, studies have clearly shown tree farming by individuals to be far more productive, and far less costly, compared to government plantations). At the same time, conceding the claims of industry for captive plantations would inaugurate a new *zamindari* system, thereby violating the principle of equity. From all points of view, farm forestry appears the most feasible path for meeting industrial demands. As compared to state forestry, or industrial plantations, or indeed imports, it will distribute benefits far more widely among the population. Phasing out all commercial wood production from state lands would also work, all other things remaining equal, to promote the protection of natural forests.

Why then are so many environmentalists bitterly opposed to farm forestry? Morally, their opposition stems from a larger disdain for commercial trans-actions of any kind which go against the Gandhian ethic of local autonomy and self-sufficiency. Ecologically, they view eucalyptus – overwhelmingly the dominant species on the farm – as an unmitigated evil, for it is believed to poison the soil and lead to a lowering of the water table. Socially, they have indicted tree farming for increasing inequalities within the village, in particu-lar by its displacement of food crops requiring more labour (see Bandyo-padhyay and Shiva 1984).

Viewed dispassionately, the environmentalists' case against farm forestry, its strident rhetoric notwithstanding, rests on rather uncertain foundations. The Indian farmer needs cash income for a variety of purposes, from marrying off his daughters to making capital investments on his land. If tree crops are objected to in this regard, consistency requires that sugarcane, tobacco, cotton and a host of other commercial plant species (including cereals) be banished for the same reason. From an ecological perspective, the clear-felling of natural forests to replace them with eucalyptus (undertaken in many forest areas in the 1960s and 1970s) was clearly an unwise policy, but so far as farmland is concerned, eucalyptus is probably less harmful for the local environment than are other important cash crops such as sugarcane. The social criticism has more to commend it, though it too must be set against the gains in intersectoral equity (i.e. between town and country) represented by tree farming. Most substantively, farm forestry can only be assessed against the other options available to us – closing down all industrial units, harvesting wood from state forests, imports and the creation of captive plantations – among all of which it is incontrovertibly the most benign alternative.

Ironically, farm forestry has itself fallen on lean days. From the late 1970s, hundreds of thousands of cultivators, aided by the supply of free seedlings by the state, started planting eucalyptus and casuarina. According to one estimate, between 1981 and 1988 roughly 8,550 million trees were planted on farmland in India, with a survival rate of approximately 60 per cent. More than 80 per cent of these trees were eucalyptus. Yet by 1986 or so, disillusionment had set in, as yields and especially prices were well below expectations. In northwestern India and Gujarat – highly commercialized farming regions where eucalyptus appeared to have taken firm root – farmers were even uprooting saplings well ahead of maturity and replanting these areas with food crops.

Studies by N.C. Saxena of the Oxford Forestry Institute and by the Institute of Rural Management at Anand (Saxena 1990; Institute of Rural Management 1992) have highlighted several reasons behind this growing disenchantment with tree farming:

(a) lack of adequate technical information, in the absence of which farmers went in for high-density plantations with little spacing – this led to yields lower than anticipated, while the trees were too narrow for several commercial uses;

(b) depressed prices caused in part by competing sources of raw material for processing industry, especially subsidized supplies from forest land and imports;

(c) lack of market information on prices and sources of demand, leading farmers to depend heavily on middlemen, whose own margins were therefore much higher than the farmers';

(d) the existence in some states of restrictions on the harvest and transport of wood, compounded by the cost of obtaining permits from the bureaucracy; and

(e) in some cases, the lowering of yields of agricultural crops caused by declining soil fertility ascribed to the planting of eucalyptus.

These problems notwithstanding, long-time observers of farm forestry are agreed that it constitutes the most desirable option for generating raw material for industry. However, to make farm forestry remunerative and productive, a policy package with the following elements needs to be executed as soon as possible:

(a) The abolition of all laws and controls inhibiting individuals from planting or selling trees grown on private land.

(b) A moratorium on supplies of timber from state forest land and a phasing out of timber imports. A particularly poignant case is the recent eucalyptus glut in the northern state of Haryana, where prices fell during the 1980s from Rs 340 to Rs 190 per tonne. Ironically, in the neighbouring state of Himachal Pradesh excellent cedar forests are being felled for the manufacture of apple crates, which could just as easily be made from Haryana eucalyptus.

(c) These situations would be avoided when restrictions on timber harvesting from state land are coupled with efforts to link farmers directly with processing industry. Tree growers' co-operatives must be promoted to eliminate middlemen, and in support the state should stipulate that all wood-using enterprises (including those for perishable fruits like apples and oranges) be required to obtain their supplies from individual tree growers or their co-operatives.

(d) As a subsequent step, farmers' co-operatives might even set up processing units (e.g. small paper mills) to retain within the farm sector a greater proportion of value added. The milk and sugar co-operatives of western India might provide pointers in this regard (Attwood and Baviskar 1988).

(e) Scientific research needs to be undertaken on high-value, low-rotation tree species other than eucalyptus, that better complement food crop production.

(f) As is already the case for other cash crops, state bodies might generate and disseminate information on demand, supply, prices, etc., keeping farmers informed about prevailing market conditions.

(g) Finally, channels must exist for the flow of credit to tree growers, particularly to small farmers working on marginal land who might not otherwise take to long-gestation crops.

In conclusion, we should once again stress that shifting the burden of commercial wood production to private farmland makes eminent ecological sense. Owing to the greater productiveness and cost-effectiveness of farmers, the policies advocated here would require perhaps as little as one-quarter of the area required if we were to continue the traditional system of harvesting timber from state forests. By any reckoning, this constitutes an enormous potential saving of the country's natural endowment.

## THE IMPERATIVE OF EQUITY

We come finally to the subsistence function of the forests, namely, the meeting of the varied biomass needs of India's villagers. Foremost among these needs are fuelwood and fodder, though access to small timber, thatch, green manure and raw materials for artisans such as basket weavers is also important for the livelihood of numerous villagers. Acute shortages of these materials prevail all over the country. While women have to walk longer and longer distances for fuelwood, and graziers to forage over a much wider area for fodder for their livestock, many wood-working artisans have been forced to abandon their calling altogether owing to the unavailability of raw material. Indeed, it is these shortages that lay behind the forest-based conflicts of the 1970s and 1980s (see Chapter 3). Fulfilling the biomass requirements of peasants, tribals. nomads and artisans in a sustainable and efficient manner thus constitutes perhaps the most urgent task of Indian forest policy. Ironically, the forestry report of the National Commission of Agriculture, a considered official statement on the subject, rejected these demands as illegitimate, arguing that villagers should depend instead on private lands and resources for generating biomass for their use (National Commission of Agriculture 1976).

Although some forest officials (and the odd wildlife conservationist) still maintain that state forest lands have no subsistence function whatsoever, this view is simply not valid. Demands of fuel and fodder, in particular, are too huge to be met in any significant proportion from private land. In 1981, for example, aggregate fuelwood demand in India was estimated at 262 million tonnes (mt), of which only 49 mt was being met from crop residues. While total fodder demand was estimated at 560 mt, the availability of agrowastes for this purpose was only 273 mt. Moreover, the bulk of crop residues are claimed by farmers with large holdings, who can in addition avail themselves of alternative fuel sources such as kerosene and cooking gas. But a majority of poor and landless peasants, nomads and artisans must perforce meet their biomass requirements from public land. The position with respect to thatch and building materials is undoubtedly very similar, whereas for a variety of non-timber forest products, villagers must turn *in toto* to lands owned not by individuals or communities but by the state.

On grounds of equity, therefore, the recommendations of the National Commission of Agriculture can be quickly cast aside, for the varied biomass needs of hundreds of millions of villagers can only be met, in the main, from lands presently constituted as 'reserved forests'. One could of course continue the present system, under which these demands are indeed fulfilled from state lands, but in a haphazard and unregulated fashion, and often by stealth. Another option, seriously advocated by some environmentalists (see Agarwal and Narain 1990) is to abolish state control totally, and let these lands revert to village communities to manage as best they can.

But where state control has lamentably failed to meet the subsistence imperative of forest policy, transferring reserved forests to the nearest hamlet would only, in the vast majority of cases, substitute one unregulated, inequitable and unsustainable system for another. This transfer might have the claims of justice – righting a historical wrong by returning forests confiscated by the state decades ago to those who previously owned them –

but it is only in exceptional cases, and under outstanding leadership, that village communities today can rise above divisions of caste and class, and the growth of individualism, to act consistently in their collective interest. (Their capacity to do so is also greatly undermined by the overall centralization of political authority that obtains over most of India.) All the same, we believe that a network of lands, constituted in the main from what are at present reserved forests, should indeed be totally devoted to meeting the diverse biomass needs of local people. Yet much more thought needs to be given to the appropriate management system, or systems, for these lands, in particular to the precise rights and responsibilities of the state and villagers respectively.

Valuable clues in this regard may be found in the experiences of two large experiments in subsistence forestry. One is the network of *van panchayats* (village forests) in the central Himalaya, the other the dynamic joint forest management programme of the West Bengal Forest Department. Although occurring in two widely separated areas with very different social structures, and originating in quite different historical processes, both systems offer us sharp pointers to the future direction forest management might take.

The *van panchayats* of the hill districts of Uttar Pradesh constitute the only major network of village forests, mandated by law, in India. They originated as a consequence of massive popular unrest at the state reservation of forests in the British-ruled Kumaun division. These forests were, for the colonial state, an important source of pine resin and timber; at the same time, they were situated in a region of great strategic significance, bordering both Nepal and Tibet and home to some of the bravest soldiers of the (British) Indian Army. Smouldering resentment at the state usurpation of forests culminated in a widespread peasant movement in 1921 that virtually paralysed the administration (see Guha 1989a). In its aftermath, the state decided to withdraw control over less valuable (from a commercial point of view) forest areas. Where in most other parts of the subcontinent the British were implacably opposed to promoting village-owned and -managed forests, in Kumaun they worked swiftly to allay discontent by allowing peasants effective control over large areas of woodland.

Under the Kumaun *panchayat* forest rules of 1931 (amended in 1976) a *van panchayat* can be formed, out of non-private land within the settlement boundaries of a village, on application to the deputy commissioner of one-third or more of its residents. Once formed, the *van panchayat* must elect its own managing committee of five to nine members, which then assumes control of the forest. Under the act, *van panchayats* have powers to regulate the use of the forest by villagers – e.g. the closure to grazing in certain seasons, restrictions on fuelwood extraction by individual households, and regulation of lopping – and levy fines on offenders. They can also prevent people from other villages from using their forest and prevent encroachments for dwellings or cultivation. Usually, a full-time watchman is appointed, paid from contributions by villagers. However, no timber trees may be felled without the permission of the Forest Department. The department is also empowered to give technical advice on timber and resin extraction. Of the receipts of *van panchayats*, 40 per cent are assigned to the Forest Department (for technical advice that is in fact very rarely given), 20 per cent is allotted to the *zilla parishat* (district council) while 40 per cent is kept with the deputy commissioner on account of the *van panchayat* itself, which with official permission may use this money for community services and local improvements.

As of 1985, there were 4,058 *van panchayats* in the Kumaun and Garhwal divisions, covering an area of 469,326 ha. Several studies have commented on the relatively healthy state of *panchayat* forests – usually in as good a condition as, and sometimes in even better shape than, reserved forests supposedly managed on 'scientific' lines – and this is corroborated by our own field experience. Oak forests in particular are invariably well maintained, though in some areas pine forests are also coveted by villagers for their grass. This is not to say that the functioning of *panchayat* forests might not be considerably improved. *Van panchayats* presently lack powers to collect fines directly from offenders (who may appeal to the deputy commissioner or go to court) or the ability fruitfully to spend money accumulated in their account on village improvements (thus in Ranikhet subdivision, Rs 3.8 million had accumulated unspent in the *van panchayat* account kept with the district magistrate). For its part, the government's *van panchayat* inspectors are too few and too poorly paid to fulfil their supervisory role properly, while the Forest Department does not feel obliged to provide technical advice as required under the act (Somanathan 1991; Ballabh and Singh 1988; Shah 1989).

These limitations notwithstanding, as an ecologically viable and socially equitable system of resource management, the network of *van panchayats* is a salutary reminder of the potential force of what often appears to be a tired cliché, namely 'popular participation'. A quite different system of popular participation in forest management has been more recently crafted in the state of West Bengal, far distant from the Kumaun Himalaya. In 1972, the West Bengal Forest Department recognized its failures in reviving degraded sal forests in the southwestern districts of the state. Traditional methods of surveillance and policing had led to a 'complete alienation of the people from the administration', resulting in frequent clashes between forest officials and villagers. Forest- and land-related conflicts in the region were also a major factor in fuelling the militant peasant movements led by the Naxalites. Accordingly, the department changed its strategy, making a beginning in the Arabari forest range of Midnapore district. Here, at the instance of a far-seeing forest officer, A.K. Bannerjee, villagers were involved in the protection of 1,272 hectares of badly degraded sal forest. In return for help in protection, villagers were given employment in both silvicultural and harvesting operations, 25 per cent of the final harvest, and allowed fuelwood and fodder collection on payment of a nominal fee. With the active and willing participation of the local community, the sal forests of Arabari underwent a remarkable recovery – by 1983, a previously worthless forest was valued at Rs 125 million.

Following the success of the Arabari scheme, village forest protection committees (FPCs) were started by the Forest Department in other areas. As of July 1990, there were 1,611 FPCs protecting 191,756 ha, primarily of degraded sal coppice forests, in the districts of Midnapore, Bankura and Purulia. This accounts for almost 47 per cent of the forest area in these three districts. The FPCs collectively have about 150,000 participating members. They have been most successful where the forest to household ratio is high; that is, where the dependence on forests for livelihood security is the greatest. The functioning of the FPCs has also been facilitated by prompt action by the Forest Department on complaints by villagers regarding illegal harvesting by outsiders. While the regrowth of these predominantly sal forests has been impressive, other trees such as mahua, kusum, amla, neem and karanj have also benefited from villager–Forest Department protection and co-operation. Where villagers were earlier forced to look for work elsewhere, the cumulative benefits of joint forest management have resulted in a significant reduction in seasonal migration out of these areas. Restored to effective control over their environment, ecosystem people are no longer forced to become ecological refugees (Malhotra and Poffenberger 1989; Malhotra et al. 1991; West Bengal Forest Department 1988; Steward 1988).

We may highlight five key features of the success of the FPCs in southwestern Bengal:

(a) No additional funding has been required for these schemes. In fact, Forest Department employees have devoted considerably less time and effort to organizing FPCs than they previously devoted to policing. This has important implications for Forest Department–local community relations elsewhere in the country.

(b) The benefits of FPCs have cut across both ethnic and political boundaries. FPCs have been successfully formed in villages with tribal, non-tribal and mixed populations, as well as in villages owing allegiance to the Congress, Communist Party of India (Marxist) and Jharkhand Parties. Interestingly, a study of forty-two FPCs in one part of Midnapore district revealed that all-tribal FPCs and mixed FPCs with a higher proportion of tribals were performing best of all – this being ascribed to higher tribal dependence on forest produce and tribals' more intimate knowledge of forest ecology (Malhotra et al. 1991).

(c) As the scheme has been successfully extended to more than a thousand villages, it cannot be said to be an isolated case or a flash in the pan. Rather, it has been tried and successfully tested in a variety of situations.

(d) The experience of the FPCs has further undermined the conventional wisdom, on which much forest management has been hitherto based, that timber is the main produce of the forests. Thus the regrowth of sal has facilitated the emergence of a large diversity of plants in the understorey that have an incredible variety of local uses as food, fuel, fodder, medicinal plants and sundry processing materials for household use and sale. One study documents 155 different species of plants and animals as being harvested by villagers from sal forests (Malhotra *et al.* 1991). Of these, a flora of at least seventy plant species were identified as being used regularly and in substantial quantities. These included sal seeds and leaves, mahua flowers and seeds, kendu leaves, tubers, mushrooms and cocoons. On a conservative estimate, 17 per cent of household income, on the average, came from the collection of non-timber forest produce. Estimated over a ten-year period, this income would be seven times more than the realization of the 25 per cent share of the harvest of sal poles which the Forest Department initially offered villagers as the main economic benefit of joint forest management.

(e) Finally, and more than anything else, the experience of the FPCs points to a qualitatively new relationship between the Forest Department and local people. While the West Bengal Forest Department is still in effective control of these areas, it has shown a greater willingness than its counterparts in other states to share power, authority and economic benefits with the villagers. At the same time, the successful functioning of the FPCs has restored a sense of dignity and self-worth among these communities, now confirmed as joint managers of the forest.

Two independent but close observers of the evolution of joint forest management offer this evaluation of its progress:

> The lesson we draw from the forest protection committees of West Bengal is one of hope for the state, for the nation, and for the world. The process of forest degeneration can be reversed. Through partnerships between foresters and forest communities, effective protection can be established with ecological and economic benefits for the community and the larger society. However, the task is not an easy one. It requires the political will of the state to delegate responsibilities to the forest communities, and changes in policies and procedures that may have been in effect for over a century. Ultimately, it requires a transition from management practices developed during the 19th century, to a management system that can respond to the social, economic and ecological needs of the 21st century. This requires dynamic leadership within the forest department to allow for a transition from traditional modes of timber production and forest policing, to a cooperative, responsive ability to work with rural communities.
>
> (Malhotra and Poffenberger 1989)

The experience of FPCs in West Bengal over the past two decades offers a fascinating comparison to the rather longer history of *van panchayats* in the Uttar Pradesh Himalaya. There are some major differences between the two schemes. As regards the respective social structures, hill villages are marked by comparatively little differentiation, while the villages of West Bengal are characterized by considerable heterogeneity with respect to caste, class and ethnic group. From an economic point of view, Himalayan forests are valuable to local communities in large part because of the close integration of agriculture and animal husbandry – with grass in pine forests and oak leaves as fodder being particularly prized. By contrast, the tribe and caste groups of the West Bengal FPCs cherish a much greater variety of non-timber forest produce, collected both for consumption and sale. Finally, *van panchayats* have to contend with an indifferent and even hostile official environment: nowhere are bureaucrats and politicians as authoritarian and corrupt as in Uttar Pradesh, while in recent years the state has tried hard to bring *van panchayats* more closely under its control. The FPCs of West Bengal have the inestimable advantage of functioning in a more congenial environment – with sympathetic politicians committed to decentralization and rural development, and bureaucrats providing inspirational leadership in the spread and functioning of protection committees.

These divergences apart, there are some notable similarities as well. In both cases, forests are in relatively good condition – ecologically speaking, certainly in better shape than under previous management systems. Common, too, is the diminution of conflict between peasants and the state over forest resources – though in the one case (West Bengal) this has been achieved through genuine partnership, in the other (Kumaun) by marking out distinct, non-overlapping areas for the Forest Department and village communities to manage. Third, in both schemes women are among the main beneficiaries of forest management. As primary gatherers and collectors of forest produce, women stand to gain most from forest protection and regeneration. Admittedly, it is only in the exceptional case – for example, where motivated voluntary organizations are active (Burra 1991) – that women have a formal leadership role in *van panchayats* or forest protection committees. But they do play a key informal role in the detection of intruders and those who violate the rules; in the forest much of the time, they are in effect the eyes and ears of the village community. Fourth, both schemes are genuinely decentralized, with benefits flowing largely to the village instead of (as is always the case with commercial forestry) to more powerful omnivore groups outside.

In their own, very different, ways, *van panchayats* and FPCs offer alternative paradigms to the authoritarian system of state forestry that has prevailed over much of India for well over a century. Where the latter rests on the exclusion (through policing) of rural communities, these systems work on the principle of inclusion. By restoring local control over management and use, they have overcome, with some success, the alienation of villagers from forests that has almost everywhere been the unhappy consequence of state forestry.

Notably, neither system has exhausted its potential. Thus *van panchayats* constitute just over 7 per cent of the forest area of the districts in which they operate – and there is no reason why, as concerned local scientists and activists have demanded, the large areas of 'civil and soyam' forests, presently open-access, heavily degraded forests under the nominal control of the district administration, should not be brought under the more effective and equitable *panchayat* system (see Shah 1989). In other areas with comparable, relatively undifferentiated, social structures, the constitution of a network of village forests, with the administration playing an advisory and supporting role, might be a worthy aim of state policy. In the majority of forest regions, however, it is more likely that a system based on the West Bengal experience of joint village community–Forest Department management is able to succeed. Sadly, despite the repeated urgings of sympathetic officials in the central Ministry of Environment and Forests, and although some individual forest officials have, on their own, started FPCs in their division, other state forest departments are lagging far behind in this regard. Political corruption, the territorial instincts of forest officials and the sloth and inertia of Indian administrations generally constitute formidable obstacles to the spread of joint forest management outside the state of West Bengal. Yet these obstacles must be overcome if India's forests are to be managed equitably, sustainably and without friction between state and citizen.

## CONCLUSION

Let us now recapitulate our suggestions. The structure for the administration of public (including forest) lands remains essentially colonial in nature. While reform of agricultural land was pressed forward following independence, the management of public lands has remained frozen. Obviously it too needs a radical reorientation. These lands should be divided into two categories: (a) lands devoted to ecological security, and (b) community-managed lands devoted to providing livelihood security through a production system compatible with ecological security. The commercial plant production function should be fully shifted to private agricultural lands. Given such an outlook, the foresters would play the role of joint managers with people of lands devoted to ecological security or to livelihood security and an extension machinery serving tree farmers. Notably, the assignation in the past of overlapping, often conflicting, functions to a unitary forest department as well as the same patch of forest, without a clear priority being assigned among these functions, has in many areas been a prime factor behind deforestation. Therefore, it would be best, as we suggest, to separate these functions. This management system should be worked out and implemented on the basis of a detailed decentralized land use planning exercise which would start afresh with land capability rather than the nature of bureaucratic control of land as its starting-point. Once an appropriate land use plan, with emphasis on the urgency of ecological security and livelihood security, is worked out, then its proper implementation could be organized not as a centralized bureaucratic exercise but as a location-specific, people-oriented exercise. This calls for strengthening of the village- and district-level planning and administration machinery with higher-level controls primarily geared to ensure that the twin considerations of ecological security and livelihood security are given due weight.

This separation of objectives, functions and management systems for the three main categories of land use outlined above – nature reserves, community forests and farm forests respectively – must be the starting-point of forest administration. A shift away from state monopoly is an essential precondition for both ecological security and livelihood security. Indeed, active involvement of the people is also necessary for alleviating the bitter, wasteful and often violent conflicts between the state machinery and the rural population over access to forest produce. That healthy forest cover can be brought about only through a close co-operation between government and the villagers was well realized by one of our early nationalist organizations, the Pune Sarvajanik Sabha. Contesting the colonial Forest Act of 1878 for its excessive reliance on state control, the Sabha pointed out that the maintenance of forest cover could more easily be brought about by

> taking the Indian villagers into confidence of the Indian government. If the villagers be rewarded and commended for conserving their patches of forest lands, or for making plantations on the same, instead of ejecting them from the forest lands which they possess, or in which they are interested, emulation might be evoked between neighbouring villages. Thus more effective conservation and development of forests in India might be secured, and when the villagers have their own patches of forests to attend to government forests might not be molested. Thus the interests of the villagers as well as the government can be secured without causing any unnecessary irritation in the minds of the masses of the Indian population.

More than a century on, these sentiments remain strikingly relevant. For we are yet in search of a truly democratic and participatory system of forest management; one founded not on mutual antagonism, but on a genuine partnership between the state and its citizens.

# 8

# IS THERE SAFETY IN
# NUMBERS?

## NO MORE INDIAS?

A decade after India attained independence, the writer Aldous Huxley was
deeply pessimistic about the future of a culture he had closely studied and
long admired. As he wrote to a friend:

> India is almost infinitely depressing; for there seems to be no solution
> to its problems in any way that any of us [would] regard as acceptable,
> the prospect of overpopulation, underemployment, growing unrest,
> social breakdown, followed, I suppose, by the imposition of a military
> or communist dictatorship.

(Grover Smith 1969: 926)

It is noteworthy that Huxley puts 'overpopulation' at the top on his list of
Indian problems (in another letter from that 1961 trip, he wrote to his brother,
the biologist Julian Huxley, of 'the impossibility of [India] keeping up with
the population increase'). Indeed, he implies that there is a direct causal link
between growing numbers and the other problems he alludes to. When the
environmental debate acquired force in the West a few years later, one of the
most vocal strands in the debate likewise held overpopulation to be the prime
reason for ecological degradation. Inevitably, the discussion focused heavily
on India, which came to be regarded by many Western environmentalists as
the classic basket case, unable adequately to feed, clothe or house its growing
population.

Modern environmentalists who focus on the 'population problem' are
usually termed 'neo-Malthusians' after the English parson, Thomas Malthus,
who in the late eighteenth century first predicted that the growth in human
numbers would outstrip the growth in food supply. Indisputably the best-
known neo-Malthusian is the Stanford biologist, Paul Ehrlich. This is how
Ehrlich begins Chapter 1 of his 1969 book, *The Population Bomb*:

> I have understood the population explosion intellectually for a long
> time. I came to understand it emotionally one stinking hot night in
> Delhi a couple of years ago. My wife and daughter and I were returning

176

to our hotel in an ancient taxi. The seats were hopping with fleas. The only functional gear was third. As we crawled through the city, we entered a crowded slum area. The temperature was well over 100, and the air was a haze of dust and smoke. The streets seemed alive with people. People eating, people washing, people sleeping. People visiting, people arguing and screaming. People thrusting their hands through the taxi window, begging. People defecating and urinating. People clinging to buses. People herding animals. People, people, people, people.

(Ehrlich 1969: 15)

In this highly charged description, exploding numbers are held guilty of pollution, stinking hot air, and even technological obsolescence (the 'ancient taxi'). If the Malthusians are right we might as well tear up our wills and wait for doomsday. In the grim scenario painted by them, purposive human action might almost seem pointless – for no development strategy, whether 'business-as-usual' or conservative-liberal-socialism, could succeed in the face of exploding numbers. But of course the case of the Malthusian ecologists, whose motto is 'No More Indias', is flawed in many respects. For one thing, it relies heavily on a mistaken analogy taken from the living world. This is the concept of 'carrying capacity', used by biologists to define the maximum number of individuals of a species that can be supported by a particular habitat. But when applied to human populations, the concept is a slippery one. It fails to take account of the ingenuity of people, their abilities through technical change to squeeze more out of a given habitat. Moreover, the needs of humans, as distinct from animals, are culturally determined, while (again unlike animals) they might have access through transport and communication to resources from far-flung localities (see Hartman 1987).

In the West itself, the arguments of the neo-Malthusians were quickly subject to critical scrutiny. From the right, liberal economists accused them of gravely underestimating both human ingenuity and the abundance of natural resources: one scholar even claimed that as intelligent and creative individuals were the 'ultimate resource', rapid population growth was in fact a good thing (Simon 1981). (These claims draw strength from countries such as the Netherlands, with a higher population density than Bangladesh but hugely prosperous none the less.) From the left, socialists disputed the contention of Malthusians that growing numbers were the prime cause of environmental degradation in the West, arguing instead that it was imperfect technology and the workings of the market that were responsible for pollution and resource exhaustion (see Commoner 1971). The Malthusians have also been taken to task for their apparent preference for coercive measures to bring about family planning and birth control.

Not surprisingly, within India left-leaning environmentalists have been quick to take offence at Malthusian prognoses, the more so as India's own indigenous omnivores, when called upon to explain environmental degradation or social strife, invariably point an accusing finger at the propensity of ecosystem people to breed in large numbers. Most commonly, the rebuttal by environmentalists of these arguments invokes the question of consumption. Whether judged in energy or material terms, an average American's demands on the earth's resources are at least an order of magnitude greater than those of an average ecosystem dweller of the Third World. The birth of one American child will thus have an environmental impact equal to that of (say) the birth of several dozen Bangladeshi children. By this reckoning, if there is a population problem at all, it exists in affluent consumer societies such as the United States.

This is a line of argument that is difficult to dispute, but some Indian environmentalists have gone on to dismiss altogether the role of population in ecological degradation. Our own line of analysis also suggests that there are more important structural factors behind the massive degradation of India's land and water resources. But surely it is going too far to reject the need for family planning, as the otherwise estimable second citizens' report does while claiming that 'If India's people were to go hungry, it can be said with authority that it would not have *anything* to do with their number but with the callous mismanagement of the country's natural resources' (Centre for Science and Environment 1985: 162; emphasis added).

Where one school, characteristically Western in origin, thus regards population growth as the main contributory factor to environmental abuse, another school, quite dominant in India, tends to treat the question of human numbers as irrelevant to the development debate. Both positions are obviously too extreme. Without in any way underestimating the part played by inequity and inefficiency in causing environmental degradation, one must recognize that in many local situations rapid population growth is indeed exerting unsustainable pressures on grazing land, water resources or food supply (Jodha 1986). Yet the way out does not lie in either of the two solutions favoured by omnivores, global and national, namely, coercive family planning, or leaving the ecosystem people (who are primarily responsible for population growth) to their own fate. What then is the outlook of the conservative-liberal-socialist in relation to the population–environment nexus?

We believe that people are motivated above all by their perceived self-interest and the interest of their family members, or to a lesser degree that of some small, homogeneous social group to which they belong. They are rarely moved by the interest of a large heterogeneous group, especially if that group be characterized by high levels of inequality. Human populations will, then, tend to grow when people perceive their interests to lie in producing several offspring. Such is likely to have been the case over much of human history. After all, the entire present human population probably derives from a parental population of a few thousand *Homo sapiens sapiens* who might have developed the current linguistic abilities around fifty thousand years ago (Cavalli-Sforza *et al.* 1994). On this assumption human populations have on average been doubling every 350 years. With mortality rates reported for pre-industrial populations of Europe, an average woman reaching adulthood would need to bear slightly over six offspring for populations to grow at this rate. Evolution seems to have led all animals to tend to want to rear a goodish number of offspring; it is quite plausible that humans too tend to identify self-interest with bearing six or seven children (Dawkins 1976).

Human population growth may, then, be decelerated only when humans come to view their self-interest as consisting in a much smaller number of offspring; or when the mortality rate is particularly high owing to war, disease or starvation. Leaving out the second possibility, there seem to be three scenarios under which humans might voluntarily produce a smaller number of progeny:

(a) Among highly mobile hunter-gatherers women might find it impossible to be on the move carrying more than one offspring at a time. If so, they must pace children about four years apart, perhaps by suckling each child over a long period.

(b) Among highly sedentary hunter-gatherer-horticulturists with limited territories and in acute conflict with neighbouring tribes, groups may be at a disadvantage when growing too large in size, exceeding the carrying capacity of the territory and occasionally facing severe food shortages. They may therefore restrain population growth in the interest of their relatively small, egalitarian groups.

(c) Among modern industrial societies the young may be obliged to spend a long time acquiring the necessary training to handle the complex artefacts and information necessary for gainful and status-worthy employment. Under these circumstances the young cannot help their parents by adding to the family income; instead, parents have to invest heavily in enhancing the quality of their offspring. The parents may then identify their self-interest, as well as that of the offspring, in producing a small number, perhaps one or two of them (Caldwell 1982).

The world over, human societies are rapidly moving away from the first two scenarios. A number of industrial countries have, however, undergone a demographic transition over the past century corresponding to the third scenario. In India too there is evidence of such a transition taking place among certain strata of the population. Perhaps the best set of such evidence comes from the People of India Project of the Anthropological Survey of India (K.S. Singh 1992). This project attempts to map the entire human surface of India based on interviews with over 25,000 individuals belonging to the 4,635 communities to which the entire population of the country has been assigned. These interviews, in conjunction with the collation of other available information, permit each community to be characterized by some 500 traits relating to ecology, food habits, social organization, occupation, social and economic status, and response to modernization. Information has also been gathered on the number of offspring perceived to be desired by most members of any given community – recorded as 1 or 2, 3, 4 or more. The data thus lend themselves well to an analysis of how a variety of traits characterizing any particular community relate to the number of offspring desired by members of that community.

The data clearly show that communities that report a desire for a small number of offspring display the following features:

(a) They are engaged in occupations in the organized industry–services sectors, or are owners of land under intensive agriculture.
(b) They tend to have access to modern amenities such as electricity and tap water, television and agrochemicals.
(c) Both boys and girls in the community receive high levels of education.
(d) Most of the communities are urban, or have combined rural–urban distribution.
(e) Most of the communities involved belong to the upper-caste groups.

This is a picture consistent with the third scenario we outlined above, that of the demographic transition in industrial societies. In such communities parents have an interest in imparting high levels of education to the offspring over many years. Indeed, to have both boys and girls studying up to postgraduate level – that is, to the age of 22 or more – is a feature of many communities wanting few children. This education is essential in equipping them to find opportunities in the modern sector, a sector involved in the use of complex artefacts that require extensive learning for effective handling. Since activities in the modern sector are largely concentrated in urban localities, and towns and cities have been provided with modern amenities, it is to be expected that most such communities would have access to these amenities. It is the upper castes that dominate the modern industries–services sector for a variety of historical reasons, and it is these who have largely come to desire few children.

Communities desiring four or more offspring on the other hand exhibit a contrasting set of attributes:

(a) They are engaged in occupations primarily involving agricultural labour, herding, hunting-gathering or fishing.
(b) They have limited access to modern amenities. In most such communities women must fetch water and fuelwood, and help in fishing or looking after livestock.
(c) Neither boys nor girls in these communities receive much education; many drop out at an early age to help out their families.
(d) Children in many such communities are engaged in paid labour.
(e) Numerous such communities are becoming progressively impoverished; they are largely landless; they indicate recent involvement in or increased

dependence on wage labour, and decreased consumption of luxury food such as fruit.

(f) Many such communities are from tribal or lower backward castes; all are exclusively rural or forest dwellers.

For these communities there are no opportunities to invest in children's education to equip them to enter the organized industries–services sector. Since both men and women are engaged throughout life in unskilled labour, children can very quickly achieve the status of earning members of the family. They thus become an economic asset at an early age and members of these otherwise mostly assetless communities are motivated to produce a large number of offspring. Notably enough these are also communities most directly dependent on natural resources such as fuelwood. Their continually increasing population does imply an ever increasing pressure on the dwindling stock of the country's natural resources.

One may visualize five routes towards a deceleration of population growth:

(a) Compulsion to limit the number of offspring, practised with some success in China, but which turned out to be a disastrous failure in India during the Emergency years of 1975–7.

(b) Provision of specific, one-time incentives, such as cash rewards for undergoing vasectomies or tubectomies.

(c) Exhortation, advertisements on television and so on.

(d) Improved access to information and the hardware of family planning.

(e) A restructuring of the economy and the society leading to widespread motivation to produce a smaller number of offspring.

It is our contention that the first three routes can have only limited positive impact; indeed, the attempts at coercion during the Emergency were greatly counterproductive in the long run. Improving access to information and drugs, implements or surgical operations for family planning is undoubtedly of value. But in itself this will not be enough. Indeed, the Anthropological Survey of India data reveal that in all the states and union territories of India many communities employ modern family planning methods; but only a fraction of these favour one or two children. A significant number of communities employing modern family planning methods still desire three, four or more children. Unless there is a change in motivation, access to family planning methods by itself will not work.

What form of restructuring of society and the economy will motivate most of India's people to produce a small number of offspring? We believe it to be neither the ongoing inequitable, resource-exhausting pattern of development, nor the return to an agrarian-pastoral society favoured by a section of environmentalists. Instead, the path of equitable and resource-efficient development advocated by us would be most compatible with finally bringing the growth of India's population to a halt. The key factor is the value attached

by society to skilled manpower – skills that require years of patient investment by parents and specialist teachers. The skills required for low-input
subsistence agriculture and animal husbandry are rapidly absorbed at home
by boys and girls; parents can therefore produce large numbers of them
without interfering with their training. Indeed, the traditional blessing to a
newly married girl in India, 'bear eight sons over a long married life',
represents the value conferred on large families by an agrarian society.

Large investments in the training of offspring are called for when they must
learn to handle complex artefacts: machinery, chemicals and sophisticated
technical information. The industrial societies of the United States, Europe
and Japan have created conditions under which the majority of people have
jobs in either industry, services or highly mechanized/chemicalized agriculture requiring such training. But this is achieved through investing large
amounts of energy, material and informational resources to build up these
sectors. That is why per capita consumption of resources in these countries
is more than ten times that in countries like India (Durning 1992). India
simply does not have at its disposal sufficient levels of energy or material
resources to support the employment of every one of its citizens in the
modern sector. This is in part why organized industry has failed to create a
demand for trained personnel at rates that could substantially reduce the
population pressure in the Indian countryside. Irrigated, intensive agriculture
has done better, and substantial new job opportunities have opened up,
although mostly for unskilled labour, in pockets of productive agriculture.
But resources have been used extremely inefficiently in both industry and
intensive agriculture, so that resource investments have yielded very inadequate returns in terms of employment for skilled personnel in both these
sectors. In part this has been compensated for by the explosion of job
opportunities in the services sector – especially in governmental and quasi-
governmental organizations. But this is counterproductive because this
sector, above all, promotes wastage of resources. Faced with a limited
resource base given its large population, and quite unable to bring in resources
from outside, India has further failed to use what it has in an effective fashion
to create new job opportunities for trained personnel. This is the root cause
of India's failure to halt the explosion of its large population.

The only way, then, is to try to make the most of India's natural resources,
to conserve them, to use them ever more effectively. Obviously Indians must
come to adopt the Japanese philosophy of patiently getting more and more
out of the natural resource base, year in and year out, without of course
emulating Japan's appetite for natural resources from outside its borders. This
will go hand in hand with a broader-based development, doing away with the
current pattern of enclaves of industry and intensive agriculture prospering
parasitically by guzzling on the resources of hinterlands that in turn become
progressively impoverished. This will be a strategy of many small, timely
interventions, focusing on a rebuilding, not exhaustion, of the natural

resource base. Such a process would be put in place only with local communities being handed back control over the local resource base, sustainably developing these resources using local labour as well as folk knowledge and wisdom suitably married to modern scientific understanding. Only such a process can enhance the quality of life of the masses of the Indian population, lifting them out of the compulsion to liquidate whatever resources they can lay hands on to eke out a living. Only such a process has the potential to create job opportunities for large numbers of trained people to participate in a process of carefully planned development of the local natural resource base. Only such a process holds the promise of motivating the Indian people on a broad enough scale to bring the ongoing explosion of population eventually under control.

# 9

# RESOURCES OF HOPE

## THINK GLOBALLY, ACT LOCALLY

Ecologists, the American conservationist Michael Soule has remarked, live in a world of wounds: wounds to both nature and society, as witness the intensification of social conflict that has almost everywhere accompanied the devastation of the Indian countryside. In this book, we have documented the range of factors behind the tearing up of the social and natural fabric of India. But we have also tried to document the myriad attempts to heal these wounds, the quiet, persistent and often unhonoured struggles to mitigate social inequalities and restore the health of the environment.

Ours is a counsel not of despair but of hope, and it is with a recapitulation of these positive initiatives that we wish to end our narrative. We focus first of all on the district of Uttara Kannada in the southern state of Karnataka – a district that captures in microcosm many of the processes of ecological change described in this study. This is a region of low, undulating hills that run right into the Arabian Sea to the west and merge with the Deccan plateau to the east. The hills are extensively wooded, with over 60 per cent of the land under forest cover. The region has excellent rainfall, with annual precipitation exceeding 5,000 mm near the crestline of the Western Ghats. From this tract of heavy rainfall originate short west-flowing rivers that descend steeply to the coastline, as well as tributaries of the great river Krishna which flows eastwards. The district has a relatively low density of human population at 104 persons per sq km in 1991. Fishing, rice, coconut, betelnut, pepper, cardamom and cotton cultivation and manganese mining are the mainstay of the economy. The district also supports fish processing, paper, plywood, tiles and caustic soda industries. The people of the district have traditionally depended heavily on the forests for fuelwood, grazing, leaf manure, construction material and a wide range of non-wood forest produce.

Hand in hand with this intimate link with forest resources go numerous traditions of conservation and sustainable use. As illustrated in Chapter 2, despite official hostility some villages of Uttara Kannada have managed successfully to protect and manage community forests that through regulated harvests continue to meet their requirements of fuel, fodder and small timber. Other patches of forest are preserved as sacred groves, and given total protection by peasants. One of the most notable of these is the grove of Karikanamma (= mother goddess of the dark forest), a magnificent stand of dipterocarp forest perched on a high hill looking far out to the Arabian Sea. The Western Ghats have two species of *Dipterocarpus*, the flagship genus of the tropical rain forests of Asia. Uttara Kannada is the northernmost limit of· the geographical range of this genus on the Western Ghats; and *Dipterocarpus* has nearly vanished from everywhere else in the district. It has disappeared elsewhere because it was greatly valued by and overexploited for the plywood industry. It persists in the grove because the long-held beliefs of ecosystem people have prevented its felling in this one sacred forest.

Studies suggest that as much as 6 per cent of the land in Uttara Kannada was once under sacred habitats: groves, ponds, pools in river courses. Today this proportion has come down to just 0.3 per cent in a 5 km x 5 km area we studied intensively in the Siddapur *taluk* of the district. This area still retains fifty-two sacred groves, some as large as 2–3 ha scattered throughout the countryside. These groves currently harbour remnant patches of evergreen rain forest, refuges for many species that have become scarce over the rest of the district. The lofty trees represent large amounts of cash, cash that cannot but be a great temptation to the poor peasants who are continuing to protect the groves, sometimes to the extent of not even removing fallen fruit. Foresters have been quite unmindful of the value of these traditional systems of conservation till recent times; indeed, some groves of over 100 ha in size in this district were clear-felled by the Forest Department to raise eucalyptus plantations in the 1970s. There are pressures on the groves from some segments of the village community as well, from people who are hungry for cash or land for cultivation. Yet recent years have seen a growing recognition of the significance of these wholly indigenous systems of conserving bio-diversity – and scientists, social activists and environmental organizations have all joined hands in studying these practices and campaigned for their continuation (Subash Chandran and Gadgil 1993).

Whereas sacred groves testify to the persistence of age-old conservation practices, the district of Uttara Kannada has also contributed substantially to the modern environmental movement, beginning with the opposition to the Bedthi dam project in 1979–80. The Bedthi is one of the four major west-flowing rivers of Uttara Kannada, the others being the Kali, Aghanashini and Sharavathy. These rivers originate in the Western Ghats at altitudes of around 600 m and drop abruptly to sea level. Just on the southern border of Uttara Kannada, in the district of Shimoga, are the Sharavathy waterfalls. The story goes that Sir M. Visvesvarayya, the famous engineer-statesman, exclaimed, 'Oh, what a waste of energy!' when he first saw these falls. The waters of the Sharavathy were duly harnessed through two hydroelectric projects in the early years of independence. This was followed by a series of dams on the

185

Kali initiated in the 1960s. Plans were then set in motion to tap the hydroelectric power of the Bedthi and Aghanashini.

Environmental impact assessment of major projects became mandatory in 1978, and a hydroelectric project on the Bedthi river came up for review. One of the present writers was a member of a committee set up to prepare such an assessment. The committee met one morning in the state capital of Bangalore, was briefed by the state power corporation and then drove the 500 km to the Bedthi valley. We reached the actual project location next morning, spent just three hours touring around the dam site and a small part of the submersion area and then sat down to write the report. This was clearly being treated as a formality to be got over, with no serious intention of any careful assessment of the environmental impact of the project. Indeed, the region was quite unfamiliar to the other committee members. But a report recommending that the project be awarded an environmental clearance was pushed through despite minority protest.

But a good proportion of the farmers whose lands were slated for submergence in the Bedthi project were well-educated betelnut, pepper and cardamom producers belonging to the upper-caste group of *Haviks*. Roused against this charade of an environmental clearance, their well-organized co-operative society funded an alternative assessment. This was helped along by the fact that a dissenting member of the committee could pass on to them the basic information on project design, information that is normally withheld from all ordinary citizens. This alternative assessment showed that the original official exercise was seriously flawed, for it deliberately under-estimated adverse impacts. Not only were a variety of environmental consequences ignored, but proper computations suggested that the economic benefit : cost ratio was less than 1.5, the standard set by the Planning Commission for acceptance of a power project. This fresh assessment was followed by an open seminar presided over by the doyen of Kannada literature, Dr Shivaram Karanth, and addressed by, among others, the *Chipko andolan* leaders Sunderlal Bahuguna and Chandiprasad Bhatt, and by Professor V.K. Damodaran, a stalwart of the Kerala Sastra Sahitya Parishat. Officials of the Karnataka government's Forest and Planning Departments also participated, although the Power Corporation pulled out at the last moment. Held in January 1980, this was the first-ever public hearing on a development project in India. After this, the project was shelved (Sharma and Sharma 1981).

In his valedictory address to the Bedthi seminar, Dr Shivaram Karanth had stressed that the betelnut gardeners too were guilty of the wasteful use of natural resources. Some sensitive individuals among them rose to the challenge of setting their own house in order and through a co-operative society launched a project for the good management of the soil, water, vegetation and livestock in and around their villages. Beginning in the early 1980s the project brought together farmers, scientists from agricultural and technical institutions, voluntary agencies and government officials in a co-operative effort. Around the same time the government of Karnataka had launched its own progressive programme of integrated development of watersheds, bringing together specialists from its various agencies under the umbrella of a dryland development board. This programme also visualized involvement of local villagers as partners in the good management of private and public lands. The programme, initially focusing on the drier areas of the state, provided an appropriate framework for carrying forward the project initiated by the betelnut gardeners of Uttara Kannada.

Over a decade now the betelnut gardeners have been pursuing attempts at an integrated development of land, water, vegetation and livestock in their hilly terrain. At times there have been frustrations with different interests among the village population pulling in different directions, as well as irritation with the operation of the government machinery that has been funding these efforts. But there have been some most encouraging successes as well. Two of these include development of fodder resources and diffusion of fuel-efficient wood stoves. The youth club in the village of Bellikeri in Sirsi *taluk*, for example, has organized a co-operative fodder farm on forest land assigned to one of its members. Gradually the production of fodder is becoming self-supporting on the basis of the sale of hand-harvested fodder grass to the local villagers (Prasad *et al.* 1985).

Another successful experiment has involved propagation of fuel-efficient wood stoves based on a design developed by highly qualified chemical engineers of the Indian Institute of Science in Bangalore. In the rest of the state this programme, dependent on government subsidies, has run into problems, but among the betelnut gardeners there is real concern about the erosion of forest resources in their vicinity. The horticulturists are also interested in using newer sources of fuel such as betelnut husk. The new design of their bathwater heating stove permits the use of this rather abundant and otherwise unused resource as fuel. In consequence, new fuel-efficient wood stoves have diffused quite rapidly in the community.

Even as the farmers of the district were coming together for more responsible resource use, the Bedthi project was revived by the government of Karnataka in 1991. Expectedly, the revived project has met with opposition. Strikingly, this time protests have been accompanied by a bold altern-ative design offered by the people of Uttara Kannada. With the help of a retired engineer who once headed the Karnataka Power Corporation, the betelnut gardeners have been able to propose an alternative scheme to generate power, involving sixteen smaller reservoirs in place of the single large reservoir planned by the government. The sixteen smaller reservoirs would submerge much less land under forest, cultivation and human habitation, and are therefore an option more acceptable to the local people. This alternative design could in fact generate a larger total amount of hydroelectric power over the year, although the single large reservoir could produce more power during the dry season. More importantly, the sixteen-reservoir alternative would generate power at a substantially lower cost than the power corporation's design. The alternative design also visualizes possibilities of using tidal flow as well as irrigated energy plantations to further augment power production.

Competent engineers have been involved in working out this alternative design, and it is quite likely that their claims would hold up under close scrutiny. What is crucial here is that the government monopoly is being broken – and through an initiative of the local people. Indeed, the betelnut gardeners are exploring the possibility of floating a public company to execute the project, which they are confident is economically far more viable than the State Power Corporation's project. At all levels, then, from sacred groves and village forests to major hydroelectric projects the resourceful people of Uttara Kannada are giving us hope that the tide of environmental degradation might yet be turned around; and in ways that would benefit the broader masses of the population.

## A TRADITION TO LIVE UP TO

These initiatives in Uttara Kannada exemplify a strain of constructive social activism and critical enquiry that runs deep in Indian culture. We might thus situate these initiatives in the context of the traditions upheld by the remarkable individuals to whom we have dedicated this book. Like Shivaram Karanth (the grand old man of the environmental movement in Uttara Kannada), the great ornithologist Salim Ali was tirelessly active well into his eighties, mapping India's rich biological diversity. As one who was always alert to the need to integrate conservation with local needs and traditions (see Ali 1977), Salim Ali would have greatly approved of the efforts under way to protect sacred groves. The social reformer Jotiba Phule, himself from the stock of ecosystem people, was one of the first to bring to wider notice the close integration of agriculture and forestry, a link he saw being broken by the takeover of forests and common lands by the state (see Phule 1883). The rehabilitation of the social and natural world of rural India was also the lifelong concern of those two great Gandhians, J.C. Kumarappa and Mira Behn. It was Kumarappa (1946) who first wrote of the 'economy of permanence', and both he and Mira Behn spent many years understanding the workings of Indian agriculture and trying to restore it to a sound ecological footing.

Outside Uttara Kannada, indeed in almost all other parts of India, social activists are upholding the traditions associated with those such as Phule, Kumarappa, Mira Behn and Salim Ali. Some of these initiatives focus more on resistance to environmental destruction or resource capture; others on ecological restoration and the adoption of environment-friendly technologies. Some have tried to adapt modern technology to the rural context;

others have sought instead to draw upon and revive the rich traditions of prudent use among ecosystem people – such as community wood-lots, sacred groves or rain-fed tanks. These efforts have been guided by Gandhians, Marxists and wholly apolitical social workers. Some initiatives, like the work of Chandi Prasad Bhatt and the Chipko movement, or the Baliraja dam in drought-prone Sangli, have been extensively written about; others are unknown outside their immediate locality, though none the less important for that.

There has even been the odd initiative emerging from the omnivore sector. Thus the Tatas, long the most progressive of India's industrial houses, were found in a recent study of magnesite mining in the Himalaya to be exceptional in their concern for minimizing the environmental impact of opencast mining, while at the same time providing infrastructure and employment to local villagers (Institute of Social Studies Trust 1991). Nor have the communists been lagging in this regard. Thus one of the most remarkable environmental success stories of recent years comes from the Marxist-ruled state of West Bengal, the village forest protection committees studied in Chapter 7.

With regard to this particular initiative, there is little doubt that the West Bengal Forest Department has been far more willing than its counterparts in other states to share power with ecosystem people. Here the success of the village forest committees cannot be isolated from the wider processes of political change in the state. Since coming to power in 1977, the alliance of left-wing parties which has ruled West Bengal has crafted local level institutions that have real decision-making power, real control over financial resources. Moreover, the elected representatives at village and district levels have real authority over government officials. This decentralization of power has been accompanied by a substantial degree of land reform, so that rural society in Bengal is less inequitable than in the rest of the Gangetic plain (see Kohli 1987). Rising expectations of ecosystem people, and growing powers to enforce these expectations, have been coupled here with a move towards a less class-ridden society. This is the context in which local-level political leadership, and a more responsive bureaucracy, have worked together to bring about a broad-based and markedly successful programme of natural resource management in the sal forests of Bengal.

The forest protection committees of West Bengal are a perfect vindication of the core message of this book, the need to blend ecology with equity. They provide further testimony of the need for a genuinely decentralized political system country-wide, where powers to use natural resources lie not with insensitive and corrupt bureaucracies but with the people who most deeply depend on these resources. Today those in power, the omnivores of India, can successfully pass on the costs of resource abuse and environmental degradation to the masses of ecosystem people, and to ecological refugees. So long as this situation persists there is little real hope. But the self-interest of India's ecosystem people is congruent for the most part with the good husbanding of natural resources, at least in their own localities. The real solution for the long-term health of the environment thus lies in passing effective political power to the people.

The state of Karnataka, of which Uttara Kannada is part, itself flirted briefly with decentralized institutions on the West Bengal pattern between 1986 and 1990. A prime mover behind the constitution of *mandal panchayats* and *zilla parishats* in the state was the late socialist leader Abdul Nazir Saab, one of the most outstanding politicians in the history of free India, and chronologically the last of the exemplars to whom this book is dedicated. As Rural Development Minister, Nazir Saab was able to motivate a lethargic bureaucracy enough to provide drinking water to every village in the state, an act which earned him the appellation, richly deserved, of 'Neer Saab' (the man who brought water). Despite the fierce antipathy of the same bureaucracy, Nazir Saab and his colleagues in the Janata Dal Party – which came to power in Karnataka in 1983, after long years of Congress rule – were able to push through the *panchayati raj* scheme. In its brief life, the system underlined one cardinal truth: namely, that the interest of politicians in environmental issues is, by and large, inversely proportional to the size of their constituency. In our experience, Members of Parliament appear to care the least, members of state legislatures a little more, members of district councils a little more still, and members of *mandal panchayats* emphatically the most. A quarter of the seats at the two lower levels were reserved for women, who were often most vocal in calling, for example, for people-oriented forest management. Yet the system was not to the liking of omnivores, whose own interests it seriously threatened. When Congress returned to power in Karnataka in 1990, it moved quickly to suspend and eventually emasculate *zilla parishats* and *mandal panchayats*. Supporters of this move claimed that the decentralization of power only meant the decentralization of corruption. Yet the state Congress government that abolished the *panchayats* has since been acknowledged to have been the most corrupt administration ever to rule Karnataka. In December 1994 the Janata Dal returned to power in the state, promising to restore the *panchayat* system.

There is, of course, a real danger that an elite of upper-caste landlords would come to dominate lower-level political institutions. This fear prompted Dr B.R. Ambedkar, the great leader of India's lower castes, to oppose the devolution of power to village institutions at the time of national independence. He believed that only a disinterested elite, manning the agencies of the state, would be able to protect the vulnerable rural poor from exploitation by the dominant upper castes. This was also the hope of classical socialists, who likewise pinned their faith in the ability of the state to bring about a just and equitable society. But forty-five years of growing state power in India have painfully belied these hopes. The record of state-guided development is abysmal, when reckoned by the criteria of economic growth, environmental stability or social equity. It is this failure that lies behind our plea for a decentralized political system in preference to the continuing centralization of power in the hands of a narrow elite, the iron triangle of omnivores. But one must recognize that mere decentralization without redistribution would be self-defeating: hence the significance of land reform, education and health care, along with equitable access to common property resources. This would lead to the institution of a political system best summed up by the word 'empowerment', much as the colonial regime could be summed up by the word 'subjugation', and the system we presently live under as 'patronage'.

There are vast resources we can draw upon in moving towards such a political system. These include a rich history of social activism that stretches from Jotiba Phule to Abdul Nazir Saab and beyond, as well as the traditions of ecological prudence that still persist among ecosystem people in far-flung parts of India. In this book we have outlined a philosophical approach for bringing under one roof these varied traditions, recasting what may appear to be contesting ideologies in one unifying framework. Thus our emphasis on strong local communities borrows from the Gandhian tradition, that on democratic institutions and private enterprise from liberal capitalism, and that on equity from Marxism. This is an eclecticism that comes out of our reading of the ecological history of modern India, but we believe it also to be in the larger Indian tradition of tolerating, assimilating and synthesizing diverse strands. What we have called conservative-liberal-socialism might just provide the springboard for moving India towards a politics of empowerment, and towards an economy of permanence.

# GLOSSARY OF WORDS IN INDIAN LANGUAGES

*aankhbandi* (allowance): literally, closing the eye. A bribe paid to an official not to take action against some violation.

*andolan*: a social movement.

*ashram*: traditionally abode of sages; currently used especially as a place to conduct constructive social activities.

*astraole*: a fuel-efficient wood stove designed by a group of engineers at the Indian Institute of Science in Bangalore.

*ayurveda*: Indian medicine.

*baandh*: dam.

*babu*: a white-collar worker.

*bachao*: save.

*bandh*: shutdown.

*Bhoodan*: literally: gift of land. A well-known social movement that urged landowners to gift land for redistribution favouring smallholders and landless peasants.

*bidi*: an Indian cheroot where tobacco is rolled in leaves of *Diospyros* trees.

*chawls*: multistoreyed tenements housing poorer families in urban areas.

*Chipko*: literally: to hug. A well-known environmental movement against forest destruction which started with peasants hugging trees to prevent their being cut.

*dharna*: a form of sit-down strike.

*dhoti*: a piece of cloth traditionally used as a lower garment by men.

*gherao*: a form of protest involving surrounding a person and preventing him/her from moving at will.

*Gramdan*: literally: gift of a village. A well-known social movement in which all landowners in a village gifted their lands for redistribution among the entire village population.

*hartal*: a strike involving closure of shops, factories, offices, transport facilities.

*Jharkhand Mukti Morcha*: a political movement for creation of a separate state of Jharkhand in tribal areas of central India.

*jail bharo andolan*: a form of protest involving courting arrest to fill up jails.

*jal samadhi*: a form of protest involving immolation in rising waters of a dam.

*Jan Sangharsh*: people's struggle.

*Loot-mar-ka-mahina*: literally: the month of plunder.

*mandal*: village cluster

*mandal panchayat*: literally: council of five members for a group of villages. In practice the council may include many more members.

*mukti sangharsh*: literally: a struggle for liberation.

*nagarpalika*: system of decentralized political institutions for the urban parts of India.

*nallah*: stream; *bandh*: stoppage.

*Naxalites*: groups working for armed revolution.

*Naxalbari*: a village in West Bengal where the activities of ultra-left Naxalites began.

*padayatra*: foot march; a form of public demonstration.

*panchayat*: a council of five members.

*panchayati raj*: system of decentralized political institutions for the rural parts of India involving *mandal panchayats* and *zilla parishats*.

*pani panchayat*: a council of villagers set up to manage common-property water resources.

*panidari*: a feudal system of ownership rights over a stretch of river.

*parishat*: assembly council.

*pathashalas*: schools in the traditional Indian system of education.

*phad*: a traditional system of irrigation in northern Maharashtra.

*rasta roko*: an agitation involving disruption of road traffic.

*ryotwari*: a system of assigning land where the rights of tillers were better recognized by the British authorities in southern and western India.

*sangharsh samiti*: struggle committee.

*satyagraha*: literally: insistence on truth. A form of non-violent protest popularized by Mahatma Gandhi.

*shikar*: hunting, particularly of bigger animals.

*taluk*: county.

*targetbaji*: unproductive pursuit of paper targets.

*van panchayat*: village forest council.

*zamindari*: a system of assigning land, introduced by the British authorities in eastern and northern India, where the local chieftains were made owners of very large tracts of land .

*zilla*: districts.

*zilla parishat*: an assembly of elected members governing a district.

# GLOSSARY OF WORDS
# REFERRING TO INDIAN
# COMMUNITIES

*Ambiga*: a fisherfolk community of Uttara Kannada district in the state of Karnataka.

*Badaga*: a community of cultivators in the Nilgiris district of Tamil Nadu.

*Brahman*: a member of an upper-caste priestly community.

*Halakki Vakkals*: a community of lower-caste peasant cultivators in Uttara Kannada district in the state of Karnataka.

*Haviks*: a community of upper-caste priests-cum-horticulturists in coastal and hill areas of the state of Karnataka.

*Kshatriya*: a member of an upper-caste warrior community.

*Sudra*: a member of lower-caste peasant, herder or artisanal communities.

*Vaisya*: a member of an upper-caste trading/artisanal community.

# BIBLIOGRAPHY

Achaya, K.T. (1993) *A Companion to Indian Food and Food Materials*, Delhi: Oxford University Press.

Agarwal, A. (1986) 'Human–nature interactions in a Third World country', *The Environmentalist*, 6.

—— and Narain, S. (1990) *Towards Green Villages*, New Delhi: Centre for Science and Environment.

Agnihotri, I. (1993) *Ecology, Land Use and Colonization: the Canal Colonies of Punjab*, mimeo, New Delhi: Nehru Memorial Museum and Library.

Ali, S. (1977) 'Wildlife conservation and the cultivator. Presidential letter', *Hornbill*, April–June, pp. 6, 36.

Alvares, C. (1989) 'No!', *The Illustrated Weekly of India*, 15 October.

—— (1992) *Science, Development and Violence: The Twilight of Modernity*, Delhi: Oxford University Press.

Anklesaria Aiyer, S. (1988) 'Narmada project: government–opposition alliance against Amte', *Indian Express*, 30 October.

Anon. (1991) 'The sangarsh yatra: a first hand account', *Narmada: A Campaign Newsletter*, Nos 7 and 8, April.

Areeparampil, M. (1987) 'The impact of Subarnarekha multipurpose project on the indigenous people of Singhbhum', in *People and Dams*, New Delhi: Society for Participatory Research in Asia.

Attwood, D.W. and Baviskar, B.S. (eds) (1988) *Who Shares?*, Delhi: Oxford University Press.

Bahuguna, S. (1983) *Walking with the Chipko Message*, Silyara (Tehri Garhwal district): Navjivan Ashram.

Bajaj, J.K. (1982) 'Green revolution: a historical perspective', *PPST Bulletin*, 2: 87–113.

Ballabh, N. and Singh, K. (1988) *Van (Forest) Panchayats in Uttar Pradesh Hills: A Critical Analysis*, Mimeo, Anand: Institute of Rural Management.

Bandyopadhyay, J. (1987) 'Political economy of drought and water scarcity', *Economic and Political Weekly*, 12 December.

—— (1989) *Natural Resource Management in the Mountain Environment: Experiences from the Doon Valley, India*, Kathmandu: International Centre for Integrated Mountain Development.

—— and Shiva, V. (1984) *Ecological Audit of Eucalyptus Cultivation*, New Delhi: Natraj Publishers.

Bhaskaran, S.T. (1990) 'The rise of the environmental movement in India', *Media Journal*, 37(2).

Bhatia, B. (1992) 'Lush fields and parched throats: political economy of ground water in Gujarat', *Economic and Political Weekly*, 19–26 December.

Bhatt, C.P. (1984) *Himalaya Kshetra ka Niyojan* (Renewal of the Himalaya), Gopeshwar: DGSM.
—— (1992) *The Future of Large Dam Projects in the Himalaya*, Nainital: Pahar.
Bhuskute, V.M. (1968) *Mulshi Satyagraha*, Dastane: Pune.
Brandis, D. (1884) *Progress of Forestry in India*, Edinburgh: McFarlane & Erskine.
Burra, N. (1991) *Women and Wasteland Development: A Review of NGO Experience*, mimeo, New Delhi: International Labour Office.
Caldwell, J.C. (1982) *Theory of Fertility Decline*, London and New York: Academic Press.
Calman, L. (1985) *Protest in Democratic India*, Boulder: Westview Press.
Cavalli-Sforza, L.L., Menozzi, P. and Piazza, A. (1994) *The History and Geography of Human Genes*, Princeton, NJ: Princeton University.
Centre for Science and Environment (CSE) (1985) *The State of India's Environment 1984–85: A Second Citizens' Report*, New Delhi: Centre for Science and Environment.
—— (1987) *The Wrath of Nature: The Impact of Environmental Destruction on Floods and Droughts*, New Delhi: Centre for Science and Environment.
Collins, G.F.S. (1921). *Modifications in the Forest Settlements: Kanara Coastal Tract*, Part 1, Karwar: Mahomedan Press.
Commoner, B. (1971) *The Closing Circle*, New York: Alfred Knopf.
Concerned Scholars (1986) *Bharat Aluminium Company: Gandhamardan Hills and People's Agitation*, Sambalpur: from the authors.
Conquest, R. (1968) *The Great Terror*, London: Weidenfeld & Nicolson.
Dalal, N. (1983) 'Bring back my valley', *The Times of India*, 10 July.
Dasgupta, P. (1982) *The Control of Resources*, Delhi: Oxford University Press.
Dasmann, R.F. (1988) 'Towards a biosphere consciousness', in D. Worstèr (ed.) *The Ends of the Earth: Perspectives on Modern Environmental History*, Cambridge: Cambridge University Press.
Dawkins, R. (1976) *The Selfish Gene*, Oxford: Oxford University Press.
Department of Science and Technology, Government of India (1990) 'Fertilizer use', in *Perspectives in Science and Technology*, Vol. 2, Science Advisory Council to the Prime Minister, Department of Science and Technology, Government of India, New Delhi: Har-Anand Publications and Vikas Publishing House.
Desai, A.R. (ed.) (1979) *Agrarian Struggles in India*, Delhi: Oxford University Press.
—— (ed.) (1986) *Agrarian Struggles in India since Independence*, Delhi: Oxford University Press.
Desmond, R. (1992) *The European Discovery of Indian Flora*, Delhi: Oxford University Press.
Devalle, S.B.C. (1992) *Discourses of Ethnicity*, Delhi: Sage Publishers.
Dhara, R. (1992) 'Health effects of the Bhopal gas leak: a review', *Epidemiologia e Prevenzione*, 52: 22–31.
Dietrich, G. (1989) 'Kanyakumari march: breakthrough despite breakup', *Economic and Political Weekly*, 20 May.
D'Monte, D. (1981) 'Time up for Tehri', *Indian Express*, 30 May.
—— (1985) *Temples or Tombs? Industry versus Environment: Three Controversies*, New Delhi: Centre for Science and Environment.
Dogra, B. (1992) *Chilika Lake Controversy: Dollars versus Livelihood*, New Delhi: from the author.
—— , Nautiyal, N. and Prasun, K. (1983) *Victims of Ecological Ruin*, Dehra Dun: from the authors.
Durning, A. (1992) *How Much Is Enough?*, The Worldwatch Environmental Alert Series, New York and London: W.W. Norton.

*Economic and Political Weekly* (1991) 'Gujarat: an advertiser's supplement', 5–12 January.

Ehrlich, P. (1969) *The Population Bomb*, New York: Ballantine Books.

Elwin, V. (1964) *The Tribal World of Verrier Elwin: An Autobiography*, Bombay: Oxford University Press.

Fernandes, W. and Kulkarni, S. (eds) (1983) *Towards a New Forest Policy*, New Delhi: Indian Social Institute.

—— and Ganguly-Thukral, E. (1988) *Development and Rehabilitation*, New Delhi: Indian Social Institute.

Food and Agricultural Organization, United Nations (1984) *Intensive Multiple-Use Forest Management in Kerala*, FAO Forestry Paper 53, FAO, Rome.

Gadgil, M. (1979) 'Hills, dams and forests: some field observations from Karnataka Western Ghats', *Proceedings of the Indian Academy of Sciences*, 2(3): 291–303.

—— (1989) 'Deforestation: problems and prospects', *Wastelands News*, Supplement to SPWD Newsletter, 4(4), May–July.

—— (1993) 'Biodiversity and India's degraded lands', *Ambio*, 22 (2–3): 167–72.

—— and Guha, R. (1992) *This Fissured Land: An Ecological History of India*, Delhi: Oxford University Press, and Berkeley, Calif.: University of California Press.

—— and Iyer, P. (1989) 'On the diversification of common property resource use by the Indian society', in F. Berkes (ed.) *Common Property Resources: Ecology and Community Based Sustainable Development*, London: Belhaven Press.

—— and Malhotra, K.C. (1982) 'Ecology of a pastoral caste: Gavli Dhangars of peninsular India', *Human Ecology*, 10: 107–43.

—— and Prasad, S.N. (1978) 'Vanishing bamboo stocks', *Commerce*, 136(3497): 1000–4.

—— and Rao, P.R.S. (1994) 'A system of positive incentives to conserve biodiversity', *Economic and Political Weekly*, August 6.

—— and Subash Chandran, M.D. (1988) 'On the history of Uttara Kannada forests', in J. Dargavel, K. Dixon and N. Semple (eds) *Changing Tropical Forests*, Canberra: Australian National University.

—— and Subash Chandran, M.D. (1992) 'Sacred groves', in G. Sen (ed.) *Indigenous Vision: People of India. Attitudes to the Environment*, Delhi: Sage Publications and Delhi: India International Centre, New Delhi.

—— and Vartak, V.D. (1975) 'Sacred groves of India: a plea for continued conservation', *Journal of Bombay Natural History Society*, 72: 314–20.

——, Pillai, J. and Sinha, M. (1989) 'Report of a study conducted on behalf of fuelwood and fodder study group', Planning Commission, Government of India.

——, Subash Chandran, M.D., Hegde, K.M., Hegde, N.S., Naik, P.V. and Bhat, P.K. (1990) *Report on Management of Ecosystem to the Development of Karnataka's Coastal Region*, New Delhi: The Times Research Foundation.

Galeano, E. (1989) 'The other wall', *New Internationalist*, November.

Ganguly-Thukral, E. (ed.) (1992) *Big Dams, Displaced People*, New Delhi: Sage Publishers.

Ghosh, A. (1991) 'Probing the Jharkhand question', *Economic and Political Weekly*, 4 May.

Grover Smith (ed.) (1969) *Letters of Aldous Huxley*, London: Chatto & Windus.

Guha, R. (1983) 'Forestry in British and post-British India: a historical analysis', *Economic and Political Weekly*, 29 October and 5–12 November.

—— (1989a) *The Unquiet Woods: Ecological Change and Peasant Resistance in the Himalaya*, Delhi: Oxford University Press, Berkeley: University of California Press.

—— (1989b) 'Radical American environmentalism and wilderness preservation: a Third World critique', *Environmental Ethics*, 11(1).

Guha, S. (ed.) (1992) *Agricultural Productivity in British India*, Delhi: Oxford University Press.

Hart, H.C. (1956) *New India's Rivers*, Bombay: Orient Longman.

Hartman, B. (1987) *Reproductive Rights and Wrongs*, New York: Harper & Row.

Hays, S.P. (1957) *Conservation as the Gospel of Efficiency: The Progressive Conservation Movement, 1880–1920*, Cambridge, Mass.: Harvard University Press.

—— (1987) *Beauty, Health and Permanence: Environmental Politics in the United States, 1955–85*, New York: Cambridge University Press.

Herring, R.J. (1983) *Land to the Tiller*, New Haven: Yale University Press.

Hiremath, S.R. (1987) 'How to fight a corporate giant', in A. Agarwal, D. D'Monte and U. Samarth (eds) *The Fight for Survival*, New Delhi: Centre for Science and Environment.

—— (1988) 'Western Ghats march was an education', *Deccan Herald*, 14 February.

Howard, A. (1940) *An Agricultural Testament*, Oxford: Oxford University Press.

Illich, I. (1978) *Towards a History of Needs*, Berkeley: Heyday Books.

Institute of Rural Management (1992) *Farm Forestry: A Review of Issues and Pragmatic Approaches*, mimeo, Anand: Institute of Rural Management.

Institute of Social Studies Trust (1991) *Mining in the Himalayas: Report on a Field Study in the Almora and Pithoragarh Districts*, New Delhi: Institute of Social Studies Trust.

Jain, L.C. (1983) *Textile Policy Set to Annihilate Employment in the Woollen Cottage Sector*, Delhi: Industrial Development Services.

Jain, S.C. (1989) *Paper Industry: Raw Material Scenario*, mimeo, New Delhi: Straw Products Ltd.

Jan Vikas Andolan (1990) *Jan Vikas Andolan: A Working Perspective*, mimeo.

Jeffrey, R. (1992) *Politics, Women and Well Being: How Kerala Became a 'Model'*, London: Macmillan.

Jodha, N.S. (1986) 'Common property resources and rural poor in dry regions of India', *Economic and Political Weekly*, 5 July.

—— (1990) *Rural Common Property Resources: Contributions and Crisis*, Foundation Day Lecture, New Delhi: Society for Promotion of Wastelands Development.

Joshi, D. (1983a) 'Magnesite udyog: vinash ki ankahi kahani' (Magnesite mining: destruction's untold story), New Delhi: *Parvatiya Times*.

—— (1983b) 'Magnesite kanan se Himalaya ko bhaari kshati' (Grievous injuries to the Himalaya caused by magnesite mining), *Parvatiya Times*, 31 November.

Joshi, P.C. (1975) *Land Reform in India*, New Delhi: Allied Publishers.

Kalpavriksh (1988) *The Narmada Valley Project: A Critique*, New Delhi: Kalpavriksh.

Kanvalli, S. (1991) *Quest for Justice*, Dharwad: Samaja Parivartana Samudaya.

Kerala Sastra Sahitya Parishat, (1984) *Science as Social Activism: Reports and Papers on the Peoples Science Movement in India*, Trivandrum: Kerala Sastra Sahitya Parishat.

Kohli, A. (1987) *The State and Poverty in India*, Cambridge: Cambridge University Press.

Kothari, A., Pande, P., Singh, S. and Variava, D. (1989) *Management of National Parks and Sanctuaries in India. A Status Report*, New Delhi: Indian Institute of Public Administration.

Kothari, R. (1984) 'The non party political process', *Economic and Political Weekly*, 2 February.

Krishnan, M. (1975) *India's Wildlife in 1959–70: An Ecological Survey of the Larger Mammals of Peninsular India*, Bombay: Bombay Natural History Society.

Kumar, K.G. (1989) 'Police brutality besieges ecology: national fishermen's march', *Economic and Political Weekly*, 27 May.

## BIBLIOGRAPHY

Kumarappa, J.C. (1938) *Why the Village Movement?*, Wardha: All India Village Industries Association.

—— (1946) *Economy of Permanence*, Varanasi: Sarva Seva Sangh Prakashan.

Kurien, J. (1978) 'Entry of big business into fishing', *Economic and Political Weekly*, 10 October.

—— (1993) 'Ruining the commons: overfishing and fishworkers' actions in south India', *The Ecologist*, 23(1): 5–12.

—— and Achari, T. (1990) 'Overfishing along Kerala coast: causes and consequences', *Economic and Political Weekly*, 1–8 September.

Lal, J.B. (1989) *India's Forests: Myth and Reality*, Dehra Dun: Natraj Publishers.

Ludden, D. (1985) *Peasant History in South India*, Princeton, NJ: Princeton University Press.

Maitra, S. (1992) 'Provision of basic services in the cities and towns of the national capital region: issues concerning management and finance', Ph.D. dissertation, Jawaharlal Nehru University, New Delhi.

Malhotra, K.C. and Poffenberger, M. (eds) (1989) 'Forest regeneration through community protection: the West Bengal experience', *Proceedings of the Working Group Meeting on Forest Protection Committees, Calcutta, June 21–22*, Calcutta: West Bengal Forest Department.

Malhotra, K.C., Deb, D., Dutta, M., Vasulu, T.S., Yadav, G. and Adhikari, M. (1991) *Role of Non-Timber Forest Produce in Village Economy: A Household Survey in Jamboni Range, Midnapore District, West Bengal*, mimeo, Calcutta: Institute for Biosocial Research and Development.

Martinez-Alier, J. (1990) *The Environmentalism of the Poor*, research proposal, New York: Social Science Research Council.

Misra, D.N. (1984) 'Unjust blame on foresters', *The Times of India*, 28 January.

Misra, T.P. (1946) *Halt Hirakud Dam*, Sambalpur: Anti-Hirakud Dam Committee.

Mitra, A. (1979) 'Integrated strategies for economic and demographic development', *Economic and Political Weekly*, 3 February.

Modi, L.N. (1988) 'Whose forests are they anyway?', *Blitz*, 10 December.

Mukul (1993) 'Villages of Chipko movement', *Economic and Political Weekly*, 10 March.

Mundle, S. and Rao, M.G. (1991) 'Volume and composition of government subsidies in India 1977–78 to 1987–88', *Economic and Political Weekly*, 4 May.

Naidu, N.Y. (1972) 'Tribal revolt in Parvatipuram agency', *Economic and Political Weekly*, 25 November.

Nandy, A. (1987) *Traditions, Tyrannies and Utopias*, Delhi: Oxford University Press.

—— (ed.) (1989) *Science, Hegemony and Violence: A Requiem for Modernity*, Delhi: Oxford University Press.

Narain, S. (1983) 'Brutal oppression of fisherfolk', *The Times of India*, 11 October.

*Narmada* (1989–90) *Narmada: A Campaign Newsletter* (various issues), distributed by the Narmada Bachao Andolan, New Delhi.

National Commission of Agriculture (1976) *Report of the National Commission of Agriculture*, Vol. 9, Delhi: Ministry of Agriculture, Government of India.

National Fisherfolk Forum (1989) *Kanyakumari March: Breakthrough despite Breakup*, Cochin: National Fisherfolk Forum.

Norton, B.G. (1987) *Why Preserve Natural Variety?*, Princeton, NJ: Princeton University Press.

Paranjpye, V. (1981) 'Dam: are we damned?' in L.T. Sharma and R. Sharma (eds) *Major Dams: A Second Look*, New Delhi: Gandhi Peace Foundation, and Sirsi: Totgars Co-operative Sale Society.

Pathak, S. (1987) *Uttarakhand mein Kuli Begar Pratha* (The forced labour system in Uttarakhand), New Delhi: Radhakrishnan Prakashan.

# BIBLIOGRAPHY

Pattanaik, S.K., Das, B. and Mishra, A. (1987) 'Hirakud dam project: expectations and realities', in *People and Dams*, New Delhi: Society for Participatory Research in Asia.

Peoples Union for Civil Liberties (1985) *Bastar: Ek Mutbhed ki Jaanch* (Bastar: an enquiry report), New Delhi: Peoples Union for Civil Liberties.

Peoples Union for Democratic Rights (1982) *Undeclared Civil War*, New Delhi: Peoples Union for Democratic Rights.

—— (1986) *Gandhamardhan Mines: A Report on Environment and People*, New Delhi: Peoples Union for Democratic Rights.

Phule, J. (1883) (1969) *Shetkaryacha Asud: The Whipcord of the Farmer (1882–83)*, Pune: Maharashtra Sahitya and Sanskriti Mandal.

Prasad, S.N. (1984) 'Productivity of eucalyptus plantations in Karnataka', in J.K. Sharma, C.T.S. Nair, S. Kedarnath and S. Kondas (eds) *Eucalyptus in India: Past, Present and Future*, Peechi: Forest Research Institute.

—— ,Hegde, M.S., Gadgil, M. and Hegde, K.M. (1985) 'An experiment in eco-development in Uttara Kannada district of Karnataka', *South Asian Anthropologist*, 6: 73–83.

PUDR (1986) *Simlipal Report*, New Delhi: Peoples Union for Democratic Rights.

Raghunandan, D. (1987) 'Ecology and consciousness', *Economic and Political Weekly*, 5 May.

Rai, U., Mukul, S.B. and Kumar, D. (1991) *Call of the Commons: People versus Corruption*, New Delhi: Centre for Science and Environment.

Rangarajan, M. (1992) 'Forest policy in the central provinces, 1860–1914', unpublished D.Phil. dissertation, Faculty of Modern History, Oxford: University of Oxford.

Ravindranath, N.H., Shailaja, R. and Revankar, A. (1989) 'Dissemination and evaluation of fuel-efficient and smokeless Astra stove in Karnataka', *Energy Environment Monitor*, 5: 48–60.

Reddy, A.K.N. (1982) 'An alternative pattern of Indian industrialization', in A.K. Bagchee and N. Bannerjee (eds) *Change and Choice in Indian Industry*, Calcutta: K.P. Bagchee.

Roy, A.K., Seshadri, S., Ghotge, S., Deshpande, A., and Gupta, A. (1982) *Planning the Environment*, Anuppur: Vidushak Karkhand.

Sainath, P. (1993) 'Palamau tribals in army's firing range', *The Times of India*, 12 November.

Sangwan, S. (1991) *Science, Technology and Colonialism: An Indian Experience, 1757–1857*, Delhi: Anamika Prakashan.

Saraf, S. (1989) 'An odyssey to save the Sivaliks', *The Hindu*, 26 February.

Sarkar, S. (1983) *Modern India, 1885–1947*, New Delhi: Macmillan.

Saxena, N.C. (1990) *Farm Forestry in North-West India*, Studies in Sustainable Forest Management, No. 4, New Delhi: Ford Foundation.

Sengupta, N. (ed.) (1982) *Jharkhand: Fourth World Dynamics*, New Delhi: Authors' Guild.

—— (1985) 'Irrigation: traditional versus modern', *Economic and Political Weekly*, special number, August.

Shah, S.L. (1989) *Functioning of Van Panchayats in Eight Hill Districts of Uttar Pradesh: An Analysis of Present Malaise and Lessons for Future in the Context of the Proposed Van Panchayat Niyamwali 1989*, mimeo from the author, Almora.

Shankari, U. (1991) 'Major problems with minor irrigation', *Contributions to Indian Sociology*, 25(1).

Sharma, J.K., Nair, C.T.S., Kedarnath, S. and Kondas, S. (eds) (1984) *Eucalypts in India: Past, Present and Future*, Peechi: Kerala Forest Research Institute.

Sharma, L.T. and Sharma, R. (1981) *Major Dams: A Second Look*, New Delhi: Gandhi Peace Foundation and Sirsi: Totgars Co-operative Sale Society Ltd.

Shiva, V. (1988) *Staying Alive: Women, Ecology and Survival in India*. New Delhi: Kali for Women, and London: Zed Press.

—— , Anderson, P., Schucking, H., Gray, A., Lohmann, L. and Cooper, D. (1991a) *Biodiversity: Social and Ecological Perspectives*, Penang, Malaysia: World Rainforest Movement.

—— , Bandyopadhyay, J., Hegde, P., Krishnamurthy, B.V., Kurien, J., Narendranath, G., Ramprasad, V. and Reddy, S.T.S. (1991b) *Ecology and the Politics of Survival*, Tokyo: United Nations University Press, and New Delhi: Sage Publications.

Simon, J. (1981) *The Ultimate Resource*, Princeton, NJ: Princeton University Press.

Simon, J.L. and Kahn, H. (ed.) (1984) *The Resourceful Earth: A Response to Global 2000*, Oxford: Basil Blackwell.

Singh, K.S. (1992) *People of India: An Introduction*, Calcutta: Anthropological Survey of India.

Singh, N.J. (1992) 'Salt affected soils in India', in T.N. Khoshoo and B.L. Deekshatulu (eds) *Land and Soils*, New Delhi: Har-Anand Publications.

Singh, S. (1994) 'Environment, class and state in India: a perspective on sustainable irrigation', unpublished Ph.D. dissertation, Department of Political Science, Delhi University.

Somanathan, E. (1991) 'Deforestation, property rights and incentives in central Himalaya', *Economic and Political Weekly*, 29 June.

Somashekara Reddy, S.T. (1988) 'Tank irrigation in Karnataka', *Swayam Gramabhydaya*, 6(4): 1–5.

Space Applications Centre (1993) 'Environmental appraisal for sustainable eco-development through remote sensing data: a case study for a few watersheds of Chamoli district, central Himalaya', Scientific Note, Ahmedabad: Space Applications Centre.

State Watershed Development Cell (1989) *Watershed Development Programme for Rainfed Agriculture*, Bangalore: State Watershed Development Cell, Government of Karnataka.

Steward, W. (1988) *Common Property Resource Management: Status and Role in India*, mimeo, New Delhi: The World Bank.

Subash Chandran, M.D. and Gadgil, M. (1993) '"Kans" – safety forests of Uttara Kannada', in H. Brandl (ed.) *Proceedings of IUFRO Forest History Group Meeting on Peasant Forestry*, Freiburg, Germany, No. 40.

Subbarayappa, B.V. (1992) *In Pursuit of Excellence: A History of the Indian Institute of Science*, New Delhi: Tata McGraw-Hill.

Sukumar, R. (1989) *The Asian Elephant: Ecology and Management*, Cambridge: Cambridge University Press.

—— (1994) *Elephant Days and Nights: Ten Years with the Indian Elephant*, New Delhi: Oxford University Press.

Thakurdas, P. *et al.* (1944) *Memorandum Outlining a Plan of Economic Development for India*, London: Penguin.

Thurow, L. (1980) *The Zero Sum Society: Distribution and the Possibilities for Economic Change*, New York: Basic Books.

Valdiya, K.S. (1992) 'Must we have high dams in the geodynamically active Himalayan domain?', *Current Science*, 63(6): 289–96.

—— (1993) 'Uplift and geomorphic rejuvenation of the Himalaya in the Quaternary period', *Current Science*, 64(11, 12): 873–84.

Vijayan, V.S. (1987) *Keoldeo National Park Ecology Study*, Bombay: Bombay Natural History Society.

Vijaypurkar, M. (1988) 'Lessons from a march', *Frontline*, 20 February–4 March.

Vinayak, A. (1990) 'Tribals try to save Eastern Ghats', *Free Press Journal*, 19 December.

Viswanathan, S. (1984) *Organizing for Science: The Making of an Industrial Research Laboratory*, Oxford: Oxford University Press.

Voelcker, J.A. (1893) *Report on the Improvement of Indian Agriculture*, Calcutta: Government Press.

Vohra, B.B. (1973) 'A charter for the land', *Economic and Political Weekly*, 31 March.

—— (1980) *A Land and Water Policy for India* (Sardar Patel Memorial Lectures), New Delhi: Publications Division.

—— (1982) 'Proper land management', *Indian Express*, 27 December.

Von Oppen, M. and Subba Rao, K.V. (1980) 'Tank irrigation in semi-arid tropical India', mimeo, Patancheru: International Crop Research Institute for Semi-Arid Tropics.

West Bengal Forest Department (1988) *Project Report on Resuscitation of Sal Forests of South-West Bengal through People's Participation*, Calcutta: West Bengal Forest Department.

Whitcombe, E. (1971) *Agrarian Conditions in Northern India*, Vol. 1: *The United Provinces under British Rule, 1860–1900*, Berkeley, CA: University of California Press.

—— (1982) 'Irrigation', in D.Kumar (ed.) *The Cambridge Economic History of India*, Vol. 2, Cambridge: Cambridge University Press.

—— (1993) 'The costs of irrigation in British India: waterlogging, salinity, malaria' in D. Arnold and R. Guha (eds) *Nature, Culture, Imperialism: Essays on the Environmental History of South Asia*, Delhi: Oxford University Press.

Whyte, R.O. (1968) *Land, Livestock and Human Nutrition in India*, New York: F. Praeger.

Zachariah, M. and Sooryamurthy, R. (1994) *Science for Social Revolution*, New Delhi: Sage Publications.

# INDEX